Y0-EII-489

INTERNATIONAL SERIES OF MONOGRAPHS IN
NATURAL PHILOSOPHY
GENERAL EDITOR: D. TER HAAR

VOLUME 57

MACROSCOPIC ELECTROMAGNETISM

OTHER TITLES IN THE SERIES IN NATURAL PHILOSOPHY

MACROSCOPIC ELECTROMAGNETISM

BY

F. N. H. Robinson

Clarendon Laboratory, Oxford

PERGAMON PRESS

OXFORD · NEW YORK · TORONTO
SYDNEY · BRAUNSCHWEIG

Pergamon Press Ltd., Headington Hill Hall, Oxford
Pergamon Press Inc., Maxwell House, Fairview Park, Elmsford, New York 10523
Pergamon of Canada Ltd., 207 Queen's Quay West, Toronto 1
Pergamon Press (Aust.) Pty. Ltd., 19a Boundary Street, Rushcutters Bay, N.S.W. 2011, Australia
Vieweg & Sohn GmbH, Burgplatz 1, Braunschweig

All Rights Reserved. No part of this publication may be reproduced, stored in a retrieval system, or transmitted in any form or by any means, electronic, mechanical, photocopying, recording or otherwise, without the prior permission of Pergamon Press Ltd.

First edition 1973

Library of Congress Cataloging in Publication Data

Robinson, Frank Neville Hosband.
Macroscopic electromagnetism.

(International series of monographs in natural philosophy, v. 57)
Bibliography: p.
1. Electromagnetic theory. I. Title.
QC670.R62 1973 537 73–4280
ISBN 0–08–017647–X

Printed in Great Britain by Bell and Bain Ltd., Glasgow

Contents

Contents

Contents

Contents

Preface

ADVANCED texts on electromagnetic topics are usually concerned with either the mathematical development of the theory, or with applications to specific types of system. Thus any extended discussion of the basic principles and laws is only to be found in elementary introductory texts, designed for readers who have neither the background of physical knowledge nor the mathematical equipment to benefit from a formal and rigorous treatment. As a result it is not uncommon to find a considerable skill in solving particular classes of electromagnetic problem allied with a vague sense of generalized confusion about the basic principles. This book is an attempt to bridge this gap.

Nowhere is this confusion more evident than at the point where contact is established between the laws of macroscopic electromagnetism and their interpretation in microscopic or atomic terms. The most notorious symptom is the disquiet with which many physicists and engineers approach any discussion involving the two magnetic vectors **B** and **H**, but, were esoteric dielectric media as common as permanent magnets, no doubt the two electric vectors **E** and **D** would be regarded with equal distrust. This topic, in one guise or another, therefore occupies a substantial fraction of the book. Three other topics which are given prominence are the basic properties of the six electromagnetic vectors, **E**, **P**, **D**, **B**, **M** and **H**, the question of force in electromagnetic systems and the role of energy in electromagnetic calculations especially in relation to thermodynamics.

The book is written throughout in S.I. units, but it must be admitted that the use of these units in elementary teaching, whatever their practical advantages, is a potent source of confusion. If a distinction, wholly artificial, is made between the vectors **B** and **H** in vacuum and this is allied to an undue emphasis on extended, isotropic, homogeneous linear media, it is not surprising that the very real distinction between **B** and **H** or, for that matter, **E** and **D** in material media is made to seem illogical and incomprehensible.

The laws of macroscopic electromagnetism and the properties of the macroscopic electromagnetic variables are independent of any microscopic interpretation; they were after all known to Maxwell, yet at the present time it is almost impossible to present a treatment of macroscopic electromagnetic phenomena without referring to their connection with the atomic structure of matter. The nature of this connection was outlined by H. A. Lorentz at the beginning of the century and involves a number of subtle and difficult problems which Lorentz was careful to leave as open questions. Later authors have, however, by and large, not only assumed that these questions were trivial but unfortunately also that they could be answered by crude, and generally inconsistent, arguments. Of these we now consider three common examples. The first, and most sophisticated, is that the averaging procedure implied by Lorentz, which leads from a microscopic theory in terms of electrons to a macroscopic theory in terms of continuous variables, is connected with the statistical averaging processes of quantum mechanics. This can easily be seen to be false for, if it were true that all macroscopically observable variables were quantal averages, we should, contrary to experience, be unable to observe macroscopic fluctuations arising from the statistical nature of a quantum mechanical description. The second is that the relation can be demonstrated by simple diagrams. This is generally first presented in connection with the relation between dielectric polarization **P**, regarded as the electric dipole moment density, and the effective body or surface charge. Critical inspection of these diagrams, however, invariably leads to the conclusion either that the net charge in any macroscopic volume is identically zero or that, in the case of surface charge, it is independent of the value of **P**, provided that **P** is non-zero. The third and most common explanation of Lorentz's procedure is that it represents a volume average over a region of macroscopic dimensions. Because any macroscopic region contains such a large number N of elementary charged particles fluctuations in N, which will be of order $N^{\frac{1}{2}}$, can be neglected. Now not only is it untrue that N is always large, the electron beam of a vacuum tube is a case in point, but, as we now show by a numerical example, fluctuations of even one particle, let alone $N^{\frac{1}{2}}$ particles, would have gross observable effects. A copper sphere of radius 1 mm contains some 10^{20} mobile electrons and 10^{20} immobile ionic atomic cores. Placed in a field of 1 millivolt per metre, which is a strong field by radio-frequency standards, macroscopic theory predicts, and the result can be verified experimentally, that it will acquire an induced dipole moment of 10^{-22} coulomb metre. Since the electronic charge is $1{\cdot}6 \times 10^{-19}$ coulomb this corresponds to the net displacement of no more than one electron from one hemisphere of the body to the other. In fact, as we shall see, the connection between microscopic structure and macroscopic laws is neither quantum mechanical nor statistical, nor can it be readily represented in precise form by simple diagrams. The widespread use of these erroneous notions is a major obstacle to an understanding of the macroscopic laws of electromagnetism.

A second obstacle is of a different kind. It arises from a tendency to avoid tedious formal arguments, based on the fundamental laws, by introducing plausible but unsupported *ad hoc* statements. Again it will be useful to illustrate this by an example. The majority of texts state, as a matter of definition, that the couple acting on a magnetic needle of moment **m** in a fluid medium, in which there is a uniform impressed field $\mathbf{B} = \mu\mu_0\mathbf{H}$ where $\mu > 1$, is $\mathbf{m} \wedge \mathbf{B}$ while a few texts state that it is $\mathbf{m} \wedge \mu_0\mathbf{H}$. Now, if **m** has already been defined, both these statements cannot be true. In fact, as Stopes-Roe and Whitworth (1971) have recently shown experimentally, the couple is $\mathbf{m} \wedge \mu_0\mathbf{H}$. The point of this example is that neither of these formulae is an admissible *fundamental* law. In this book, in agreement with the implicit assumptions of almost every writer on electromagnetism, the fundamental macroscopic laws are taken to be the four Maxwell equations

$$\nabla . \mathbf{B} = 0, \qquad \nabla \wedge \mathbf{E} + \dot{\mathbf{B}} = 0, \qquad \nabla . \mathbf{D} = \rho \qquad \text{and} \qquad \nabla \wedge \mathbf{H} = \mathbf{J} + \dot{\mathbf{D}}$$

and the equation for the force density in vacuum $\mathbf{f} = \rho\mathbf{E} + \mathbf{J} \wedge \mathbf{B}$, together with the explanatory rules: the current density **J** is the flux vector for charge, in vacuum $\mathbf{D} = \varepsilon_0\mathbf{E}$ and $\mathbf{B} = \mu_0\mathbf{H}$ where ε_0 and μ_0 are absolute constants, and in matter $\mathbf{D} = \varepsilon_0\mathbf{E} + \mathbf{P}$ and $\mathbf{B} = \mu_0(\mathbf{H} + \mathbf{M})$ where **P** and **M** are properties of the material medium. Once these laws have been stated it is possible to define the magnetic moment of a body, and then the couple acting on the body in any given situation is a matter for calculation, not definition. This particular example happens to be a case where intuition substituted for argument leads to an incorrect result but other cases, less clear cut, abound. Thus, for example, few texts on electromagnetism, thermodynamics, solid state physics or statistical mechanics ever explain why the work done by magnetic forces per unit volume, which is $\mathbf{H} \cdot d\mathbf{B}$, can often be replaced by the apparently quite different expression $-\mathbf{M} \cdot d\mathbf{B}$ or even $-\mu_0\mathbf{M} \cdot d\mathbf{H}$.

It will be apparent then that, in this book, we attempt to set out the basic concepts and laws of macroscopic electromagnetism, and some of their most important consequences, without ignoring their relation to the microscopic atomic structure of matter, but also without, if possible, introducing hidden assumptions which may or may not be self-consistent. Classical electromagnetism raises many problems of an almost metaphysical nature such as the nature of the elementary charged particles and these are entirely ignored, as are all questions connected with special relativity. There is also nothing about mathematical techniques for solving particular classes of problem, e.g., in potential theory or wave theory, and applications are only mentioned in passing where they illustrate or emphasize a particular point. All these topics can be found treated at length elsewhere and the Bibliography makes some suggestions for further reading.

Chapters 1 to 4 contain introductory material, not all of it presented in conventional form. The connection between macroscopic and microscopic theory is set out in chapters 5 to 8. Chapter 9 deals with the basic properties of the macroscopic fields and 10 with energy and forces. Chapter 11 contains ancillary material on waves needed in later sections. Chapters 12, 13 and 14 deal with the nature of the constitutive relations and their atomic interpretation, chapter 15 with thermodynamics and chapter 16 dealing with statistical mechanics reverts to some of the problems considered in earlier chapters, especially 13 and 14. The next two chapters deal with the closely related topics of thermal radiation and fluctuations; finally chapter 19 gives a brief summary of some of the leading ideas and contains a bibliography. Although the mathematical apparatus required of the reader is modest, a thorough knowledge of vector methods and Fourier series and a nodding acquaintance with Cartesian tensors will suffice; the text makes such extensive use of possibly unfamiliar vector relations that these have been collected together for proof in an appendix.

Obviously any discussion of the microscopic structure of matter involves quantum mechanical notions and so, although classical language is used throughout most of the discussion, quantum mechanical language is used whenever it is necessary or simpler. Also concepts drawn from statistical mechanics are used freely and so the reader innocent of any knowledge of these two subjects will be at something of a disadvantage. On the other hand, the sophisticated reader will, no doubt, find the treatment of atomic topics naïve. This is to some extent intentional, not only because a full treatment would be inappropriate in a book on electromagnetism, but also because, although simple arguments may occasionally mislead, there is little point in expressing an easily measured material property in terms of an impressive array of matrix elements few of which are calculable or accessible to independent measurement. From the point of view of macroscopic electromagnetism, atomic and solid state physics are not precise predictive theories but explanatory theories, leading to an understanding of empirical regularities in observed phenomena.

For the physicist mathematics is a concise and rigorous language which connects one set of equations, with a recognizable physical content, with a second set of equations whose physical content may be either simpler or more useful. Very often we can omit most or possibly all of a mathematical argument by relying on physical intuition. This is an economical but risky procedure, for few of us are entirely free from physical misconceptions of which we are totally unaware. Certainly in writing this book I have had to discard many notions about the properties of the electromagnetic field which I had hitherto held to be obvious. The great advantage of a mathematical argument is that we can check each step to see if it is in fact logically compelling, and results derived mathematically give us a series of fixed reference points against which we can check our physical intuition.

For this reason the argument of this book is conducted mainly, though not exclusively, in mathematical terms. A mathematical derivation may be less memorable, congenial or obvious than a series of physical arguments but it is less prone to logical error. We cannot escape the use of physical intuition in setting up the equations which govern a system, in guiding the direction of our mathematical manipulations and in interpreting our final results, but the introduction of new physical arguments in the middle of a calculation requires exceptional care. The new physical notions so introduced have usually not been subject to such close scrutiny as the original statements with which we began the calculation.

I have tacitly assumed that the reader has already been exposed to a conventional course in electromagnetism. In the last chapter, in addition to a brief bibliography there is a summary of some of the main results obtained in the text. The reader may find it helpful to read this summary after chapter 1. I hope that he will find some answers to problems that he has been unable or unwilling to resolve for himself and that he will not feel that the book should have been called "Elementary Electromagnetism from an Unnecessarily Pedantic Standpoint".

Acknowledgement. I am indebted to many of my colleagues, especially Professor Sir Rudolf Peierls, for their patience as I sought to clarify my ideas, but to Dr. Geoffrey Brooker I am indebted in a different way. His ruthless and incisive comments on the text were not only salutary but also invariably helpful. Those errors and infelicities that remain are all my own work.

CHAPTER I

Introduction

MAXWELL'S equations in S.I. units are

$$\nabla \cdot \mathbf{B} = 0 \quad (1.1a) \qquad \nabla \wedge \mathbf{E} + \dot{\mathbf{B}} = 0 \quad (1.1b)$$

$$\nabla \cdot \mathbf{D} = \rho \quad (1.1c) \qquad \nabla \wedge \mathbf{H} - \dot{\mathbf{D}} = \mathbf{J} \quad (1.1d)$$

In vacuum $\mathbf{D} = \varepsilon_0 \mathbf{E}$ and $\mathbf{H} = (1/\mu_0)\mathbf{B}$ where ε_0 and μ_0 are universal constants. In the presence of matter $\mathbf{D} = \varepsilon_0 \mathbf{E} + \mathbf{P}$ and $\mathbf{H} = (1/\mu_0)(\mathbf{B} - \mu_0 \mathbf{M})$ where $\mathbf{P}$ and $\mathbf{M}$ represent the effects of charges in matter, i.e. the polarization or magnetization. Together with the Lorentz expression for the force density

$$\mathbf{F} = \rho \mathbf{E} + \mathbf{J} \wedge \mathbf{B} \tag{1.2}$$

these equations form a succinct summary of the laws of classical continuum electromagnetism as it applies to macroscopic systems. At the time these laws were formulated the electron had not been discovered and no question arose of their possible relation to a more fundamental theory, in which the quantities ρ, $\mathbf{J}$, $\mathbf{P}$ and $\mathbf{M}$ are expressed in terms of the effects of elementary sub-microscopic charged particles and only the vectors $\mathbf{E}$ and $\mathbf{B}$ are needed to describe the fields.

A theory of this type was first developed by H. A. Lorentz in 1906 and subsequently published as *The Theory of Electrons*. This has more recently been reprinted by Dover Publications (1951) and an excellent modern account and extension of Lorentz's ideas has been given by Rosenfeld (1951). All subsequent treatments of the macroscopic field equations have been based on concepts first introduced by Lorentz, and we now give a brief summary of this theory.

We take as our starting point the field equations in vacuum

$$\nabla \cdot \mathbf{B} = 0 \quad (1.3a) \qquad \nabla \wedge \mathbf{E} + \dot{\mathbf{B}} = 0 \quad (1.3b)$$

$$\varepsilon_0 \nabla \cdot \mathbf{E} = \rho \quad (1.3c) \qquad (1/\mu_0)\nabla \wedge \mathbf{B} - \varepsilon_0 \dot{\mathbf{E}} = \mathbf{J} \quad (1.3d)$$

and the equation for the force acting on a single charged particle

$$\mathbf{F} = q\mathbf{E} + q\mathbf{v} \wedge \mathbf{B}. \tag{1.4}$$

In (1.3c) and (1.3d) ρ and $\mathbf{J}$ are to include all the charges (nuclei, electrons) present in the system. Then, following Lorentz and Rosenfeld, we assume that in a macroscopic system the quantities which appear in (1.3) and (1.4) are to be averaged over volumes which, while containing many sub-microscopic particles, are nevertheless infinitesimal on a macroscopic scale. Thus equation (1.3c) for example becomes

$$\langle \varepsilon_0 \nabla \cdot \mathbf{E} \rangle = \langle \rho \rangle.$$

This averaging procedure commutes with space and time differentiation so that we have

$$\varepsilon_0 \nabla \cdot \langle \mathbf{E} \rangle = \langle \rho \rangle.$$

The four equations of the set (1.3) are, therefore, still valid if **E**, **B**, ρ and **J** are regarded as averages in this sense. Next we investigate the nature of the averages of ρ and **J**. In doing this we introduce the notion that the charges in matter can be separated into "true" or mobile charges such as ions in an electrolyte or electrons in a metal, which can make unlimited excursions under the influence of a field, and bound charges such as nuclei and electrons in atoms whose excursions can never exceed distances of atomic dimensions. Thus we write

$$\rho = \rho_t + \rho_b \tag{1.5a}$$

$$\mathbf{J} = \mathbf{J}_t + \mathbf{J}_b. \tag{1.5b}$$

We next define the vectors $\langle \mathbf{P} \rangle$ and $\langle \mathbf{M} \rangle$ (we omit discussion of higher order moments) as the electric and magnetic dipole moment densities due to bound charges. It is then relatively easy to show that

$$\langle \rho_b \rangle = -\nabla \cdot \langle \mathbf{P} \rangle \tag{1.6a}$$

$$\langle \mathbf{J}_b \rangle = \langle \dot{\mathbf{P}} \rangle + \nabla \wedge \langle \mathbf{M} \rangle. \tag{1.6b}$$

Thus if we now set $\mathbf{D} = \varepsilon_0 \mathbf{E} + \mathbf{P}$ and $\mathbf{H} = 1/\mu_0 \mathbf{B} - \mathbf{M}$ we recover the original macroscopic equations. This, on the face of it, appears to be eminently satisfactory; however, it suffers from two fundamental defects, both associated with the use of an "infinitesimal" volume average. The first defect is seen most clearly if we consider (1.6a) and, to illustrate the nature of the defect, we consider a linear one dimensional array of alternating negative and positive ions, such as occurs in polar crystals, e.g. ZnO. This is shown in Fig. 1.1. It is fairly obvious

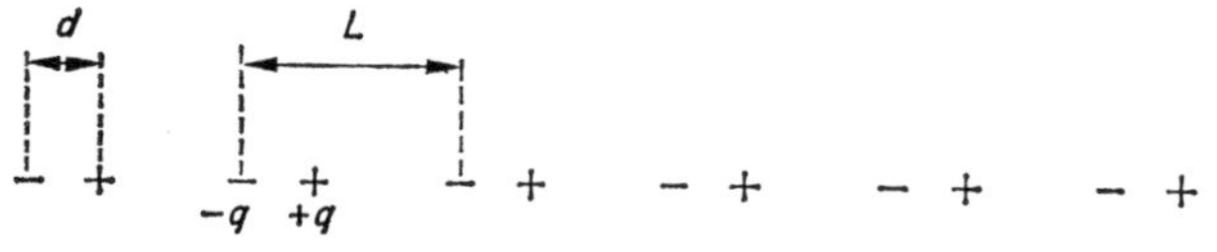

FIG. 1.1. A linear array of charges.

that the dipole moment per unit length of this array is $+(qd/L)$ so that if the length of the whole array is $NL+d$, the total moment is $(N+1)qd$. However, we could equally well describe this array as having a dipole moment of $-q[(L-d)/L]$ per unit length together with two odd charges $-q$ at one end and $+q$ at the other end. The total dipole moment is now still $q(NL+d) - Nq(L-d) = (N+1)qd$. Thus there is no unique way of defining P, the moment per unit length, unless we insist that **P** is defined over a length containing no net charge. This excludes the second choice and allows only the first choice. This, of course, is connected with the fact that the dipole moment of a region containing positive and negative charges is only unique if the region as a whole is exactly neutral. Thus, in (1.6a) the vector $\langle \mathbf{P} \rangle$ is only unique if the volume averaging is done over volumes that are exactly neutral. In this case either $\langle \rho_b \rangle \equiv 0$ or the volume average on the left of (1.6a) is taken over a different volume. Both alternatives are unsatisfactory. The one denies the possibility of using (1.6a) altogether, the other leads to ambiguous results. The second defect is that we have assumed that it is always possible to choose a volume which is "infinitesimal" in a macroscopic sense and yet contains an enormous number of sub-microscopic charges. This is in general untrue. In many applications where we treat either ρ_t or ρ_b as a macroscopically continuous variable and obtain results in excellent accord with experiment, no such volume exists. We give two examples.

In a low-noise travelling-wave tube, a macroscopically significant length of the electron beam is determined by the frequency of operation and the beam velocity. Suppose that the beam voltage is 625 volts so that its velocity is $c/20$ and we are amplifying signals at 5×10^9 Hz. The free space wavelength is 6 cm and a macroscopically insignificant length of the beam is certainly much less than 6/20 cm. Let us assume that 0·01 cm is adequate. A typical beam current is 160 μA or 10^{15} electron s^{-1}. The number of electrons in 0·01 cm is therefore $(20/c) \times 10^{13} \approx 10^4$. This is hardly large enough to treat ρ as continuous to parts in 10^6 as we do in travelling-wave tube theory in excellent agreement with experiment. Our second example is drawn from the refraction of light by solids, a phenomenon described with great accuracy by the continuum equations. Here the wavelength in a solid is perhaps 3000 Å and the largest linear dimension that we can conceivably regard as macroscopically insignificant can hardly exceed 1000 Å. A volume of 1000 Å cubed in a typical solid will normally contain no more than 10^8 particles taking part in optical effects, and adding one unneutralized electron to such a volume yields the quite absurd value of 160 coulomb m^{-3} for ρ_b.

These two considerations are enough to indicate that, whatever we mean by the averages in the Lorentz theory, they cannot be simple volume averages. Any attempt to explain the electric and magnetic properties of matter in terms of such averages is bound to lead to confusion. Since Lorentz's time it has been suggested that the averages are more properly regarded either as time averages over electronic motion in atoms or as quantum mechanical averages. Neither proposal is satisfactory. The periods of electronic motion in atoms are around 10^{-15} to 10^{-16} s, and any average which eliminates the details of this motion also eliminates optical effects which occur at frequencies up to 10^{15} Hz. In any case exactly the same averaging problem occurs in vacuum electronics and plasmas with unbound electrons. A quantum mechanical average will only affect the situation when the spread in the electron wave function is sufficient to make a localized description of atoms or free electrons impossible. This does not occur in vacuum tubes, plasmas, gases, or indeed many liquids or solids. Thus the Lorentz theory, which is in excellent agreement with experiment, cannot rely for its validity on the particular form of average suggested by Lorentz and accepted by most subsequent authors. The general assumption that this form of average is adequate is responsible for much of the confusion attendant on discussions of the relations between the macroscopic equations and microscopic theory.

It is, perhaps, surprising that, from the time of Lorentz's original work up to around 1950, the Lorentz averaging procedure was, with a few minor modifications, few of them improvements, generally accepted. The first attempt to treat the average in a different way is due to Mazur and Nijboer (1953) in a paper entitled "The Statistical Mechanics of Matter in an Electromagnetic Field". Mazur and Nijboer and later authors, notably de Groot (1969), regard the average as an ensemble average. Thus they show that, if we allow for the uncertainty in the position of the individual sub-microscopic charges, due to either quantum mechanics or thermal energy, then the equations which result, as averages of the microscopic Lorentz equations, do indeed have the form of the macroscopic Maxwell equations. The proof is quite general and can be given in either classical, quantum mechanical or relativistic terms. This approach is, by itself, insufficient. The passage from the microscopic to the macroscopic equations depends critically on the assumption that the statistical ensemble distribution function in coordinate space varies only slowly over a region of the order of the mean spacing between the sub-microscopic particles of the system. This implies that the ensemble used to describe the system must be so coarse that no meaning can be attached to any sort of material structure on an atomic scale. Thus the ensemble average approach, although appropriate to gases, cannot be applied to solids or even to a liquid with a sharply defined surface. Indeed we can also assure ourselves by detailed calculation that, except in gases, the statistical ensemble is never coarse enough to lead to the macroscopic field equations.

Mazur and Nijboer remark that the Lorentz volume average is, in some ways, similar to a special form of ensemble average. In a statistical ensemble we include all possible microscopic particle configurations compatible with the macroscopic specification of the system (i.e. the number of particles, the temperature, etc.). If now we augment this ensemble (which expresses our ignorance about the detailed microscopic state of the system) by an ensemble with new configurations, each obtained from one of the original configurations by rigid translations of all the particles together, this is equivalent to giving the origin $\mathbf{r}_0$ from which we measure the particle coordinates, a statistical distribution about the true origin from which we measure the field coordinates. Thus, if we allow $\mathbf{r}_0$ to cover a region of dimensions Λ, averages in this ensemble are equivalent to volume averages over elements of volume Λ^3. If we now choose Λ^3 so that it contains many elementary charges a straightforward application of Mazur and Nijboer's argument yields the macroscopic field equations. From this point of view the procedure is entirely satisfactory. It is, however, both conceptually and practically unsatisfactory. Conceptually, because it introduces a wholly artificial, and potentially serious, element of statistical uncertainty into the situation, and practically because it leads to incorrect average results. Thus, for example, if we apply the method to the interaction of light of wavelength λ with a dielectric medium, the fields due to the particles of the medium are underestimated by a factor of at best $1-\frac{1}{2}(2\pi\Lambda/\lambda)^2$. If therefore we were to use this method to discuss total internal reflection of light we would never find exactly total reflection, but at most a reflection coefficient $R \approx 1-\frac{1}{2}(2\pi\Lambda/\lambda)^2$. For glass we can hardly take Λ to be much smaller than 10 Å while for blue light in glass $\lambda \approx 3000$ Å so that the error in R is -2×10^{-4} which, though small, is unacceptable. The theory of ensemble averages of microscopic equations has an important part to play in electromagnetic theory and we shall discuss it in chapter 6; it is not, however, by itself sufficient to justify the use of the macroscopic equations as *exact* equations in *macroscopic* problems.

We now turn to another and simpler approach to the problem. In general in physics, and especially in statistical physics, we regard a macroscopic problem as one in which the detailed structure of the system on a microscopic scale is inaccessible to us. Thus, if we have a gram molecule of argon in a volume of 22·4 litres at 273 K this information is, by itself, insufficient to allow us to calculate the exact location and momentum of each atom. We therefore consider an ensemble of identically specified systems, with systems of the ensemble in every possible microscopic configuration compatible with the macroscopic data. We then construct expressions for the value of a macroscopic variable, say the pressure p for each such microscopic configuration. The ensemble average of these expressions yields the expectation value of the pressure, and, if we wish, we can also obtain the fluctuations in the pressure by calculating the ensemble average of p^2. We can, however, consider a macroscopic problem in a different way, as a problem characterized by a scale, so that, even if we knew the exact configuration of the system, we should not wish to make use of all this information. Some aspects of the detailed configuration would be irrelevant to the macroscopic problem. This is clearly the notion behind Mazur and Nijboer's use of the augmented ensemble. The augmented ensemble is, however, unsatisfactory because, although it simplifies the resulting equations, by eliminating irrelevant information, it does so in an uncontrolled way, and, in the process, also modifies possibly relevant information. We must therefore consider some other way of modifying the field equations which is free from this objection.

We take the attitude that a macroscopic problem is always characterized by a finite scale of length. In optics this is related to the wavelength, but in other problems it may be related to a mechanical tolerance, etc. If, therefore, we were to express all the quantities in the problem as spatial Fourier series or integrals, our interest in their spectrum would terminate at some fairly definite cut-off frequency k_0. Thus, in X-ray diffraction, we might require $k_0 \approx 10^9\ \text{cm}^{-1}$, in optics $10^5\ \text{cm}^{-1}$ and in electronics $10^4\ \text{cm}^{-1}$. The procedure we seek is therefore one which truncates or terminates the spatial Fourier spectrum of the variables at or near a definite cut-off frequency k_0. We shall show in chapter 5 that this procedure can be defined in a rigorous way and that, although it turns out to be in some ways similar to Mazur and Nijboer's ensemble average, it leads from a particle description to the macroscopic field equations as exact relations between macroscopic variables. It also corresponds in a direct way to our intuitive ideas about forming volume averages of atomic quantities. The nature of the problem under discussion, rather than the physical state of the system, determines the correct choice of k_0. The state of the system and its constitution then determines whether this value of k_0 will lead to a useful simplification of the problem. In general we find that if $k_0 < 10^{12}\ \text{cm}^{-1}$ we are free from the need to speculate about the internal structure of the elementary charges, while if $k_0 < 10^6\ \text{cm}^{-1}$ we need not discuss the internal structure of atoms. If, further, a_0 is the mean spacing of the *charges* and $k_0 a_0 \ll 1$ we can always treat electricity as a fluid, while if in addition $k_0 A_0 \ll 1$ where A_0 is the *interatomic* spacing we can also use Maxwell's equations in their conventional form (1.1).

We now turn from a consideration of the relation between microscopic and macroscopic fields to the subject matter of the next three chapters. In this book our emphasis is throughout on the fields rather than on the potentials. This does not necessarily mean that we attach any special metaphysical status to the fields that does not equally adhere to the potentials, nor that we are insensible to the elegance of a treatment of electrodynamics in terms of the potentials alone. It is, however, quite certain that had electromagnetic theory first been discovered in this form, leaving us ignorant of the existence of fields, many of the topics which now excite the attention of the quantum electrodynamicist would still be awaiting discovery. In the practical applications of electromagnetic theory to the mundane topics of electronics, optics, atomic physics and so on, the fields are the primary tools of the working physicist.

In the next chapter we give a traditional account of the origins of the electromagnetic field equations by generalization from a few simple, if rather hypothetical, observations. This serves mainly to establish operational definitions of the fields and their primary properties. The account will make rather more use of formal vector analysis than is usual in a first introduction. In particular we relate Ampère's law to the Biot–Savart law without appealing to current loops, infinite straight wires, etc.

Although we treat the potentials as subsidiary to the fields there are some topics where they are such useful auxiliary concepts that it would be perverse to ignore them and so we devote chapter 3 to a brief discussion of some of their properties in relation to the field equations.

Introduction

Since we propose to discuss the electromagnetic properties of solids in terms of the vectors **P** and **M** and the tensor Q_{ij} or dyadic $\underline{\underline{\mathbf{Q}}}$ whose microscopic interpretations are as electric dipole moment density, magnetic dipole moment density or electric quadrupole moment density, we cannot escape defining the multipole moments of a localized charge and current distribution. Fortunately we require only a few low order moments and so we can dispense with much of the elaborate mathematics required for a more general discussion.

CHAPTER 2

The Basic Field Equations

Introduction

Electromagnetism, like other branches of theoretical physics, arose from humble origins, in this case simple qualitative observations about the forces between certain curious mineral objects and the phenomena observed when other objects are rubbed together. As these observations became more systematic, rules and regularities began to appear and eventually these rules were crystallized as the laws of electromagnetism. A set of mathematical equations such as Maxwell's equations is, however, devoid of all physical significance unless we know what the variables in the equations mean in physical terms. Thus the equation

$$\delta \int_a^b T \, \mathrm{d}r = 0$$

could be a statement of the laws of any one of several branches of physics and, until we know what the letters mean, we cannot begin to guess whether it refers to, say, geometrical optics or mechanics. For this reason it is important that we see how Maxwell's equations arose as a generalization of the laws of static and quasi-static phenomena, since only in this way can we acquire a sense of their physical content.

We have some freedom of choice in choosing a starting point but we shall follow tradition and use the following basic laws:

1. The inverse square law of force between charges.
2. Conservation of charge.
3. The existence of currents.
4. The assumption that magnetism is solely due to currents and is described by the Biot–Savart law.
5. Faraday's law of electromagnetic induction.

Electrostatics

We begin with electrostatics, a science which originates in the observation that bodies can be electrified by heat, friction or chemical action and then exert forces on each other at a distance. The forces act along the line joining the bodies, may be attractive or repulsive and obey an inverse square law. Action and reaction balance and the effects of several bodies are linearly additive.

These observations allow us to assign a charge q to each body so that the force exerted and experienced by the body are both proportional to q. We then summarize the laws in the expression for the force $\mathbf{F}(\mathbf{r}_1)$ exerted on a charge q_1 at $\mathbf{r}_1$ by charges q_k at $\mathbf{r}_k$. If we let

$\mathbf{r}_{1k} = \mathbf{r}_1 - \mathbf{r}_k$ and introduce a universal constant ε_0, connected with a choice of units, we have

$$\mathbf{F}(\mathbf{r}_1) = q_1 \sum_{k \neq 1} \frac{q_k \mathbf{r}_{1k}}{4\pi\varepsilon_0 r_{1k}^3}. \tag{2.1}$$

We define the electric field $\mathbf{E}(\mathbf{r}_1)$ at $\mathbf{r}_1$ so that

$$\mathbf{F}(\mathbf{r}_1) = q_1 \mathbf{E}(\mathbf{r}_1) \tag{2.2}$$

and then the field due to the charges is

$$\mathbf{E}(\mathbf{r}_1) = \sum_{k \neq 1} \frac{q_k \mathbf{r}_{1k}}{4\pi\varepsilon_0 r_{1k}^3}. \tag{2.3}$$

Since for any closed surface S with a positive outward normal we have

$$\oint_S \frac{\mathbf{r} \cdot \mathrm{d}\mathbf{S}}{4\pi r^3} = 1$$

if the surface encloses the origin, and zero if it does not, we then obtain Gauss' theorem

$$\varepsilon_0 \int_S \mathbf{E} \cdot \mathrm{d}\mathbf{S} = \sum_k q_k = Q \tag{2.4}$$

where Q is the total charge within S.

At this stage we run into a difficulty if we consider the properties of point charges and so, for the moment, we restrict our attention to a situation in which the charge is regarded as a continuous fluid with a density ρ, which has the property

$$\int_V \rho \, \mathrm{d}^3\mathbf{r} \to \text{order } a^2 \text{ or less}$$

as the linear dimensions a of the volume of integration tend to zero. We can then write equation (2.3) as

$$\mathbf{E}(\mathbf{r}_1) = \int_V \frac{\rho(\mathbf{r}_2)\mathbf{r}_{12} \, \mathrm{d}^3\mathbf{r}_2}{4\pi\varepsilon_0 r_{12}^3} \tag{2.5}$$

and Gauss' theorem becomes

$$\varepsilon_0 \int_S \mathbf{E} \cdot \mathrm{d}\mathbf{S} = \int_V \rho \, \mathrm{d}^3\mathbf{r}. \tag{2.6}$$

The divergence of a vector $\mathbf{A}$, i.e. div $\mathbf{A}$ or $\nabla \cdot \mathbf{A}$, is defined by

$$\nabla \cdot \mathbf{A} = \lim_{V \to 0} 1/V \oint_S \mathbf{A} \cdot \mathrm{d}\mathbf{S}$$

where V is the volume within the closed surface S. Thus Gauss' theorem can be written in the differential form

$$\varepsilon_0 \nabla \cdot \mathbf{E} = \rho. \tag{2.7}$$

The identity

$$\frac{\mathbf{r}_{12}}{r_{12}^3} = -\mathbf{grad}_1 \left(\frac{1}{r_{12}}\right) = -\nabla_1 \left(\frac{1}{r_{12}}\right) \tag{2.8}$$

in conjunction with (2.5) allows us to write E as the gradient of a scalar potential

$$\mathbf{E} = -\nabla\phi \tag{2.9}$$

where

$$\phi(\mathbf{r}_1) = \int_V \frac{\rho(\mathbf{r}_2)\,d^3\mathbf{r}_2}{4\pi\varepsilon_0 r_{12}}. \tag{2.10}$$

Thus the electrostatic field is conservative or irrotational and

$$\nabla \wedge \mathbf{E} = \textbf{curl}\,\mathbf{E} = 0. \tag{2.11}$$

This obviously leads to the further relation

$$\nabla \wedge (\nabla \wedge \mathbf{E}) = 0. \tag{2.12}$$

In Cartesian co-ordinates, but not in general,

$$(\nabla \wedge (\nabla \wedge \mathbf{E}))_i = (\nabla(\nabla \cdot \mathbf{E}))_i - (\nabla \cdot \nabla)E_i, \tag{2.13}$$

thus, in a region free from charge where $\nabla \cdot \mathbf{E} = 0$ we have

$$(\nabla \cdot \nabla)E_i = \nabla^2 E_i = 0. \tag{2.14}$$

Since this refers only to the individual components E_i, one at a time, any Cartesian-like component of $\mathbf{E}$ satisfies Laplace's equation. The operator ∇^2 can be expressed in any convenient coordinate system.

The electrostatic scalar potential ϕ in general satisfies

$$\nabla^2\phi = -\nabla \cdot \mathbf{E} = -\rho/\varepsilon_0 \tag{2.15}$$

which is known as Poisson's equation. In empty space it satisfies Laplace's equation

$$\nabla^2\phi = 0. \tag{2.16}$$

As a consequence all three second derivatives of ϕ cannot simultaneously have the same sign and there are no absolute maxima or minima of ϕ in empty space (Earnshaw's theorem).

If we allow point charges the situation is rather more complicated and we should have to discuss whether by a point, in this connection, we mean the same thing as a point in a discussion of space as a continuum. We shall not pursue this question but instead take the view that a point charge is the limit of a continuous distribution. Thus, if we consider the class of functions $\delta(\mathbf{r})$ with the property that

$$\int_V \delta(\mathbf{r})\,d^3\mathbf{r}$$

is unity whenever the volume of integration includes the origin and

$$\delta(\mathbf{r}) = 0$$

whenever $r > \varepsilon \to 0$, we can express a distribution of "point" charges q_k at $\mathbf{r}_k$ as a charge density

$$\rho(\mathbf{r}) = \sum_k q_k \delta(\mathbf{r}_k - \mathbf{r}). \tag{2.17}$$

The function $\delta(\mathbf{r})$ is essentially the three dimensional Dirac delta function and its mathematical properties can be investigated with considerably more rigour than we have indicated. (For a brief discussion see Messiah, 1961.) If we adopt this view of point charges then all the results derived for a continuous distribution remain valid.

The Basic Field Equations

Conservation of Charge

Experimentally, charge can be transported from one location to another, either by the bodily motion of charged bodies, or through certain types of material, notably metals, which do not themselves move. The existence of conductors leads us to define a current, or flux of charge, I as the rate of transport of charge and a current density $\mathbf{J}$ so that the current I crossing an element of area dS is $\mathbf{J} \cdot d\mathbf{S}$.

If matter in an isolated region of space is initially uncharged no process acting within this region, e.g. friction, subdivision of the bodies, etc., leads to a change in the net charge although equal amounts of positive and negative charge may be created. This leads to a new physical law, the conservation of charge, which we can express as

$$\operatorname{div} \mathbf{J} + (\partial \rho / \partial t) = 0$$

or

$$\nabla \cdot \mathbf{J} + \dot{\rho} = 0. \tag{2.18}$$

Since $\rho = \varepsilon_0 \nabla \cdot \mathbf{E}$ this leads to the relation

$$\nabla \cdot (\mathbf{J} + \varepsilon_0 \dot{\mathbf{E}}) = 0. \tag{2.19}$$

The vector $\mathbf{J} + \varepsilon_0 \dot{\mathbf{E}}$ is solenoidal and so can be expressed as the curl of some new vector. We return to this point at a later stage. The quantity $\varepsilon_0 \dot{\mathbf{E}}$ has the dimensions of a current density and is referred to as the displacement current density or more loosely as the displacement current.

Conductors

To complete the formal structure of electrostatics we must consider briefly the notion of a conductor as a medium which cannot support an electric field, i.e., charges in a conductor always distribute themselves so that there is no net field in the conductor. This is essentially a new physical law, but it does contain our earlier definition of a conductor as a medium which transports electric charge. We may note, in passing, that it is only a useful definition of a conductor in electrostatics and that many media, e.g. silicon, which behave as conductors in weak fields do not do so in higher fields. A useful device, the field effect transistor, is based on this property.

Once we leave the realm of pure electrostatics we can have a field $\mathbf{E}$ in a conductor but it is accompanied by a current density $\mathbf{J}$. By considering the work done in taking charge around a closed path part of which lies in vacuum and part in a conductor we see that in the conductor the field does work on the charges at a rate $\mathbf{E} \cdot \mathbf{J}$ per unit volume. This work is either converted into heat or mechanical energy in the conductor and a continued current is only possible if there is a continuous source of energy, e.g., a battery, a van der Graaf machine, or a temperature difference somewhere in the system.

In many cases Ohm's law is obeyed and $\mathbf{J}$ is proportional to $\mathbf{E}$. If we let $\mathbf{J} = \sigma \mathbf{E}$ the conductivity σ in S.I. units is expressed as reciprocal ohms per metre or mho m^{-1}. For copper $\sigma \approx 10^8$ mho m^{-1} but values ranging from 10^8 to 10^{-15} or less occur in other media.

The equations

$$\varepsilon_0 \nabla \cdot \mathbf{E} = \rho, \qquad \mathbf{J} = \sigma \mathbf{E}, \qquad \nabla \cdot \mathbf{J} + \dot{\rho} = 0$$

yield, in the *interior* of a conductor,

$$\dot{\rho}+\rho/\tau_d = \dot{\rho}+\rho\sigma/\varepsilon_0 = 0. \tag{2.20}$$

The "dielectric relaxation" time τ_d varies from 10^{-16} seconds or less in metals to about one month in some plastics. Any charge density established in the interior of a conductor collapses in a time of the order of τ_d. This does not apply at the surface and indeed in a static problem charge on a conductor must reside on its surface.

The theory of direct current circuits is obviously related to electrostatics, although some concepts, e.g. a sustained current flow, lie outside electrostatics. We shall assume that the reader is so utterly familiar with this topic that any further discussion would be superfluous.

Magnetostatics

The basic phenomena of magnetostatics are associated with the forces acting between conductors carrying currents. The fundamental laws are similar in their basic structure to electrostatics though their geometrical form is somewhat more complicated. They are summarized in Biot and Savart's law for the force dF_{12} acting on a current element $\mathbf{J}_1 dV_1$ at $\mathbf{r}_1$ due to an element $\mathbf{J}_2\, dV_2$ at r_2.

$$d\mathbf{F}(\mathbf{r}_1) = \mu_0\, [\mathbf{J}_1\, dV_1 \wedge (\mathbf{J}_2\, dV_2 \wedge \mathbf{r}_{12})]/4\pi r_{12}^3. \tag{2.21}$$

The universal constant μ_0 is associated with a choice of units.

The force between two closed localized current systems is

$$\mathbf{F}_{12} = \frac{\mu_0}{4\pi}\int_{V_1} dV_1 \mathbf{J}_1 \wedge \int_{V_2} \frac{\mathbf{J}_2 \wedge \mathbf{r}_{12}}{r_{12}^3}\, dV_2. \tag{2.22}$$

We use the vector relation

$$\mathbf{A} \wedge (\mathbf{B} \wedge \mathbf{C}) = \mathbf{B}(\mathbf{A} \cdot \mathbf{C}) - \mathbf{C}(\mathbf{A} \cdot \mathbf{B}) \tag{2.23}$$

to write this as

$$\mathbf{F}_{12} = \frac{\mu_0}{4\pi}\int_{V_2} dV_2 \mathbf{J}_2 \int_{V_1} \frac{\mathbf{J}_1 \cdot \mathbf{r}_{12}}{r_{12}^3}\, dV_1 - \frac{\mu_0}{4\pi}\int_{V_1} dV_1 \int_{V_2} \mathbf{r}_{12} \frac{\mathbf{J}_1 \cdot \mathbf{J}_2}{r_{12}^3}\, dV_2. \tag{2.24}$$

Now

$$\mathbf{J}_1 \cdot \frac{\mathbf{r}_{12}}{r_{12}^3} = -\mathbf{J}_1 \cdot \nabla_1 \left(\frac{1}{r_{12}}\right) = -\nabla_1 \cdot \left(\frac{\mathbf{J}_1}{r_{12}}\right) + \frac{1}{r_{12}} \nabla_1 \cdot \mathbf{J}_1$$

and for steady currents in a stationary state $\dot{\rho} = 0$ so that $\nabla_1 \cdot \mathbf{J}_1 = 0$. Thus the integral over V_1 in the first term of (2.24) is

$$\int_{V_1} \frac{\mathbf{J}_1 \cdot \mathbf{r}_{12}}{r_{12}^3}\, dV_1 = -\int_{V_1} \nabla_1 \cdot \left(\frac{\mathbf{J}_1}{r_{12}}\right) dV_1.$$

This can be integrated to give a surface integral over the boundary of V_1 where, from our assumption that the circuits were closed, the normal component of $\mathbf{J}_1$ is zero. Thus this term is zero and so

$$\mathbf{F}_{12} = -\frac{\mu_0}{4\pi}\int_{V_1} dV_1 \int_{V_2} \mathbf{r}_{12} \frac{\mathbf{J}_1 \cdot \mathbf{J}_2}{r_{12}^3}\, dV_2. \tag{2.25}$$

This not only checks that action and reaction balance but can be used to calculate **F** in terms of the geometry of the two circuits with great accuracy. It is used in the design of a precision instrument, the current balance. For this reason (2.25) is used to define the unit of current, the ampere, in mechanical terms. The constant μ_0 is arbitrarily set equal to $4\pi \times 10^{-7}$. It is usual to attribute dimensions to μ_0 and to express charges in coulombs or ampere-seconds, thus treating the ampere as the basic electromagnetic unit. Thus we have either

$$\mu_0 = 4\pi \times 10^{-7}\ \text{joule amp}^{-2}\ \text{m}^{-1}$$

$$\mu_0 = 4\pi \times 10^{-7}\ \text{joule coulomb}^{-2}\ \text{s}^2\ \text{m}^{-1}$$

or more usually

$$\mu_0 = 4\pi \times 10^{-7}\ \text{henry m}^{-1}$$

where

$$1\ \text{henry} = 1\ \text{volt amp}^{-1}\ \text{s}$$

and

$$1\ \text{volt} = 1\ \text{joule amp}^{-1}\ \text{s}^{-1} = 1\ \text{joule coulomb}^{-1}.$$

Because electrostatic measurements are much less easy to perform the constant ε_0 is most accurately obtained from the expression for the velocity of light c. Thus

$$\mu_0 \varepsilon_0 c^2 = 1 \quad \text{and so} \quad \varepsilon_0 = \frac{10^7}{4\pi c^2} \tag{2.26}$$

where c is the latest agreed experimental value of the velocity of light in vacuum. The simplest dimensions of ε_0 are farad m^{-1} where 1 farad = 1 coulomb volt^{-1} = 1 coulomb^2 joule^{-1}. Since $c \approx 3 \times 10^8\ \text{m s}^{-1}$ we have the useful *approximate* results

$$(\mu_0 \varepsilon_0)^{-\frac{1}{2}} \approx 3 \times 10^8\ \text{m s}^{-1}, \qquad (\mu_0/\varepsilon_0)^{\frac{1}{2}} \approx 120\pi\ \text{ohms} \tag{2.27}$$

where 1 ohm = 1 volt amp^{-1}. Even more approximately $\mu_0 \approx 10^{-6}$ henry m^{-1} and $\varepsilon_0 \approx 10^{-11}$ farad m^{-1}.

If $\text{d}\mathbf{F}(\mathbf{r}_1)$ is the force acting on a current element $\mathbf{J}_1\ \text{d}V_1$ at $\mathbf{r}_1$ we can define the magnetic field $\mathbf{B}(\mathbf{r}_1)$ so that

$$\text{d}\mathbf{F}(\mathbf{r}_1) = \mathbf{J}_1\ \text{d}V_1 \wedge \mathbf{B}(\mathbf{r}_1). \tag{2.28}$$

If we have a charge distribution ρ fixed in a body which moves with a velocity $\mathbf{v}$ it is obvious that the current crossing unit area fixed in space as the body moves through the area is $\mathbf{J} = \rho\mathbf{v}$ and it is not implausible to assume that if q is the charge within an infinitesimally small volume its contribution to the current is $\rho\mathbf{v}\ \text{d}V = q\mathbf{v}$. Certainly we could use (2.28) together with the corresponding electrical relation to express the force density or body force as

$$\mathbf{f} = \rho\mathbf{E} + \mathbf{J} \wedge \mathbf{B} \tag{2.29a}$$

or

$$\mathbf{f} = \rho\mathbf{E} + \rho\mathbf{v} \wedge \mathbf{B}, \tag{2.29b}$$

but it does not necessarily follow that the force on a true point charge is

$$\mathbf{F} = q\mathbf{E} + q\mathbf{v} \wedge \mathbf{B}. \tag{2.30}$$

Fortunately, the difference between (2.29b) and (2.30) is only significant if we intend to discuss problems concerned with the internal structure and finite size of electrons. For our purposes (2.30) and (2.29b) can be treated as equivalent. For a very full discussion of just those problems that we propose to neglect the reader should consult Rohrlich (1965).

The magnetic field **B** which appears in (2.28) is related to its sources, the currents **J** (we use source here in a general sense and not in the special sense of hydrodynamics) by

$$\mathbf{B}(\mathbf{r}_1) = \mu_0 \int_{V_2} \frac{\mathbf{J}(\mathbf{r}_2) \wedge \mathbf{r}_{12}}{4\pi r_{12}^3} \, dV_2. \tag{2.31}$$

If we use the identity (2.8) together with the vector relation

$$\nabla \wedge (\psi \mathbf{F}) = \psi \nabla \wedge \mathbf{F} - \mathbf{F} \wedge (\nabla \psi)$$

and note that $\nabla_1 \wedge \mathbf{J}(\mathbf{r}_2) = 0$ since ∇_1 operates on $\mathbf{r}_1$ which does not appear in $\mathbf{J}(\mathbf{r}_2)$, we obtain

$$\mathbf{B}(\mathbf{r}_1) = \nabla_1 \wedge \mathbf{A}(\mathbf{r}_1) \tag{2.32}$$

where the magnetostatic vector potential is

$$\mathbf{A}(\mathbf{r}_1) = \mu_0 \int_{V_2} \frac{\mathbf{J}(\mathbf{r}_2) \, dV_2}{4\pi r_{12}}. \tag{2.33}$$

The vector **B** is therefore solenoidal and

$$\nabla . \mathbf{B} = 0. \tag{2.34}$$

All the flux lines of **B** form closed loops and there are no free magnetic poles equivalent to free electric charge. This is a direct consequence of our *assumption* that magnetic effects arise only from currents.

Any square integrable vector whose divergence and curl are both zero everywhere is also zero, thus $\nabla \wedge \mathbf{B}$ cannot be zero. In general, for a square integrable vector we have the identity [see Appendix (A72)]

$$\mathbf{K}(\mathbf{r}_1) = \frac{1}{4\pi} \left\{ \nabla_1 \wedge \int_{V_2} \frac{\nabla_2 \wedge \mathbf{K}(\mathbf{r}_2) \, dV_2}{r_{12}} - \nabla_1 \int_{V_2} \frac{\nabla_2 . \mathbf{K}(\mathbf{r}_2) \, dV_2}{r_{12}} \right\}. \tag{2.35}$$

We can make this relation plausible by noting that any vector can always be expressed as

$$\mathbf{K} = -\nabla \psi + \nabla \wedge \mathbf{G}$$

where $\nabla . \mathbf{G} = 0$. We then have, in Cartesian co-ordinates,

$$\nabla . \mathbf{K} = -\nabla^2 \psi$$

$$\nabla \wedge \mathbf{K} = -\nabla^2 \mathbf{G}.$$

Thus

$$\mathbf{G}(\mathbf{r}_1) = \frac{1}{4\pi} \int_{V_2} \frac{\nabla_2 \wedge \mathbf{K}(\mathbf{r}_2) \, dV_2}{r_{12}}$$

and

$$\psi(\mathbf{r}_1) = \frac{1}{4\pi} \int_{V_2} \frac{\nabla_2 . \mathbf{K}(\mathbf{r}_2) \, dV_2}{r_{12}}$$

which yields (2.35).

The Basic Field Equations

We now use the relation

$$\nabla_2\left(\frac{1}{r_{12}}\right) = -\nabla_1\left(\frac{1}{r_{12}}\right) = \frac{\mathbf{r}_{12}}{r_{12}^3} \tag{2.36}$$

in equation (2.31) and obtain

$$\mathbf{B}(\mathbf{r}_1) = \frac{\mu_0}{4\pi}\int\left\{\frac{1}{r_{12}}\nabla_2 \wedge \mathbf{J}(\mathbf{r}_2) - \nabla_2 \wedge \left(\frac{\mathbf{J}(\mathbf{r}_2)}{r_{12}}\right)\right\} \mathrm{d}V_2.$$

The integral is over all space and so the second term yields a surface integral over a region where $\mathbf{J}(\mathbf{r}_2) = 0$ and is therefore zero. If in (2.35) we now set $\mathbf{K} = \mathbf{J}$, and form $\nabla_1 \wedge \mathbf{B}(\mathbf{r}_1)$, we obtain

$$\nabla_1 \wedge \mathbf{B}(\mathbf{r}_1) = \mu_0\mathbf{J} + \mu_0\nabla_1 \int \frac{\nabla_2 \cdot \mathbf{J}(\mathbf{r}_2)}{4\pi r_{12}} \mathrm{d}V_2. \tag{2.37}$$

For a stationary system $\nabla \cdot \mathbf{J} = -\dot{\rho} = 0$ and so we obtain Ampère's law

$$\nabla \wedge \mathbf{B} = \mu_0\mathbf{J}. \tag{2.38}$$

Thus the line integral of $\mathbf{B}/\mu_0$ round a closed path is equal to the current threading the path. Equation (2.38) implies that $\nabla \cdot \mathbf{J} = 0$. This cannot, therefore, be a general result, if we wish to retain both charge conservation and the possibility that charge should ever leave a region of space. Thus we cannot in general proceed from (2.37) to (2.38).

Displacement Current

Maxwell, by considering the flow of alternating currents in capacitors, concluded that the displacement current $\varepsilon_0\dot{\mathbf{E}}$ was electrically equivalent to a current density $\mathbf{J}$ and then went on to postulate that the equivalence extended to magnetic effects. If throughout magnetism we replace $\mathbf{J}$ by $\mathbf{J}+\varepsilon_0\dot{\mathbf{E}}$ we do not alter the equation $\nabla \cdot \mathbf{B} = 0$, but in (2.37) we already have $\nabla \cdot (\mathbf{J}+\varepsilon_0\dot{\mathbf{E}}) = 0$ and so (2.38) becomes, in general,

$$\nabla \wedge \mathbf{B} = \mu_0(\mathbf{J}+\varepsilon_0\dot{\mathbf{E}}). \tag{2.39}$$

We need not dwell on the consequences of this famous hypothesis, which should be familiar to the reader, nor on the scepticism with which it was received at the time.

If we compare equation (2.39) with equation (2.37) we might be tempted to assume that the last term in (2.37), i.e. the integral, can be shown to be $\mu_0\varepsilon_0\dot{\mathbf{E}}$ with no further assumptions. Thus using only the laws of electrostatics we have

$$\nabla_1 \cdot \left\{\nabla_1 \int_{V_2} \frac{\nabla_2 \cdot \mathbf{J}(\mathbf{r}_2)}{4\pi r_{12}} \mathrm{d}V_2\right\} = -\nabla_1 \cdot \mathbf{J} = \dot{\rho} = \varepsilon_0\nabla_1 \cdot \dot{\mathbf{E}}$$

and so we obtain

$$(1/\mu_0)\nabla_1 \cdot (\nabla \wedge \mathbf{B}) = \nabla \cdot (\mathbf{J}+\varepsilon_0\dot{\mathbf{E}}).$$

This is however an identity, since both sides are zero, and has no physical content. Maxwell's hypothesis is a new physical law which lies outside static electromagnetism. It can be obtained from electrostatics by invoking special relativity but this is hardly satisfactory in view of the crucial part played in relativity by the velocity of propagation of electromagnetic phenomena in vacuum.

Electromagnetic Induction

The remaining law of electromagnetism is due to Faraday. There is an exceptionally good discussion of electromagnetic induction in vol. 2 of the *Feynman Lectures on Physics* (Feynman *et al.*, Addison-Wesley, 1964) to which the reader may refer for amplification of our remarks. One aspect of electromagnetic induction is already implicit in the Lorentz force law. If we move a conductor with velocity **v** in a field **B**, the charge carriers within it experience a force $q\mathbf{v} \wedge \mathbf{B}$ and will move under its influence, until their separation generates a field $\mathbf{E} = -\mathbf{v} \wedge \mathbf{B}$ opposing further motion. If the conductor is closed a current may flow continuously. In a closed loop of wire the electromotive force (e.m.f.) acting round the loop is

$$\psi = \oint (\mathbf{v} \wedge \mathbf{B}) \cdot d\mathbf{l} = -\oint \mathbf{B} \cdot (\mathbf{v} \wedge d\mathbf{l}). \tag{2.40}$$

If **B** is a constant uniform field,

$$\psi = -\mathbf{B} \cdot \oint \mathbf{v} \wedge d\mathbf{l},$$

and this is simply the rate at which the magnetic flux threading the loop is decreasing. We could therefore write the result as

$$\psi = -d/dt \int_S \mathbf{B} \cdot d\mathbf{S}, \tag{2.41}$$

in terms of an open surface S bounded by the loop. This result is the basis of the operation of some types of dynamo and we call it the dynamo effect.

Faraday also found that, if the loop were fixed and the sources of the field moved or alternatively their strength changed, there was a similar effect which could be expressed as

$$\psi = -\int_S \dot{\mathbf{B}} \cdot d\mathbf{S} \tag{2.42}$$

for a fixed loop. This is a new effect, unconnected with the Lorentz law (except in relativity). It can also be expressed in terms of a field **E** as

$$\oint \mathbf{E} \cdot d\mathbf{l} = -\int_S \dot{\mathbf{B}} \cdot d\mathbf{S}. \tag{2.43}$$

Note that (2.43), Faraday's law, does not require the presence of a material circuit. It relates **E** and **B** in vacuum. It is essential to distinguish this from the dynamo effect.

The curl of a vector is defined by the equation

$$(\nabla \wedge \mathbf{F})_i = \lim_{S \to 0} 1/S_i \oint \mathbf{F} \cdot d\mathbf{l}$$

where the path is round the periphery of the plane area S whose normal is along the i axis. Thus equation (2.43) yields $\nabla \wedge \mathbf{E} = -\dot{\mathbf{B}}$.

We see that the general electric field is no longer irrotational. The induced component does not affect the arguments which lead to $\varepsilon_0 \nabla \cdot \mathbf{E} = \rho$ but it does mean that **E** cannot, in general, be derived solely from a scalar potential.

The Field Equations

We have now completed our task for this chapter and arrived at the canonical set of field equations in vacuum:

$$\nabla \cdot \mathbf{B} = 0 \quad (2.44a) \qquad \nabla \wedge \mathbf{E} + \dot{\mathbf{B}} = 0 \quad (2.44b)$$

$$\varepsilon_0 \nabla \cdot \mathbf{E} = \rho \quad (2.44c) \qquad (1/\mu_0)\nabla \wedge \mathbf{B} - \varepsilon_0 \dot{\mathbf{E}} = \mathbf{J}. \quad (2.44d)$$

These equations are to be supplemented by the Lorentz force equation, either

$$\mathbf{f} = \rho\mathbf{E}+\mathbf{J} \wedge \mathbf{B}, \tag{2.45a}$$

or

$$\mathbf{f} = \rho\mathbf{E}+\rho\mathbf{v} \wedge \mathbf{B}, \tag{2.45b}$$

or

$$\mathbf{F} = q\mathbf{E}+q\mathbf{v} \wedge \mathbf{B}. \tag{2.45c}$$

We have so far taken the view that (2.45a) effectively defines **E** and **B** in operational terms. We now abandon this view, since the units of ρ and **J** in (2.45a) cannot be fixed without the use of the field equations (2.44). Henceforth we regard (2.45a, b, c) as expressing only the most important properties of **E** and **B**.

The electromagnetic equations exhibit a number of well-known invariance properties (Lorentz-invariance, inversion, time reversal, and charge conjugation invariance, reciprocity) which are extensively treated elsewhere (Jackson, 1962; Schelkunoff, 1943) but we would also like to remind the reader of the complexity of the field equations.

The vectors **E** and **J** are polar vectors whose components change sign on inversion of the co-ordinate system. **B** is an axial vector and its components are unchanged under inversion. On the other hand **E** is unchanged under time reversal whereas **B** and **J** both change sign. Symmetry arguments which apply to **E** do not always apply to **B** and vice versa.

We notice that Maxwell's equations consist of a homogeneous pair of equations $\nabla \cdot \mathbf{B} = 0$, $\nabla \wedge \mathbf{E}+\dot{\mathbf{B}} = 0$ and an inhomogeneous pair $\varepsilon_0\nabla \cdot \mathbf{E} = \rho$ and $(1/\mu_0)\nabla \wedge \mathbf{B}-\varepsilon_0\dot{\mathbf{E}} = \mathbf{J}$ with source terms ρ and **J**. Thus, however we modify the source terms to obtain a more convenient representation of the effects of matter, it will only influence the second pair of equations. This trivial observation is often a useful indication of whether an unfamiliar equation in an unfamiliar context is a general result or implies a special treatment of the source terms. In vacuum, **B** and the vector $\mu_0\mathbf{H}$ coincide and, since it also happens that in many applications of electromagnetic theory e.g. to optics, electronics, and radio this equality is also valid, the set of equations

$$\nabla \cdot \mathbf{H} = 0 \qquad \nabla \wedge \mathbf{E}+\mu_0\dot{\mathbf{H}} = 0$$

$$\varepsilon_0\nabla \cdot \mathbf{E} = \rho \qquad \nabla \wedge \mathbf{H}-\varepsilon_0\dot{\mathbf{E}} = \mathbf{J}$$

is very frequently used. This has some practical advantages but can lead to confusion. Several texts, for example, never make it entirely clear that $\nabla \cdot \mu\mu_0\mathbf{H} = 0$ is always correct whereas $\mu\mu_0\nabla \cdot \mathbf{H} = 0$ is not.

We have chosen to present the basic laws in terms of continuous charge and current distributions. As far as we are concerned it is, however, a matter of indifference whether we regard (2.45a, b) as the basic force law or (2.45c) and whether we regard

$$\rho(\mathbf{r}) = \sum_k q_k\delta(\mathbf{r}_k-\mathbf{r})$$

and

$$\mathbf{J}(\mathbf{r}) = \sum_k q_k\mathbf{v}_k\delta(\mathbf{r}_k-\mathbf{r})$$

as defining ρ and **J** in terms of point charges, or whether we regard point charges as the limits of continuous distributions. The choice of a definite point of view does, however, have consequences in any theory of the elementary charged particles and for this the reader is referred to Rohrlich's book quoted earlier.

Problems

1. In vacuum $\nabla \wedge (\nabla \wedge \mathbf{E}) = 0$ and $\nabla \cdot \mathbf{E} = 0$ and, in cylindrical coordinates, $\nabla^2 E_z = 0$. Does E_r satisfy any simple equation?

2. A system of three separate conducting electrodes at different potentials is set up. Is there any configuration of the electrodes for which a charged particle will remain at rest in the vicinity of the electrodes?

3. Use equations (2.11) and (2.18) to derive Kirchhoff's circuit laws.

4. One plate of a parallel plate capacitor consists of a thick metal plate, the other plate of a thin layer of n type silicon. Why is the charge–voltage characteristic of this system symmetric for low voltages but asymmetric for high voltages?

5. How would you attempt, in principle, to determine the value of a resistance in absolute units?

CHAPTER 3

Potentials

Introduction

The fields **E** and **B** form a six-component system, but the two homogeneous Maxwell equations

$$\nabla \cdot \mathbf{B} = 0 \quad (3.1a) \qquad \nabla \wedge \mathbf{E}+\dot{\mathbf{B}} = 0 \quad (3.1b)$$

suggest that not all these components are entirely independent or at least that we might find a more compact description of the field with fewer components. We have already seen that in electrostatics **E** can be expressed as the gradient of a single scalar potential and that **B** can be expressed as the curl of a vector potential. It is easy to see that the substitutions

$$\mathbf{B} = \nabla \wedge \mathbf{A} \quad (3.2a) \qquad \mathbf{E} = -\dot{\mathbf{A}}-\nabla\phi \quad (3.2b)$$

yield (3.1a) and (3.1b) as identities. The remaining Maxwell equations now impose conditions on the potentials **A** and ϕ. Thus

$$\varepsilon_0\nabla \cdot \mathbf{E} = \rho \quad (3.3a) \quad \text{and} \quad \nabla \wedge \mathbf{B}-\mu_0\varepsilon_0\dot{\mathbf{E}} = \mu_0\mathbf{J} \qquad (3.3b)$$

yield the equations

$$\Box^2\phi = -\rho/\varepsilon_0-(\partial/\partial t)(\nabla \cdot \mathbf{A}+\mu_0\varepsilon_0\dot{\phi}) \qquad (3.4a)$$

and

$$\Box^2\mathbf{A} = -\mu_0\mathbf{J}+\nabla(\nabla \cdot \mathbf{A}+\mu_0\varepsilon_0\dot{\phi}) \qquad (3.4b)$$

where the d'Alembertian

$$\Box^2 = \nabla^2-(1/c^2)(\partial^2/\partial t^2) = \nabla^2-\mu_0\varepsilon_0(\partial^2/\partial t^2).$$

The potentials **A** and ϕ are to some extent arbitrary in that the addition of any gradient to **A** leaves **B** unchanged. Thus if we make the *gauge* transformation

$$\mathbf{A} \rightarrow \mathbf{A}' = \mathbf{A}+\nabla\psi \qquad (3.5a)$$

$$\phi \rightarrow \phi' = \phi-\dot{\psi}, \qquad (3.5b)$$

both **B** and **E**, the physical quantities, are unchanged. We can use this to simplify the equations (3.4a) and (3.4b) for under the gauge transformation (3.5)

$$\nabla \cdot \mathbf{A}+\mu_0\varepsilon_0\dot{\phi} \rightarrow \nabla \cdot \mathbf{A}+\mu_0\varepsilon_0\dot{\phi}+\Box^2\psi, \qquad (3.6)$$

and since the function ψ is arbitrary it can be chosen so that (3.6) vanishes. In other words, we can restrict **A** and ϕ by the further condition

$$\nabla \cdot \mathbf{A}+\mu_0\varepsilon_0\dot{\phi} = 0 \qquad (3.7)$$

without conflicting with our earlier results. With this restriction, we are working in the

Lorentz gauge and ϕ and $\mathbf{A}$ satisfy the inhomogeneous wave equations

$$\Box^2\phi = -\rho/\varepsilon_0 \tag{3.8a}$$

$$\Box^2\mathbf{A} = -\mu_0\mathbf{J}. \tag{3.8b}$$

The Retarded Potentials

The static equations

$$\nabla^2\phi = -\rho/\varepsilon_0$$

$$\nabla^2\mathbf{A} = -\mu_0\mathbf{J}$$

are satisfied by

$$\phi(\mathbf{r}_1, t_1) = \iiint \frac{\rho(\mathbf{r}_2, t_1)\,\mathrm{d}V_2}{4\pi\varepsilon_0 r_{12}}$$

$$\mathbf{A}(\mathbf{r}_1, t_1) = \iiint \frac{\mu_0\mathbf{J}(\mathbf{r}_2, t_1)\,\mathrm{d}V_2}{4\pi r_{12}}$$

and we now show that (3.8a) is satisfied by

$$\phi(\mathbf{r}_1, t_1) = \iiint \frac{\rho(\mathbf{r}_2, t_2)\,\mathrm{d}V_2}{4\pi\varepsilon_0 r_{12}} \tag{3.9a}$$

with

$$t_2 = t_1 \mp (r_{12}/c), \tag{3.10}$$

so that by inference (3.8b) is satisfied by

$$\mathbf{A}(\mathbf{r}_1, t_1) = \iiint \frac{\mu_0\mathbf{J}(\mathbf{r}_2, t_2)\,\mathrm{d}V_2}{4\pi r_{12}}. \tag{3.9b}$$

The proof is simple. The integrand in (3.9a) has a singularity at $\mathbf{r}_1 = \mathbf{r}_2$. In the immediate vicinity of this point the difference $t_2 - t_1$ can be neglected and so the singularity contributes $-[\rho(\mathbf{r}_1, t_1)/\varepsilon_0]$ to $\nabla^2\phi$. Away from this point there are no singularities and so

$$\nabla^2\phi(\mathbf{r}_1, t_1) = \iiint \nabla_1^2 \left\{ \frac{\rho[\mathbf{r}_2, t_1 \pm (r_{12}/c)]}{4\pi\varepsilon_0 r_{12}} \right\} \mathrm{d}V_2.$$

The function on which ∇_1^2 operates is, as far as its dependence on r_1 goes, a scalar function of r_{12} the distance from an "origin" at r_2 and so we can use the relation

$$\nabla^2 f(r) = \frac{1}{r^2}\frac{\partial}{\partial r}\left(r^2\frac{\partial f}{\partial r}\right) = \frac{1}{r}\frac{\partial^2}{\partial r^2}(rf(r)).$$

This yields

$$\nabla_1^2\phi = \iiint \frac{1}{4\pi\varepsilon_0 r_{12}}\frac{\partial^2}{\partial r_{12}^2}\rho\left(\mathbf{r}_2, t_1 \mp \frac{r_{12}}{c}\right)\mathrm{d}V_2$$

and so

$$\nabla_1^2\phi = \frac{1}{c^2}\frac{\partial^2}{\partial t_1^2}\iiint \frac{\rho(\mathbf{r}_2, t_1 \mp r_{12}/c)\,\mathrm{d}V_2}{4\pi\varepsilon_0 r_{12}}.$$

Potentials

The charge in the infinitesimally immediate vicinity $\mathbf{r}_2 \to \mathbf{r}_1$ of the field point $\mathbf{r}_1$ does not make a finite contribution to ϕ or to $\partial^2\phi/\partial t^2$ and so this last result is

$$\nabla^2\phi = \frac{1}{c^2}\frac{\partial^2\phi}{\partial t^2}.$$

When we add the contribution to $\nabla^2\phi$ from the singularity at $\mathbf{r}_1$ we obtain (3.8a).

The solutions with $t_2 = t_1-(r_{12}/c)$ are known as *retarded* potentials and those with $t_2 = t_1+(r_{12}/c)$ as *advanced* potentials. Both types of potential satisfy the wave equations and are mathematically admissible. However, in the context of macroscopic or even atomic physics, incoming fields drive charges and currents but charges and currents produce only outgoing fields. Thus only the retarded potentials which refer to charges and currents at earlier times correspond to our notions of causality. It is usual, and we follow this practice, to ignore the advanced potentials.

The potentials in the Lorentz gauge (3.7) reduce for slowly varying systems to the static potentials. The fields due to a localized charge or current distribution near the origin then fall off, at great distances R, as R^{-2} or faster. If however the charge distribution varies rapidly with time, then although $\mathbf{A}$ still falls off as R^{-1} we have in $\mathbf{E}$ a term $-\dot{\mathbf{A}}$ which also decays as R^{-1}. It is possible to verify that there is a similar term in $\mathbf{B}$ but we shall not give the calculation since it is easier still to see that it is required by

$$\nabla \wedge \mathbf{B} = \mu_0\varepsilon_0\dot{\mathbf{E}}.$$

Thus in a system oscillating with a period τ the asymptotic behaviour of $\mathbf{E}$ and $\mathbf{B}$ at a distance R is as $(R\tau)^{-1}$. This part of the field is responsible for radiation.

If we assume a time dependence of the form exp $(j\omega t)$ the Lorentz gauge condition gives

$$\phi = -(1/j\omega\mu_0\varepsilon_0)\nabla\cdot\mathbf{A}$$

and so we can dispense with ϕ and work with $\mathbf{A}$ alone:

$$\mathbf{B} = \nabla \wedge \mathbf{A}, \qquad \mathbf{E} = -j\omega\mathbf{A}+(1/j\omega\mu_0\varepsilon_0)\nabla(\nabla\cdot\mathbf{A}). \tag{3.11}$$

In many radiation problems, especially those associated with radio aerials, the dominant term in $\mathbf{E}$ arises from $-j\omega\mathbf{A}$ and this is often used to simplify aerial calculations. This approximate result with $\mathbf{A}$ given by (3.9b) must not, however, be confused with an exact result which we now consider.

The Coulomb Gauge

The Lorentz gauge (3.7) is not the only possible choice for a constraint on $\mathbf{A}$ and ϕ. It is also possible to choose ψ in (3.5a) so that $\nabla\cdot\mathbf{A} = 0$. The potentials now satisfy

$$\nabla^2\phi = -\rho/\varepsilon_0 \tag{3.12a}$$

$$\Box^2\mathbf{A} = -\mu_0\mathbf{J}+\mu_0\varepsilon_0\nabla\dot{\phi}. \tag{3.12b}$$

This gauge with $\nabla\cdot\mathbf{A} = 0$ is variously known as the Coulomb gauge, the transverse gauge, or the radiation gauge.

In the Coulomb gauge it is useful to divide the current $\mathbf{J}$ into a transverse part $\mathbf{J}_t$ and a longitudinal part $\mathbf{J}_l$ so that

$$\mathbf{J} = \mathbf{J}_t+\mathbf{J}_l, \qquad \nabla\cdot\mathbf{J}_t = 0, \qquad \nabla\wedge\mathbf{J}_l = 0. \tag{3.13}$$

By expressing ϕ in terms of ρ using the integral of (3.12a) and then using $\nabla \cdot \mathbf{J}+\dot{\rho} = 0$, the continuity equation, we obtain

$$\varepsilon_0 \nabla \dot{\phi} = \mathbf{J}_l, \tag{3.14}$$

so that (3.12b) yields

$$\Box^2 \mathbf{A} = -\mu_0 \mathbf{J}_t \tag{3.15}$$

and therefore the vector potential in the Coulomb gauge is

$$\mathbf{A}(\mathbf{r}_1, t_1) = \iiint \frac{\mu_0 \mathbf{J}_t(\mathbf{r}_2, t_2)\, \mathrm{d}V_2}{4\pi r_{12}}. \tag{3.16}$$

Equation (3.12a) yields an unretarded scalar potential whose contribution to **E**, taken alone, would lead to instantaneous action at a distance. If, however, in a localized distribution, $\dot{\rho} = -\nabla \cdot \mathbf{J}$ is non-zero, **J** cannot (see Appendix, section 9) be decomposed into parts $\mathbf{J}_l$ and $\mathbf{J}_t$ which vanish outside the distribution. Thus **A** given by (3.16) is not solely a retarded potential but, everywhere, contains an unretarded component related to the local value of $\mathbf{J}_t$. Initially the contribution this makes to **E** just cancels the unretarded term derived from ϕ. In practice, calculations in the Coulomb gauge are very involved unless $\nabla \cdot \mathbf{J}$ is identically zero. Nevertheless, because it makes a formal separation between the electrostatic and the radiation fields, it is widely used in quantum electrodynamics even though the gauge condition itself is not Lorentz invariant.

From our point of view the most useful result of this chapter is the demonstration that the fields at a distance fall off as $1/R^2$ in the static case and $1/R\tau$ or ω/R in the time-dependent case. We shall, however, in the next chapter use the potentials to discuss the multipole moments of a localized charge distribution.

The general significance of the potentials, that they lead to simpler equations and also lend themselves better than the fields **E** and **B** to a Lagrangian or Hamiltonian formulation of electrodynamics, does not concern us. The greater convenience of this formulation is not, by itself, a sufficient reason for regarding the potentials as more than useful adjuncts in the development of the theory of the field.

Problems

1. Find the vector potential in the Coulomb gauge for the magnetic field produced by a circular loop of wire carrying a current I.

2. Find the fields described by a vector potential with the single component

$$A_x(\mathbf{r}, t) = A \exp j(\omega t - \beta z).$$

What is the time average of the z component of $\mathbf{E} \wedge \mathbf{B}$?

3. A charge is situated at the origin for $t < 0$. Between $t = 0$ and $t = \tau$ it moves at constant velocity to the point $\mathbf{r} = (0, 0, a)$. For $t > \tau$ it remains at rest at this point. Discuss the behaviour of the distant potentials **A** and ϕ in the Coulomb gauge.

4. Use equation (3.16) to derive Neumann's formula for the mutual inductance of two loops of wire.

5. Show that a localized distribution of charge and current, in which $\nabla \wedge \mathbf{J} \equiv 0$, cannot radiate.

CHAPTER 4

Multipole Moments

Introduction

Many of the most important problems in electromagnetism are concerned with small localized systems such as radio transmitters and receivers, permanent magnets, and atoms. We are interested both in the fields these systems produce at a distance and in the effect on these systems of fields due to distant charges. Situations of this type are most conveniently discussed in terms of the moments of the charge and current distributions within the system. If the charge distribution is stationary so that $\nabla \cdot \mathbf{J} = -\dot{\rho} = 0$, the moments can be unambiguously classified as electric or magnetic and all electric effects are due to electric moments and magnetic effects to magnetic moments. In the general case of a time-dependent system a formal classification into electric and magnetic moments is still possible but it is not unique and obviously magnetic fields may now be due both to magnetic moments and to changing electric moments. A general and elegant way of discussing and classifying the moments in terms of the behaviour and symmetry of the fields at a distance is given by Jackson (1962) but for our purposes this is unnecessary since we shall only be concerned with the first three electric moments (charge, dipole, and quadrupole) and the first non-vanishing magnetic moment, the dipole. We can therefore proceed in a rather more *ad hoc* fashion, beginning with the static electric moments.

The Electric Dipole

We suppose that we have a localized static distribution of charge density ρ in a static field $\mathbf{E} = -\nabla\phi$ due to distant charges. The energy of the distribution in the field is

$$W = \int \rho\phi \, dV. \tag{4.1}$$

We now assume that ϕ varies slowly over the distribution so that if we expand ϕ in a Taylor series about a fixed point $\mathbf{R}$ in the distribution (see Fig. 4.1) only the first few terms need

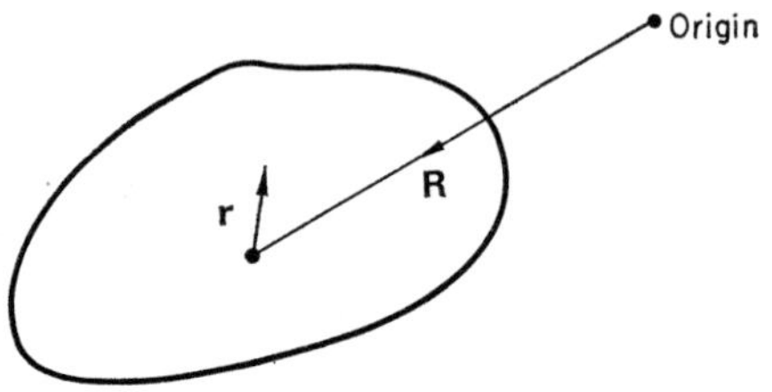

FIG. 4.1. The vectors used in equation (4.2).

be retained. We thus have

$$\phi(\mathbf{R}+\mathbf{r}) = \phi(\mathbf{R})+\mathbf{r}\,.\,\nabla_R\phi(\mathbf{R})+\tfrac{1}{2}r_ir_j\frac{\partial^2\phi(\mathbf{R})}{\partial R_i\partial R_j}+\text{etc.,} \tag{4.2}$$

or

$$\phi(\mathbf{R}+\mathbf{r}) = \phi(\mathbf{R})-\mathbf{r}\,.\,\mathbf{E}(\mathbf{R})-\tfrac{1}{2}r_ir_j\frac{\partial E_j(\mathbf{R})}{\partial R_i}+\text{etc.,} \tag{4.3}$$

where a double suffix summation is implied in the last terms of (4.2) and (4.3).

The Electric Multipoles

If we now define the multipole moments of the distribution as the charge,

$$q = \int \rho\,\mathrm{d}V, \tag{4.4}$$

the dipole moment,

$$\mathbf{p} = \int \rho\mathbf{r}\,\mathrm{d}V, \tag{4.5}$$

and the quadrupole moment,

$$Q_{ij} = \tfrac{1}{2}\int r_ir_j\rho\,\mathrm{d}V, \tag{4.6}$$

which in dyadic notation is

$$\underline{\underline{\mathbf{Q}}} = \tfrac{1}{2}\int \mathbf{rr}\rho\,\mathrm{d}V, \tag{4.7}$$

we can express the energy as

$$W = q\phi(\mathbf{R})-\mathbf{p}\,.\,\mathbf{E}(\mathbf{R})-Q_{ij}\frac{\partial E_j(\mathbf{R})}{\partial R_i}+\text{etc.} \tag{4.8}$$

We note that if q vanishes $\mathbf{p}$ is independent of the origin $\mathbf{R}$ of the expansion and if q and $\mathbf{p}$ vanish Q_{ij} is independent of $\mathbf{R}$. In general only the lowest non-vanishing moment is unique unless $\mathbf{R}$ is specified. Thus, unless there is a natural origin for the system, e.g. the nucleus in an atom, a multipole moment is only unique if all lower moments vanish.

In a neutral system with $q = 0$, the leading term in W is

$$W = -\mathbf{p}\,.\,\mathbf{E} \tag{4.9}$$

and if $\mathbf{p}$ is independent of $\mathbf{E}$ the force acting on the system is

$$\mathbf{F} = \nabla(\mathbf{p}\,.\,\mathbf{E}) = (\mathbf{p}\,.\,\nabla)\mathbf{E}. \tag{4.10}$$

In a uniform field the force vanishes but there is a couple tending to orient $\mathbf{p}$ parallel to $\mathbf{E}$. If θ is the angle between $\mathbf{p}$ and $\mathbf{E}$ the magnitude of the couple is $\Gamma = pE\sin\theta$. If $\mathbf{E}$ is along the y axis and $\mathbf{p}$ along the x axis the couple acts in a positive sense about the z axis and so we have

$$\mathbf{\Gamma} = \mathbf{p} \wedge \mathbf{E}. \tag{4.11a}$$

If q and $\mathbf{p}$ vanish the simplest mechanical effect due to Q_{ij} is a couple in a non-uniform field. We note that (4.9) is valid only if $\mathbf{p}$ is independent of $\mathbf{E}$ and we shall take up this question at a later stage.

The Magnetic Dipole

If we have a stationary system of currents of density $\mathbf{J}$ in a uniform magnetic field $\mathbf{B}$, then using the coordinate system of Fig. 4.1, the couple acting on the system is

$$\mathbf{\Gamma} = \int \mathbf{r} \wedge (\mathbf{J} \wedge \mathbf{B})\,\mathrm{d}V.$$

Multipole Moments

Since $\mathbf{V} \cdot \mathbf{J} = 0$ we can express $\mathbf{J}$ as $\mathbf{V} \wedge \mathbf{M}$ and put $\mathbf{M} = 0$ outside the distribution (see Appendix, sections 7 and 9). We then find that

$$\mathbf{\Gamma} = -\mathbf{B} \wedge \int \mathbf{M}\, \mathrm{d}V,$$

and further that

$$\int \mathbf{M}\, \mathrm{d}V = \tfrac{1}{2} \int \mathbf{r} \wedge \mathbf{J}\, \mathrm{d}V.$$

Thus if we define the magnetic dipole moment of the system as

$$\mathbf{m} = \tfrac{1}{2} \int \mathbf{r} \wedge \mathbf{J}\, \mathrm{d}V, \tag{4.12}$$

we have

$$\mathbf{\Gamma} = \mathbf{m} \wedge \mathbf{B} \tag{4.11b}$$

which is of the same form as the electric result.

The result (Appendix, section 6)

$$\int \mathbf{J}\, \mathrm{d}V = -\int \mathbf{r}(\mathbf{V} \cdot \mathbf{J})\, \mathrm{d}V + \oint \mathbf{r}(\mathbf{J} \cdot \mathrm{d}\mathbf{S})$$

applied to any volume with a closed surface S outside the distribution yields with $\mathbf{V} \cdot \mathbf{J} = -\dot{\rho}$, since the surface integral vanishes

$$\int \mathbf{J}\, \mathrm{d}V = \int \mathbf{r}\dot{\rho}\, \mathrm{d}V = \dot{\mathbf{p}}$$

so that any lower moment we might attempt to form would be zero for a static distribution and equal to $\dot{\mathbf{p}}$ for a time-dependent distribution. We note in passing that the scalar $\int \mathbf{r} \cdot \mathbf{J}\, \mathrm{d}V$ vanishes if $\mathbf{V} \cdot \mathbf{J} = 0$.

The Multipole Potentials

If we now change our coordinate system to that shown in Fig. 4.2 with an origin in the

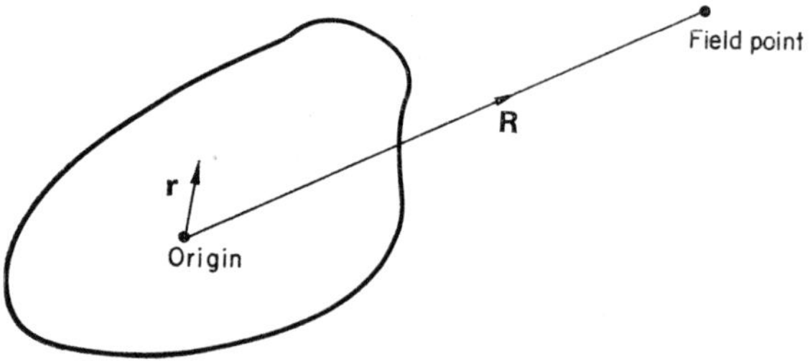

FIG. 4.2. The vectors used in equation (4.13).

distribution, the electric potential at the field point $\mathbf{R}$ due to static charges is

$$\phi(\mathbf{R}) = \frac{1}{4\pi\varepsilon_0} \int \frac{\rho(\mathbf{r})\, \mathrm{d}V}{|\mathbf{R}-\mathbf{r}|} \tag{4.13}$$

and, if we expand the denominator of the integrand as a Taylor series, i.e.

$$|\mathbf{R}-\mathbf{r}|^{-1} = \frac{1}{R} - \mathbf{r} \cdot \mathbf{V}\left(\frac{1}{R}\right) + \tfrac{1}{2} r_i r_j \frac{\partial^2}{\partial R_i \partial R_j}\left(\frac{1}{R}\right) + \text{etc.}, \tag{4.14}$$

we can express $\phi(\mathbf{R})$ as

$$\phi(\mathbf{R}) = \frac{1}{4\pi\varepsilon_0}\left\{\frac{q}{R} - \mathbf{p}\,.\,\nabla\left(\frac{1}{R}\right) + Q_{ij}\frac{\partial^2}{\partial R_i \partial R_j}\left(\frac{1}{R}\right) + \text{etc.}\right\} \tag{4.15}$$

in terms of the same electric multipole moments. The lth moment is called a 2^l pole with 2^l expressed in classical Greek. The potential due to the lth moment decreases with distance as $R^{-(l+1)}$ and varies with the linear dimensions of the distribution as r^l. At great distances $R \gg r$ the potential is dominated by the first non-vanishing term. The static electric field falls off as $(1/R^2)(r/R)^l$, but, as we shall see in a time-dependent system of frequency f, period $\tau = 1/f$ associated with a wavelength $\lambda = c/f$, the fields decay as $(1/R)(r/\lambda)^l$ with $l \geqslant 1$.

Since $\nabla^2(1/R) = 0$ except at $R = 0$, the field outside the distribution is derived from a potential $\phi(R)$ expressed as a sum of terms

$$\frac{q}{R}, \qquad -\mathbf{p}\,.\,\nabla\left(\frac{1}{R}\right), \qquad Q_{ij}\frac{\partial}{\partial R_i}\frac{\partial}{\partial R_j}\left(\frac{1}{R}\right) \ldots,$$

each of which satisfies Laplace's equation. In most texts the electric quadrupole moment is defined not by (4.6) but by

$$Q'_{ij} = \tfrac{1}{2}\int (r_i r_j - \tfrac{1}{3}r^2\delta_{ij})\rho \, \mathrm{d}V;$$

this has the advantage that Q_{ij} can be directly related to the spherical harmonics $Y_{2m}(\cos\theta, \phi)$ and satisfies $Q'_{11} + Q'_{22} + Q'_{33} = 0$. Both these two different definitions of the quadrupole moment, as either Q_{ij} or Q'_{ij}, are, for our purposes, entirely equivalent. We shall continue to use the definition (4.6) although we note that a quadrupole moment of the form $Q_{ij} = k\delta_{ij}$ does not lead to an external field, and we can use in equation (4.15) whichever definition of the quadrupole moment is most convenient for a particular problem.

The magnetic vector potential, using the coordinate system of Fig. 4.2, due to a solenoidal current with $\nabla\,.\,\mathbf{J} = 0$ so that $\mathbf{J} = \nabla \wedge \mathbf{M}$ with $\mathbf{M} = 0$ outside the distribution, is

$$\mathbf{A}(\mathbf{R}) = \frac{\mu_0}{4\pi}\int \frac{\mathbf{J}(\mathbf{r})\,\mathrm{d}V}{|\mathbf{R}-\mathbf{r}|} = \frac{\mu_0}{4\pi}\int \frac{\nabla \wedge \mathbf{M}(\mathbf{r})\,\mathrm{d}V}{|\mathbf{R}-\mathbf{r}|}.$$

A partial integration yields a vanishing surface term and leaves

$$\mathbf{A}(\mathbf{R}) = \frac{\mu_0}{4\pi}\int \mathbf{M}(\mathbf{r}) \wedge \nabla_r \frac{1}{|\mathbf{R}-\mathbf{r}|}\,\mathrm{d}V.$$

The leading term in this expression, since

$$\nabla_r \frac{1}{|\mathbf{R}-\mathbf{r}|} = -\nabla_R \frac{1}{|\mathbf{R}-\mathbf{r}|},$$

is

$$\mathbf{A}(\mathbf{R}) = \frac{\mu_0}{4\pi}\nabla_R\left(\frac{1}{R}\right) \wedge \int \mathbf{M}\,\mathrm{d}V$$

or

$$\mathbf{A}(\mathbf{R}) = \frac{\mu_0}{4\pi}\nabla_R\left(\frac{1}{R}\right) \wedge \mathbf{m}. \tag{4.16}$$

Multipole Moments

The Dipole Fields

We now look at the fields due to electric and magnetic dipoles, beginning with the electric dipole.

The electrostatic potential due to an electric dipole at the origin is from (4.15)

$$\phi = -\frac{1}{4\pi\varepsilon_0}\,\mathbf{p}\,.\,\nabla\left(\frac{1}{R}\right) = \frac{\mathbf{p}\,.\,\mathbf{R}}{4\pi\varepsilon_0 R^3}\,. \tag{4.17}$$

Notice that it can be written as $\mathbf{p}\,.\,\mathbf{E}_0$ where $\mathbf{E}_0$ is the field due to a unit charge at the origin, a result which often simplifies electrostatic calculations. The corresponding electric field is

$$\mathbf{E} = -\nabla\phi = -\nabla\left(\frac{\mathbf{p}\,.\,\mathbf{R}}{4\pi\varepsilon_0 R^3}\right). \tag{4.18}$$

If we write this as $\mathbf{E} = (1/4\pi\varepsilon_0)\nabla\{\mathbf{p}\,.\,\nabla(1/R)\}$ and note that $\nabla \wedge \mathbf{p} = 0$, $\nabla\mathbf{p} = 0$, $\nabla \wedge [\nabla(1/R)] = 0$, the identity (see Appendix)

$$\nabla(\mathbf{A}\,.\,\mathbf{B}) = (\mathbf{A}\,.\,\nabla)\mathbf{B}+(\mathbf{B}\,.\,\nabla)\mathbf{A}+\mathbf{A}\wedge(\nabla\wedge\mathbf{B})+\mathbf{B}\wedge(\nabla\wedge\mathbf{A})$$

yields

$$\mathbf{E} = \frac{1}{4\pi\varepsilon_0}(\mathbf{p}\,.\,\nabla)\nabla\left(\frac{1}{R}\right) = \frac{3\mathbf{R}(\mathbf{p}\,.\,\mathbf{R})-R^2\mathbf{p}}{4\pi\varepsilon_0 R^5}\,. \tag{4.19}$$

It is sometimes convenient to write this as

$$E_i = \Gamma_{ij}p_j, \tag{4.20}$$

where

$$\Gamma_{ij}(\mathbf{R}) = \frac{3R_iR_j-R^2\delta_{ij}}{4\pi\varepsilon_0 R^5}\,. \tag{4.21}$$

The interaction energy between two dipoles $\mathbf{p}$ at the origin and $\boldsymbol{\pi}$ at $\mathbf{R}$ is

$$W = -p_i\Gamma_{ij}(\mathbf{R})\pi_j, \tag{4.22}$$

which is symmetric in the coordinates of the dipoles since $\Gamma(\mathbf{R}) = \Gamma(-\mathbf{R})$. If we have a set of equal dipoles distributed isotropically, or on a simple cubic lattice, in a sphere centred on the origin which is a lattice point, the field at the origin due to all dipoles except that at the origin is

$$E_i = p_j \sum_{n\neq 0} \Gamma_{ij}\{\mathbf{R}(n)\}$$

and is zero. This result is important in the theory of dielectrics.

The magnetic field $\mathbf{B}$ due to a dipole $\mathbf{m}$ is, from (4.16),

$$\mathbf{B} = \nabla\wedge\mathbf{A} = \frac{\mu_0}{4\pi}\nabla\wedge[\nabla(1/R)\wedge\mathbf{m}].$$

We use the identity

$$\nabla\wedge(\mathbf{F}\wedge\mathbf{G}) = \mathbf{F}(\nabla\,.\,\mathbf{G})-\mathbf{G}(\nabla\,.\,\mathbf{F})+(\mathbf{G}\,.\,\nabla)\mathbf{F}-(\mathbf{F}\,.\,\nabla)\mathbf{G}$$

and, noting that $\nabla\,.\,\nabla(1/R) = 0$ and derivatives of $\mathbf{m}$ are zero, obtain

$$\mathbf{B} = \frac{\mu_0}{4\pi}(\mathbf{m}\,.\,\nabla)\nabla\left(\frac{1}{R}\right). \tag{4.23}$$

This is of the same form as the electric dipole field (4.19) and so

$$\mathbf{B} = \frac{\mu_0}{4\pi}\frac{3\mathbf{R}(\mathbf{m}\,.\,\mathbf{R})-R^2\mathbf{m}}{R^5}. \tag{4.24}$$

Notice that either type of dipole field can be represented as either the gradient of a scalar or the curl of a vector. Thus, in general terms, with a dipole **f** producing a field **F** we have

$$\mathbf{F} = -\nabla\psi = \nabla\left\{\frac{1}{4\pi}\mathbf{f}\,.\,\nabla\left(\frac{1}{R}\right)\right\} \tag{4.25a}$$

or

$$\mathbf{F} = \nabla\wedge\mathbf{G} = -\nabla\wedge\left\{\frac{1}{4\pi}\mathbf{f}\wedge\nabla\left(\frac{1}{R}\right)\right\}. \tag{4.25b}$$

In deriving (4.24) we have however used the relation $\nabla^2(1/R) = 0$ which was not required for (4.19), thus the behaviour at $R \to 0$ may not be the same. Further if we replace **f** by

$$\mathbf{f} = \int \mathbf{f}'\gamma(\mathbf{R})\,\mathrm{d}V$$

where $\mathbf{f}'$ is a constant and $\gamma(\mathbf{R})$ is a sharply peaked function at $\mathbf{R} = 0$ we shall have further problems at the origin.

We have now defined a set of moments, which in component form are

$$q = \int \rho\,\mathrm{d}V \tag{4.26a}$$

$$p_i = \int \rho r_i\,\mathrm{d}V \tag{4.26b}$$

$$Q_{ij} = \tfrac{1}{2}\int \rho r_i r_j\,\mathrm{d}V \tag{4.26c}$$

$$m_i = \tfrac{1}{2}\int \varepsilon_{ijk} r_j J_k\,\mathrm{d}V, \tag{4.26d}$$

and up to order $r^2\rho$ or rJ there are only two more independent moments possible; these are $\int J_i\,\mathrm{d}V$ and $\frac{1}{2}\int (J_i r_j + J_j r_i)\,\mathrm{d}V$. We have, however, the relations

$$\int \frac{\partial}{\partial r_j}(r_i J_j)\,\mathrm{d}V = \int r_i J_j\,\mathrm{d}S_j = 0$$

and

$$\int \frac{\partial}{\partial r_k}(r_i r_j J_k)\,\mathrm{d}V = \int r_i r_j J_k\,\mathrm{d}S_k = 0,$$

and so

$$\int\left(J_i + r_i\frac{\partial J_j}{\partial r_j}\right)\mathrm{d}V = \int (J_i - r_i\dot{\rho})\,\mathrm{d}V = 0$$

and

$$\int\left(r_i J_j + r_j J_i + r_i r_j\frac{\partial J_k}{\partial r_k}\right)\mathrm{d}V = \int (r_i J_j + r_j J_i - r_i r_j\dot{\rho})\,\mathrm{d}V = 0.$$

Thus the remaining moments can be expressed as time derivatives of moments in the set (4.26), i.e.

$$\int J_i\,\mathrm{d}V = \dot{p}_i \tag{4.27}$$

$$\tfrac{1}{2}\int (r_i J_j + r_j J_i)\,\mathrm{d}V = \dot{Q}_{ij}. \tag{4.28}$$

Time-dependent Potentials

The moments q, $\mathbf{p}$, $\mathbf{m}$ and Q_{ij} therefore, up to this order, form a complete set for static problems and, with their time derivatives, for time-dependent problems. Thus, even though when we go to time-dependent problems we might feel impelled to re-examine our definitions of the moments we should in fact not discover any intrinsically new moments. We are therefore justified in using the definitions of the static moments in all problems.

In time-dependent problems it is convenient, indeed almost essential, to assume that the time dependence is harmonic as $\exp(j\omega t)$. We now have three lengths to compare, r the extent of the distribution, R the distance to the field point and λ the wavelength. We shall always assume that r/R and r/λ are small but not necessarily that R/λ is small or vice versa. Since, with a harmonic time dependence, we have $j\omega\mu_0\varepsilon_0\mathbf{E} = \nabla \wedge \mathbf{B}$ and $\mathbf{B} = \nabla \wedge \mathbf{A}$ we need not consider the scalar potential. The retarded vector potential in the Lorentz gauge is

$$\mathbf{A}(\mathbf{R}, t) = \frac{\mu_0}{4\pi}\int \mathbf{J}(\mathbf{r}, t)\frac{\exp(-jk|\mathbf{R}-\mathbf{r}|)}{|\mathbf{R}-\mathbf{r}|}\,\mathrm{d}V \tag{4.29}$$

and we can expand this as a power series in both $kr = 2\pi r/\lambda$ and r/R, retaining only terms of zero or first order in r. The result is

$$\mathbf{A}(\mathbf{R}, t) = \frac{\mu_0}{4\pi}\frac{\exp(-jkR)}{R}\left\{\int \mathbf{J}\,\mathrm{d}V + \left(\frac{1}{R^2}+\frac{jk}{R}\right)\int \mathbf{J}(\mathbf{R}\,.\,\mathbf{r})\,\mathrm{d}V\right\}. \tag{4.30}$$

We have

$$\int \mathbf{J}\,\mathrm{d}V = \dot{\mathbf{p}} = j\omega\mathbf{p}$$

and

$$\int J_i(\mathbf{R}\,.\,\mathbf{r})\,\mathrm{d}V = R_j\int J_i r_j\,\mathrm{d}V.$$

If we refer to (4.26d) and (4.28) we see that this term is simply $R_j\dot{Q}_{ij}+(\mathbf{m} \wedge \mathbf{R})_i$ and so the vector potential is, dropping the explicit time dependence,

$$\mathbf{A}(\mathbf{R}) = \frac{\mu_0}{4\pi}\frac{\exp(-jkR)}{R}\left\{j\omega\mathbf{p}+\left(\frac{1}{R^2}+\frac{jk}{R}\right)(\mathbf{m} \wedge \mathbf{R}+j\omega R_j Q_{ij})\right\}. \tag{4.31}$$

The static fields are obtained by taking the limit $\omega = kc \to 0$ remembering that

$$\mathbf{E} = \frac{1}{j\omega\mu_0\varepsilon_0}\nabla \wedge \mathbf{B}$$

and we then obtain

$$\mathbf{B} = \nabla \wedge \left(\frac{\mu_0\mathbf{m} \wedge \mathbf{R}}{4\pi R^3}\right), \qquad \mathbf{E} = \frac{1}{4\pi\varepsilon_0}\nabla \wedge \left(\nabla \wedge \frac{\mathbf{p}}{R}\right),$$

dropping the term in Q for brevity. We leave it as an exercise for the reader to show that these are the normal static dipole fields.

The radiation field, when $R \gg \lambda$ or $kR \gg 1$, again dropping the quadrupole term, which we will deal with separately, is obtained from

$$\mathbf{A}(\mathbf{R}) = \frac{j\omega\mu_0}{4\pi}\frac{\exp(-jkR)}{R}\left\{\mathbf{p}+\frac{\mathbf{m} \wedge \mathbf{R}}{cR}\right\}. \tag{4.32}$$

We note that in the radiation field $E \sim cB$ and so we need only consider **B**. This is obtained from $\nabla \wedge \mathbf{A}$ as

$$\mathbf{B}(\mathbf{R}) = \frac{\mu_0 ck^2 \exp(-jkR)}{4\pi R}\left\{\mathbf{n} \wedge \mathbf{p} + \frac{(\mathbf{n} \wedge \mathbf{m}) \wedge \mathbf{n}}{c}\right\}, \tag{4.33}$$

where **n** is a unit vector $\mathbf{n} = \mathbf{R}/R$ in the direction of **R**. The contribution of the quadrupole term to the radiation field is

$$B_i = \frac{jk^3 \mu_0 c \exp(-jkR)}{4\pi R} \varepsilon_{ijk} Q_{kl} n_j n_l. \tag{4.34}$$

We see that the relative effects of **p**, **m** and Q are in the ratio p to m/c to kQ and that all the contributions give fields falling off as R^{-1}. Whether, for a given system, **p**, **m** or Q_{ij} is non-zero depends primarily on the symmetry of the system but, if we suppose that we are dealing with atoms of size r in which the electrons have velocities v, the allowed moments have magnitudes $p \sim er$, $m \sim evr$ and $Q \sim er^2$. Thus relative to the electric dipole term the magnetic dipole term is less by a factor v/c and the electric quadrupole term by a factor $kr \sim r/\lambda$. Notice also that if ω_0 is an atomic resonance frequency $v \sim \omega_0 r$ and so $v/c \sim \omega_0 r/c$, thus the magnetic dipole and electric quadrupole terms are in the ratio ω_0/ω. This is not surprising since m and $\dot{Q}$ are the anti-symmetric and symmetric parts of the same tensor $r_j J_i$.

The final case that we shall consider is the near field, i.e., the field in a region $r \ll R \ll \lambda$. In this region the main term in **B** due to **m** is of the order $\mu_0 m/R^3$ which we may compare with the near fields due to **p** and Q_{ij} omitting the radiation terms from these moments. These near fields are of order $(\mu_0 ck/R^3)p$ and $(\mu_0 ck/R^3)Q$, thus the effect of **m** will only dominate **B** if $kRcp < m$ and $kcQ < m$. In terms of an atom with $p \sim er$, $m \sim evr$, $Q \sim er^2$, $v \sim \omega_0 r$ we have $R < (v/c)\lambda \sim (\omega/\omega_0)r$ and $\omega < \omega_0$.

We may summarize these results as follows. The static magnetic field is solely due to **m**, but whenever $\dot{\mathbf{p}}$ and $\dot{Q}_{ij}$ are non-zero the magnetic field will be dominated by the effects of a changing **p** unless either $v/c \sim 1$ or $R < (v/c)\lambda$. At frequencies approaching a natural resonance frequency of the atom if we retain **m** we should also retain $\dot{Q}_{ij}$. These results apply to the non-resonant response of atoms to fields and are the reason why, in elementary treatments of optical properties, we almost always disregard all except electric dipole effects. They should not be confused with selection rules in atomic spectra. If an atomic transition involves no change in parity and a change in total angular momentum from 1 to 0 (triplet to singlet transitions for example) it will produce a radiation field of the same symmetry as the field due to a magnetic dipole and be classified as a magnetic dipole transition. The fact that magnetic dipole transitions are most readily observed at either very high frequencies in the u.v. or very low frequencies (r.f.) is more a manifestation of the laws of quantum mechanics and the sophistication of electronic apparatus than of fundamental electromagnetic laws. For a much fuller discussion the reader should consult Pershan (1967).

The Energy of a Dipole in a Field

We conclude this chapter by reverting to the question of the energy of a dipole **p** in a field **E** when **p** depends on **E**. The force acting on the dipole is undoubtedly given by $F_i = p_j(\partial/\partial r_j)E_i$ as we can see by considering the component charges of **p**. If we imagine first setting up the impressed field **E** by doing work W_0 to assemble the distant charges and then introducing the dipole along some path dr_i the total energy expended by external sources in assembling the system is

$$W = W_0 - \int F_i \, dr_i$$

or

$$W = W_0 - \int p_j \left(\frac{\partial}{\partial r_j} E_i \right) dr_i.$$

Now we may bring the dipole in along any path we choose as long as it comes from a region so far away that $\mathbf{E} = 0$. We choose a path parallel to the r_1 axis and so

$$W = W_0 - \int p_1 \frac{\partial E_1}{\partial r_1} dr_1 - \int p_2 \frac{\partial E_1}{\partial r_2} dr_1 - \int p_3 \frac{\partial E_1}{\partial r_3} dr_1.$$

Since $\nabla \wedge \mathbf{E} = 0$ we have

$$\frac{\partial E_1}{\partial r_2} = \frac{\partial E_2}{\partial r_1} \quad \text{and} \quad \frac{\partial E_1}{\partial r_3} = \frac{\partial E_3}{\partial r_1}$$

so that finally

$$W = W_0 - \int_0^E \mathbf{p} \cdot d\mathbf{E}. \tag{4.35}$$

If **p** is independent of **E** this reduces to $W_0 - \mathbf{p} \cdot \mathbf{E}$.

The expression $-\mathbf{p} \cdot \mathbf{E}$ gives the energy of a rigid distribution of charge in an impressed field **E** due to the interaction of the charges with the electric field. It also gives the interaction energy of a flexible charge distribution whose moment is **p** when the field is **E**. However as **E** is increased from zero to its final value, or the distribution is moved from one region to another, the distortion which occurs in the charge distribution results in the storage of energy in forms which are not included in the electrostatic calculation; these forms may actually be wholly or partly electrostatic on an atomic scale but, from the point of view of the calculation, can be regarded as mechanical energy. Thus, for example, if the actual system under discussion consists of charged bodies mounted on springs, although the energy stored in the distortion of a spring is ultimately related to the electrostatic forces between electrons and nuclei we should have no hesitation in describing it as mechanical energy. Equation (4.35) gives the total electrostatic and mechanical energy of the system as a function of **E** and therefore should be used to calculate the force on the system.

$$\mathbf{F} = -\nabla W = (\mathbf{p} \cdot \nabla)\mathbf{E}. \tag{4.36}$$

Problems

1. The first non-vanishing moment of an atom is the electric quadrupole moment defined by equation (4.6). In a principal axis system this is a diagonal tensor with elements Q_{ij}. Show that the force acting on the atom in a field **E** is

$$F_i = \sum_j Q_{jj} \frac{\partial^2 E_i}{\partial r_j^2}.$$

Show also that if the electric charge distribution in the atom has spherical symmetry the force vanishes.

2. Charges $+q$ are placed at the vertices of a regular tetrahedron and a charge $-4q$ at the centre. Show that the electric quadrupole moment is zero.

3. A set of electrodes produces a potential $\phi = Axy$ near the origin. What force acts on an atom placed at $x = a$ if its induced dipole moment is $\mathbf{p} = \alpha\varepsilon_0\mathbf{E}$?

4. A current I circulates in an anchor ring of radius r. What is the magnetic dipole moment of this system?

5. The induced dipole moment of an atom is given by $\mathbf{p} = \alpha\varepsilon_0\mathbf{E}$. Find the couple acting on the atom in a field $\mathbf{E}$.

6. Two charges $+q$ and $-q$ are held at an equilibrium separation r_0 by a spring. The internal potential energy of the system varies with separation as $\frac{1}{2}k(r-r_0)^2$. The system is introduced into a region of uniform field E parallel to the length of the system. Calculate the work done in introducing the system and the change in the internal potential energy of the system.

CHAPTER 5

Microscopic and Macroscopic Fields

Introduction

In this and the next three chapters we shall be concerned with the nature of the relation between the continuum equations of macroscopic electromagnetism and the atomic picture of matter as an assembly of interacting discrete point-like charges. We need to investigate this relation for two reasons, firstly because it is an essential step in relating empirical macroscopic electromagnetic phenomena to atomic theory and secondly because a misunderstanding of this relation can make it almost impossible for us to achieve a lively and intuitive comprehension of the significance of the vectors **H** and **D** or **M** and **P** which play a central role in macroscopic theory. Our central task, then, is to show how the macroscopic equations arise from a particle approach to electromagnetism and in the course of this we shall also discover the criteria which determine when a particular form of the macroscopic equations is appropriate to a particular problem.

It is often stated that the macroscopic equations are an average of the microscopic particle equations, and that this average is virtually exact because of the exceeding smallness of the elementary charge and the vast number of elementary charges in any macroscopically significant region of space. We have already given examples in the preface and chapter 1, which indicate that this view is inadequate. It is true that in many cases macroscopic results are obtained as average results, the averaging being over a statistical or quantum mechanical specification of the state of a macroscopic system, but we also know that in some cases the macroscopic results are sufficiently well resolved for us to observe fluctuations and noise of statistical or quantum mechanical origin. In other words different states of a system belonging to the same statistical ensemble lead to distinct macroscopic results. Thus, whatever process of averaging leads from the microscopic or atomic equations to the macroscopic or continuum equations cannot depend in any fundamental way on the statistical description of the system under discussion. The process must be inherent in the very notion of a macroscopic calculation and be related to the use we propose to make of the calculation.

Now, by a macroscopic calculation, we mean a calculation of finite resolution yielding results which are precise to this resolution and no more. Thus in a macroscopic calculation there will be features of the system, on a scale related to the resolution, which cannot appear explicitly in the final answer. To make this clear we give one example. If we consider the optical properties of a crystal with atoms of polarizability α on a cubic lattice of spacing a_0 at a wavelength λ, then the properties of the crystal will be determined by α/a_0^3 and, as long as $a_0 \ll \lambda$, it will make no difference if we decrease a_0 by a factor $\frac{1}{2}$ and α by a factor $\frac{1}{8}$. It is in this sense that the macroscopic results are insensitive to the microscopic details of the structure of the system. Our problem then is to describe the macroscopic averaging process in such a way that the macroscopic equations which result retain only information of macroscopic relevance about the microscopic structure. At the same time we have to ensure that the process retains all the information that we can conceivably use in the macroscopic problem under discussion. In particular if, because the system is macroscopically and incompletely specified, the actual microscopic configuration is only known to be one of the many possible configurations of a statistical ensemble, we must also ensure that our macroscopic average does not eliminate possible real statistical fluctuations in the final output of the calculation. We have, in addition, to make sure that our macroscopic averaging process is compatible with any other statistical averages that we may have to form at later stages. Finally, of course, we should like to devise an explicit procedure which leads directly to the conventional macroscopic continuum equations and provides us with a simple physical interpretation of the symbols in these equations expressed in atomic terms.

In this chapter we describe a formal procedure for smoothing the particle equations and thus obtaining continuum equations which in any defined macroscopic context can be treated as exact equations. In chapter 6, we consider how this process is related to the formation of true statistical or ensemble averages, and in chapter 7, we see how these notions are related to the use of hydrodynamic continuum equations to describe the macroscopic flow of electricity. Finally in chapter 8, we consider the effects of bound charge in matter and show how these effects can be represented in terms of polarization or magnetization vectors **P** or **M** which, in atomic terms, can be understood as dipole moment densities.

The Microscopic Fields

If we regard all electromagnetic phenomena to be due to the effects of point charges, the appropriate field equations will be

$$\begin{gathered}\nabla_R.\mathbf{b}(\mathbf{R}) = 0, \qquad \nabla_R \wedge \mathbf{e}(\mathbf{R})+\frac{\partial}{\partial t}\mathbf{b}(\mathbf{R}) = 0, \\ \nabla_R.\mathbf{e}(\mathbf{R}) = \sum_n q(n)\delta(\mathbf{r}(n)-\mathbf{R}), \\ \nabla_R \wedge \mathbf{b}(\mathbf{R})-\frac{\partial}{\partial t}\mathbf{e}(\mathbf{R}) = \sum_n q(n)\dot{\mathbf{r}}(n)\delta(\mathbf{r}(n)-\mathbf{R})\end{gathered} \tag{5.1}$$

where the point charges $q(n)$ are located at positions $\mathbf{r}(n)$ and, to save unnecessary symbols, we have temporarily set $\mu_0 = \varepsilon_0 = 1$. In this chapter we shall be considering the relation between these equations and the equations

$$\begin{gathered}\nabla.\mathbf{B} = 0, \qquad \nabla \wedge \mathbf{E}+\frac{\partial}{\partial t}\mathbf{B} = 0, \\ \nabla.\mathbf{E} = \rho, \qquad \nabla \wedge \mathbf{B}-\frac{\partial}{\partial t}\mathbf{E} = \mathbf{J}\end{gathered} \tag{5.2}$$

in which electricity is regarded as a continuous fluid. Not until chapter 8 will we need to introduce the auxiliary fields **D** and **H** or the polarization vectors **P** and **M**.

Although discussion of the relation between the microscopic field equations and the macroscopic, or continuum, equations is traditionally associated with the problem of reconciling an atomic picture of the constitution of matter with the form of the equations used in discussing the electromagnetic field in matter regarded as a continuous medium, the problem is, in fact, more general; for we often wish to use the continuum equations in systems such as electron beam tubes or plasmas where the polarization vectors **P** and **M** are not required. Thus an investigation of the conditions under which the two sets of equations (5.1) and (5.2) are equivalent is important even if we are not concerned with dielectric or magnetic media.

Our aim is not to assign logical priority to either set (5.1) or set (5.2), but to show that if we take (5.1) as the fundamental equations, then, in any macroscopic calculation, they are equivalent to (5.2). In this chapter we shall only consider the formal process which leads from (5.1) to (5.2). The relation between (5.2) and an ensemble average of (5.1) will be dealt with in the next chapter.

Equations (5.1) have one very unpleasing feature in that, if the particles of the system are given an infinitesimal displacement, the source terms on the right of the last pair of equations change discontinuously in multiples of e, the electronic charge. Certainly one of our first aims must be to eliminate this property, so that in (5.2) we can treat ρ and $\mathbf{J}$ as smooth, continuously variable functions.

Truncation

If we let

$$\begin{aligned} \rho(\mathbf{R}) &= \sum_n q(n)\delta(\mathbf{r}(n)-\mathbf{R}) \\ \mathbf{j}(\mathbf{R}) &= \sum_n q(n)\dot{\mathbf{r}}(n)\delta(\mathbf{r}(n)-\mathbf{R}) \end{aligned} \tag{5.3}$$

then we can express the variables $\mathbf{b}(\mathbf{R})$, $\mathbf{e}(\mathbf{R})$, $\rho(\mathbf{R})$ and $\mathbf{j}(\mathbf{R})$ in (5.1) as Fourier integrals so that for example,

$$\mathbf{b}(\mathbf{R}) = \int_{-\infty}^{\infty} \mathbf{b}(\mathbf{k}) \exp(2\pi i\mathbf{k}\,.\,\mathbf{R})\, \mathrm{d}^3\mathbf{k} \tag{5.4}$$

with

$$\mathbf{b}(\mathbf{k}) = \int_{-\infty}^{\infty} \mathbf{b}(\mathbf{R}) \exp(-2\pi i\mathbf{R}\,.\,\mathbf{k})\, \mathrm{d}^3\mathbf{R}. \tag{5.5}$$

In any macroscopic problem there will be a natural scale of length, and we will only be interested in components of the Fourier spectrum of the variables up to some limiting cut-off wave number or frequency k_0. For example in optics we might take k_0 to be somewhat larger than $1/\lambda$ where λ is the shortest wavelength in the problem; equally in electrostatics we might take k_0 to be somewhat greater than $1/r$ where r is the smallest radius of curvature of any portion of an electrode. Thus only those Fourier components with $k < k_0$ are relevant to the macroscopic problem. If we were simply to drop, from $\mathbf{b}$, $\mathbf{e}$, ρ and $\mathbf{j}$, all terms with $k > k_0$, not only should we retain all the necessary details of the problem but also ρ and $\mathbf{j}$ would become continuous variables. We now give a simple formal procedure for attaining this objective. The form we choose is designed to simplify later developments.

Truncation

Let $f(\mathbf{s})$ be a continuous scalar function of the three dimensional vector $\mathbf{s}$ with the integral

$$\int_{-\infty}^{\infty} f(\mathbf{s})\,\mathrm{d}^3\mathbf{s} = 1 \tag{5.6}$$

and let the Fourier transform of $f(\mathbf{s})$ be

$$F(\mathbf{k}) = \int_{-\infty}^{\infty} f(\mathbf{s}) \exp(-2\pi i \mathbf{s}\,.\,\mathbf{k})\,\mathrm{d}^3\mathbf{s}. \tag{5.7}$$

We then define the quantity $[\alpha(\mathbf{R})]$ by

$$[\alpha(\mathbf{R})] = \int_{-\infty}^{\infty} \alpha(\mathbf{R}-\mathbf{s}) f(\mathbf{s})\,\mathrm{d}^3\mathbf{s}, \tag{5.8}$$

in which α stands for any of the electromagnetic variables, e.g. $b_x(R)$, $e_z(\mathbf{R})$ or $\rho(\mathbf{R})$. The Fourier transform of $[\alpha(\mathbf{R})]$ is $\overline{\alpha(\mathbf{k})}$ and we obviously have

$$\overline{\alpha(\mathbf{k})} = \alpha(\mathbf{k})\, F(\mathbf{k}).$$

If, therefore, we choose $f(\mathbf{s})$ so that $F(\mathbf{k})$ tends rapidly to zero for $|k| > k_0$ we shall have eliminated the unwanted high frequency components of the spectrum of $\alpha(\mathbf{R})$ by forming $[\alpha(\mathbf{R})]$. However we also want to ensure that this process, which we call "truncation", does not alter the low frequency components $|k| \ll k_0$. Thus we also require

$$F(\mathbf{k}) = 1-\varepsilon, \qquad |k| \ll k_0 \tag{5.9}$$

where ε is as small as we please. This to some extent restricts the behaviour of the sampling function $f(\mathbf{s})$, for, if we consider

$$\frac{\partial^2 F}{\partial k_i^2} = \int (-2\pi i s_i)^2 \exp(-2\pi i \mathbf{k}\,.\,\mathbf{s}) f(\mathbf{s})\,\mathrm{d}^3\mathbf{s},$$

we see that

$$\left(\frac{\partial^2 F}{\partial k_i^2}\right)_{k_i \to 0} = -4\pi^2 \int_{\infty}^{\infty} s_i^2 f(\mathbf{s})\,\mathrm{d}^3\mathbf{s}.$$

The condition (5.9) requires that $\partial^2 F/\partial k_i^2 \to 0$ as $k \to 0$ and so we have

$$\int_{-\infty}^{\infty} s_i^2 f(\mathbf{s})\,\mathrm{d}^3\mathbf{s} = 0.$$

This is only possible if $f(\mathbf{s})$ can take on negative values. Thus $f(\mathbf{s})$ cannot be regarded as a probability distribution function. One possible form for $f(\mathbf{s})$ is

$$\left(\frac{\sin 2\pi k_0 s_x}{\pi s_x}\right)\left(\frac{\sin 2\pi k_0 s_y}{\pi s_y}\right)\left(\frac{\sin 2\pi k_0 s_z}{\pi s_z}\right)$$

but this has pathological convergence properties, associated with the discontinuity in its Fourier transform, $F(\mathbf{k}) = 1$, $k_i < k_0$, $F(\mathbf{k}) = 0$, $k_i > k_0$. A better behaved function is the

Fourier transform of

$$F(\mathbf{k}) = \exp\left\{-\frac{1}{2\nu}\left(\frac{k}{k_0}\right)^{2\nu}\right\} \tag{5.10}$$

where ν is integral and large. By taking a large enough, but finite, value of ν we can always satisfy (5.9), but if $\nu > 1, f(\mathbf{s})$ is no longer intrinsically positive.

It is obvious that the process of truncating a function, defined by (5.8), commutes with space and time differentiation, i.e. we have

$$\frac{\partial}{\partial R_i}[\alpha(\mathbf{R})] = \left[\frac{\partial}{\partial R_i}\alpha(\mathbf{R})\right]$$

$$\frac{\partial}{\partial t}[\alpha(\mathbf{R})] = \left[\frac{\partial}{\partial t}\alpha(\mathbf{R})\right],$$

thus the fields $\mathbf{e}$ and $\mathbf{b}$ after truncation satisfy the equations

$$\nabla \cdot [\mathbf{b}] = 0, \qquad \nabla \wedge [\mathbf{e}] + \frac{\partial}{\partial t}[\mathbf{b}] = 0,$$

$$\nabla \cdot [\mathbf{e}] = [\rho], \qquad \nabla \wedge [\mathbf{b}] - \frac{\partial}{\partial t}[\mathbf{e}] = [\mathbf{j}].$$

These equations are exact for all macroscopic purposes as long as k_0 has been correctly chosen. We have now only to investigate the properties of the truncated source terms $[\rho]$ and $[\mathbf{j}]$. If we insert ρ and $\mathbf{j}$ from (5.3) in (5.8) we obtain

$$\begin{aligned}[][\rho] &= \sum_n q(n) f(\mathbf{R}-\mathbf{r}(n)) \\ [\mathbf{j}] &= \sum_n q(n)\dot{\mathbf{r}}(n) f(\mathbf{R}-\mathbf{r}(n)).\end{aligned} \tag{5.11}$$

Thus, if $f(\mathbf{s})$ is a continuous function of its argument we see that an infinitesimal change in the particle coordinates $\mathbf{r}$ results in an infinitesimal, and no longer discontinuous, change in $[\rho]$ and $[\mathbf{j}]$. Equally an infinitesimal change in $\mathbf{R}$ also leads to infinitesimal changes. Thus $[\rho]$ and $[\mathbf{j}]$ are truly continuous functions.

We obviously want the function $F(k)$ to be real and this implies that $f(\mathbf{s})$ is an even function. Thus the rate at which the sampling function $f(\mathbf{s})$ changes in the vicinity of $\mathbf{s} = 0$ is mainly determined by

$$\left(\frac{\partial^2 f}{\partial s_i^2}\right)_{\mathbf{s}=0} = -4\pi^2 \int k_i^2 F(\mathbf{k})\, \mathrm{d}^3\mathbf{k},$$

or, in more appropriate terms, by

$$\frac{1}{f_0}\left(\frac{\partial^2 f}{\partial s_i^2}\right)_{\mathbf{s}=0} = -\frac{(2\pi k_0)^2 \int (k_i/k_0)^2 F(\mathbf{k})\, \mathrm{d}^3\mathbf{k}}{\int F(\mathbf{k})\, \mathrm{d}^3\mathbf{k}}.$$

Clearly the distance over which the sampling function $f(\mathbf{s})$ changes appreciably is of the order of $s_0 = 1/2\pi k_0$. In equations (5.11), therefore, if there are many particles within a volume of linear dimensions s_0 the truncated source functions $[\rho]$ and $[\mathbf{j}]$ will no longer be influenced by the positions of individual particles within a group but only by their mean position. In this case we can treat $[\rho]$ and $[\mathbf{j}]$ as effectively continuum fluid variables. We thus finally arrive at the set of equations (5.2), if we set $\mathbf{E} = [\mathbf{e}]$, etc.

Truncation and Averages

It is obviously a matter of some interest to see how the truncated terms $[\rho]$ and $[\mathbf{j}]$ are related to our naïve intuitive notions about averages. To demonstrate this we consider a simple one dimensional example, in which we have equal charges q evenly arranged along the x axis at $x = na_0$, where n is positive or negative integer. If we regard a_0 as a microscopic length relative to some macroscopic scale implied by the problem itself, we would have no hesitation in declaring that the macroscopic linear charge density was uniform and given by $\rho = q/a_0$. It is, however, instructive to see how we arrive at this conclusion. We certainly do not do it by calculating the charge q in a fixed small line segment of length l centred at x, for if $l = \nu a_0+\varepsilon$ where ν is integral and $\varepsilon < a_0$ we should find a charge of either $Q = \nu q$ or $Q = (\nu+1)q$, depending on the position of x. We might, however, be more sophisticated and note that, for a random choice of x over a range a_0, the probability of $Q = \nu q$ is $1-\varepsilon/a_0$ and of $Q = (\nu+1)q$ is ε/a_0 so that the expectation value is $(\nu+(\varepsilon/a_0))q$ giving $\rho = q/a_0$. This is equivalent to the process known as taking a sliding average and, as is well known, this tends to eliminate high frequency components from the average. It is obviously a useful way of picturing the process of forming an average. We can, however, look at this in a slightly more formal way, which will make contact with the techniques developed in this chapter.

Suppose that we expand the exact charge distribution as a Fourier series of fundamental period a_0 so that we obtain

$$\frac{q}{a_0}\sum_{m=-\infty}^{\infty}\exp\left(\frac{2\pi imx}{a_0}\right).$$

The charge density calculated for the line element $x-l/2$ to $x+l/2$ is

$$\rho = \frac{1}{l}\int_{x-l/2}^{x+l/2}\frac{q}{a_0}\sum_{m=-\infty}^{\infty}\exp\left(\frac{2\pi imx}{a_0}\right)\mathrm{d}x$$

which gives

$$\rho = \frac{q}{a_0}+\sum_{m=0}^{\infty}\frac{2q}{\pi ml}\sin\frac{\pi ml}{a_0}\cos\frac{2\pi mx}{a_0}.$$

For a random choice of either l or x the sum in this expression vanishes, i.e. we remove all Fourier components of ρ above $k = 1/a_0$. Thus, carried to a logical limit, our naïve averaging process is equivalent to truncating the Fourier spectrum of ρ.

We now see that, not only does the truncation procedure lead to field equations (5.2) of the conventional form, but also the interpretation of the continuous source terms ρ and $\mathbf{J}$ is entirely conventional. In practice we approach a problem in the following way. First of all we consider the nature of the results we hope to obtain, and especially the degree to which we will be able to attach any significance to Fourier components of the fields of spatial frequency k. This will allow us to fix the value of the cut-off k_0. At this stage we may either know *a priori* whether there are many charged particles within a length $1/k_0$ or we may be ignorant on this point. If we have *a priori* information we can then decide whether it will be appropriate to treat ρ and $\mathbf{J}$ as continuum variables or whether we shall have to retain the coordinates of individual particles in describing the charge and current densities. If we do not know the answer to this question we might instead assume that ρ could be treated as a continuum variable, solve the problem, and then check whether the result was consistent with our first assumption.

Truncation is a purely formal mathematical process which leads from the microscopic particle equations to continuum equations. If the value of k_0 is fixed by the nature of the problem, how far we can progress towards eliminating the particles from the problem depends on the number density N of the particles and the value of k_0. If N/k_0^3 is small all we succeed in doing is eliminating all reference to the internal structure and size of the particles. Provided that $k_0 \ll 10^{13}\ \text{cm}^{-1}$, electrons and nuclei, apart from their charge and mass, are equivalent. If $k_0 \ll 10^8\ \text{cm}^{-1}$, nuclei and ions can be regarded as equivalent although, as we shall see in chapter 8, we can consider properties other than the charge, i.e. the multipole moments, in more detail. If N/k_0^3 is large, electricity may be regarded as a continuous fluid.

In the next chapter we shall consider ensemble averages of the microscopic equations. These averages behave in mathematically very similar ways to truncation but, unlike truncation, they correspond to a definite aspect of the physical properties of the system. For example, if $\langle\ \rangle$ denotes an ensemble average we can consider both $\langle\rho^2\rangle$ and $\langle\rho\rangle^2$ and their difference has a physical significance. The truncation procedure attaches no meaning to $[E^2]$ other than $[E]^2$. In a truncated problem all variables, of actual or potential macroscopic significance, must be replaced in all equations by their truncations. This presents no problems in the field equations, which are linear, but as we shall see in chapter 7 requires very careful treatment in connection with terms in equations of motion, such as the expression $\rho\mathbf{E}+\mathbf{J} \wedge \mathbf{B}$ for the macroscopic force density.

Summary

Although this chapter has been rather brief and mainly concerned with simple mathematical manipulations the concept introduced here of macroscopic equations as the truncated form of the microscopic equations is of central importance in understanding the nature of macroscopic electromagnetic theory and its relation to the atomic structure of matter. It will therefore perhaps be useful if we recapitulate the argument.

Inherent in the idea of a macroscopic theory is the notion that some microscopic features of the system under discussion are either inaccessible to measurement or irrelevant to us. Thus, in discussing the refraction of light by a glass lens, we are utterly uninterested in the exact location of the individual atoms of the medium, or the elementary charges of the system. On the other hand a systematic displacement of these charges over regions of dimensions comparable with the wavelength is of interest. Loosely speaking we would express this by saying that only the average behaviour of the charges over a region of dimensions comparable with a wavelength and containing many individual charges is significant. Unfortunately if we pursue this idea to its logical conclusion and attempt to make the definition of the average more precise we find that it is impossible at one and the same time to define a unique average and produce a useful set of average equations. This difficulty arises because, instead of concentrating our attention on the nature of the macroscopic problem we have diverted it to the structure of the medium. In a macroscopic problem characterized by a scale Λ microscopic structure on a scale less than Λ is unwanted detail. We have to devise a procedure which, for example, when applied to a regular array of charges on a lattice of microscopic dimensions, leads to a uniform macroscopic charge density. There are a number of ways in which we could approach this problem but much the simplest method is to say that the nature of the macroscopic context implies that all the variables in the problem can have their spatial Fourier spectrum cut off at a spatial frequency $k_0 \approx 1/\Lambda$. This, as we have seen, leads from the microscopic field equations to the macroscopic equations in which ρ and $\mathbf{J}$ are continuous variables. How far truncation simplifies the problem depends on the relation between the macroscopic scale Λ and the scale of the microscopic structure of the problem (see problem 5). The truncation procedure has the advantage that it is mathematically similar to the process of forming a quantum mechanical or thermodynamic ensemble average. It is therefore compatible with the two most important physical averaging processes which take account of our ignorance about the detailed physical specification of the system (as distinct from our indifference to its details, inherent in the macroscopic context). In practice we never have to consider the details of the truncation procedure: it is sufficient to know that it exists and that for any specific macroscopic context there is a definite order of magnitude for the cut off frequency k_0. Once this is known we can turn to a consideration of the structure of the system and decide what form of the macroscopic equations will be appropriate for our discussion. If the mean particle spacing a_0 satisfies $k_0 a_0 \ll 1$ we can almost certainly treat the system as a continuum. In chapter 8 we shall discuss the conditions under which it will be useful to introduce the auxiliary vectors **P**, **D**, **M** and **H**. Finally we emphasize that the use of the truncation procedure leads to no new results: it merely provides a precise quantitative interpretation of the rather vague averaging notions introduced by Lorentz. It does, however, discourage us from attempting to give these notions an incorrect and inconsistent interpretation of the type discussed in the Preface. This is of some pedagogic value and for example considerably clarifies our ideas about the relative significance of the macroscopic fields **B** and **H** and the local magnetic field in matter.

Problems

1. A planar space charge limited diode has an anode to cathode spacing of 1 mm and an anode potential of 100 V. Estimate the largest value of k_0 for which the space charge can be treated as a continuum. (Neglect thermal emission velocities.)

2. Electromagnetic waves of frequency $\nu = c/\lambda$ propagate unattenuated in a neutral plasma of free electrons and heavy positive ions. Show that one necessary condition for treating the plasma as a continuum is $\lambda \gg e^2/\varepsilon_0 mc^2$. What other conditions will normally have to be fulfilled? (Neglect collisions.)

3. Show that if $F(k) = \exp\,[-1/2\nu(k/k_0)^{2\nu}]$ is the Fourier transform of $f(s)$ and ν is a positive integer $f(s)$ takes both positive and negative values when $\nu > 1$.

4. In a macroscopic problem Fourier components with $k = 10^5\ \text{cm}^{-1}$ are of interest and the particle density is $10^{18} = (10^6)^3\ \text{cm}^{-3}$. If we use the function in problem 3 to truncate the spectrum what is the least value of the integer ν that leads to an error of less than 1 part in 10^6 in the macroscopic components?

Microscopic and Macroscopic Fields

5. Conducting spheres of radius 1 mm are embedded in foam plastic on a regular cubic lattice of spacing 1 cm. Show that at low frequencies this medium can be treated as a continuum. What is its relative dielectric constant? What is the highest frequency at which it can be treated as a continuum? How would one attempt to describe the electromagnetic properties of the medium at a wavelength of 100 μm (1/10 mm)?

6. Light of wavelength λ falls at normal incidence on a grating with equal transparent and opaque rulings each of width a where $2a < \lambda$. Find the fraction of the incident intensity that is transmitted. Show that, as far as the transmitted light is concerned, the grating can be regarded as a continuous absorbing medium.

CHAPTER 6

Ensemble Averages

Ensembles

The truncation procedure discussed in the last chapter is a formal method for eliminating unwanted microscopic detail from the field equations. In practice, however, we can never achieve an exact specification of the configuration of all the particles of a system. Not only is this forbidden by quantum mechanics but, if the specification of the state of a system is only given in macroscopic terms, it will be insufficient to determine even the exact quantum state of the system. Thus, normally, there will be a large ensemble of possible microstates, or particle configurations, compatible with the macroscopic specification of the state of the system. The particular system under discussion must then be regarded as a system in a microstate chosen at random from this ensemble. The macroscopic specification of the state of the system determines the probability with which a particular microstate (ν) appears in the ensemble, or the probability $f(\nu)\,\mathrm{d}\nu$ that the system is in a microstate in a range $\mathrm{d}\nu$ near ν. Thus, if we can calculate $f(\nu)$, the expectation value of an observable quantity α whose value for the microstate ν is $\alpha(\nu)$ is

$$\langle\alpha\rangle = \int \alpha(\nu) f(\nu)\,\mathrm{d}\nu.$$

There is no guarantee that

$$\langle\alpha^2\rangle = \int \alpha^2(\nu) f(\nu)\,\mathrm{d}\nu$$

will be equal to $\langle\alpha\rangle^2$ and so the variable α has a statistical uncertainty.

We shall not consider quantum mechanical ensembles since they do not, in our context, differ significantly from classical ensembles and the language of classical ensemble theory is better adapted to our discussion. The reader who wishes to see how the quantum ensemble calculation proceeds should consult either Crowther and ter Haar (1971) or Mazur and Nijboer (1953).

The choice of the "volume" element $\mathrm{d}\nu$ in configuration space is, for our purposes, almost completely arbitrary, but in classical statistical mechanics it is usual to relate it to the coordinate and momentum components of the particles. Thus, in classical statistical mechanics, if we have a system of N particles, we consider a phase space of $6N$ dimensions with a volume element

$$\mathrm{d}\Gamma = \prod_n \mathrm{d}^3\mathbf{r}(n) \, . \prod_n \mathrm{d}^3\mathbf{p}(n) \tag{6.1}$$

where $\mathbf{r}(n)$ and $\mathbf{p}(n)$ are the position and momentum of the nth particle. This corresponds to the classical definition of an exact microstate as one where the coordinates and momenta of all the particles are known. The ensemble distribution function may then be written as a function of the $\mathbf{r}(n)$, $\mathbf{p}(n)$ and the time. This has the further advantage that the laws of classical mechanics ensure that, as a system develops in time from its initial microstate, the probability density f in the region occupied by the system in phase space remains constant (Liouville's theorem). We shall not need this result and so, although we shall use the language of classical statistical mechanics and the classical phase space we can at any time use a more general configuration space.

It is useful to picture points in phase space as representing the microstates of systems of the ensemble. Thus the function $f(\mathbf{r}(n), \mathbf{p}(n), t)$ gives the density of representative points in phase space. We now have for the expectation value of an observable α, whose value at $\mathbf{R}$ and t for one particular microstate $(\mathbf{r}(n), \mathbf{p}(n))$ is $\alpha(\mathbf{R}, t, \mathbf{r}(n), \mathbf{p}(n))$,

$$\langle \alpha(\mathbf{R}, t) \rangle = \int \alpha(\mathbf{R}, t, \mathbf{r}(n), \mathbf{p}(n)) f(\mathbf{r}(n), \mathbf{p}(n), t)\, \mathrm{d}\Gamma. \tag{6.2}$$

In this chapter we are not interested in the precise form of the distribution function f although the reader will, no doubt, be aware (see, e.g., Landau and Lifshitz, 1968) that for a system in equilibrium at a temperature T,

$$f = A \exp\left(-\frac{\mathscr{H}(\mathbf{r}(n), \mathbf{p}(n), \lambda)}{kT} \right)$$

where $\mathscr{H}$ is the Hamiltonian of the system as a function of the $\mathbf{r}(n)$, $\mathbf{p}(n)$ and external parameters λ such as an externally applied electric field, and A is a normalizing constant.

If the system is of limited spatial extent and total energy the distribution function f, as well as being integrable, must vanish for extreme values of any single particle coordinate or momentum. This allows us to set any expressions or integrals, involving f on the boundary of phase space, equal to zero.

Differentiation

It is obvious from (6.2) that spatial differentiation commutes with the ensemble average thus,

$$\frac{\partial}{\partial R_i} \langle \alpha(\mathbf{R}, t) \rangle = \left\langle \frac{\partial \alpha}{\partial R_i} \right\rangle. \tag{6.3}$$

On the other hand, since f depends on t, we have

$$\frac{\partial}{\partial t} \langle \alpha \rangle = \left\langle \frac{\partial \alpha}{\partial t} \right\rangle + \int \alpha \frac{\partial f}{\partial t}\, \mathrm{d}\Gamma. \tag{6.4}$$

Now, as time passes, representative points move out of a given element of phase space and so we have

$$\frac{\partial f}{\partial t} = -\sum_n \left\{ \frac{\partial}{\partial r_i(n)} (\dot{r}_i(n) f) + \frac{\partial}{\partial p_i(n)} (\dot{p}_i(n) f) \right\}.$$

If we insert this expression in the integral in (6.4) and do an integration by parts, noting that $\mathrm{d}\Gamma = \prod_n \mathrm{d}^3\mathbf{r}(n) \,.\, \prod_n \mathrm{d}^3\mathbf{p}(n)$ we obtain a zero surface integral and are left with

$$\frac{\partial}{\partial t} \langle \alpha \rangle = \left\langle \frac{\partial \alpha}{\partial t} \right\rangle + \int \sum_n \left\{ f \dot{r}_i(n) \frac{\partial \alpha}{\partial r_i(n)} + f \dot{p}_i(n) \frac{\partial \alpha}{\partial p_i(n)} \right\} \mathrm{d}\Gamma.$$

The total, or hydrodynamic, derivative in phase space is

$$\frac{\mathrm{d}}{\mathrm{d}t} = \frac{\partial}{\partial t} + \sum_n \left(\dot{r}_i(n) \frac{\partial}{\partial r_i(n)} + \dot{p}_i(n) \frac{\partial}{\partial p_i(n)} \right), \tag{6.5}$$

and so we have

$$\frac{\partial}{\partial t}\langle\alpha\rangle = \left\langle\frac{d\alpha}{dt}\right\rangle. \tag{6.6}$$

Now consider the microscopic field equation

$$\nabla_R \wedge \mathbf{b}(\mathbf{R}, t, \mathbf{r}(n), \mathbf{p}(n)) + \frac{\partial}{\partial t}\mathbf{e}(\mathbf{R}, t, \mathbf{r}(n), \mathbf{p}(n)) = 0.$$

The partial derivative $\partial/\partial t$ here signifies only that **R** is held constant in the differentiation, not that we are to ignore changes in **e** resulting from changes in the particle configuration. It should, therefore, be replaced by the total derivative d/dt and so the ensemble average variables satisfy the equation

$$\nabla_R \wedge \langle\mathbf{b}(\mathbf{R}, t)\rangle + \frac{\partial}{\partial t}\langle\mathbf{e}(\mathbf{R}, t)\rangle = 0.$$

Thus the ensemble average, like truncation, preserves the form of the field equations and, if we put $\mathbf{B}(\mathbf{R}, t) = \langle\mathbf{b}(\mathbf{R}, t)\rangle$, $\mathbf{E}(\mathbf{R}, t) = \langle\mathbf{e}(\mathbf{R}, t)\rangle$, and

$$\rho(\mathbf{R}, t) = \langle\sum_n q(n)\delta(\mathbf{r}(n)-\mathbf{R})\rangle \tag{6.7a}$$

$$\mathbf{J}(\mathbf{R}, t) = \langle\sum_n q(n)\dot{\mathbf{r}}(n)\delta(\mathbf{r}(n)-\mathbf{R})\rangle \tag{6.7b}$$

we have

$$\begin{aligned} \nabla.\mathbf{B} &= 0, \qquad \nabla \wedge \mathbf{E}+\dot{\mathbf{B}} = 0 \\ \nabla.\mathbf{E} &= \rho, \qquad \nabla \wedge \mathbf{B}-\dot{\mathbf{E}} = \mathbf{J}. \end{aligned} \tag{6.8}$$

Retardation

There is one flaw in this argument, for we have implicitly assumed that the microscopic fields **e** and **b** at (**R**, t) are completely determined by the particle configuration ($\mathbf{r}(n)$, $\mathbf{p}(n)$) at the same time t. As we know, the effects of distant particles are retarded. We can escape from this dilemma in the following way. If we assume that the particle trajectories are continuous, the value of $\mathbf{r}(n)$ at t can be expressed, in terms of its value at some other time t, in terms of the derivatives $\partial^k(\mathbf{r}(n))/\partial t^k$ at t up to some arbitrary order k. Thus, if we go to a more general configuration space, in which not only the $\mathbf{r}(n)$ and $\mathbf{p}(n)$ but also the time derivatives are treated as configuration space coordinates, all our previous arguments remain valid. For a complete discussion of this problem the reader should consult de Groot and Vlieger (1965) or de Groot (1969).

Charge and Current Densities

We must now investigate the properties of the source terms ρ and **J** in (6.7a) and (6.7b). To do this we define a new function $g_n(\mathbf{r}(n), t)$ obtained by integrating f over all the microscopic variables except $\mathbf{r}(n)$, i.e.

$$g_n(\mathbf{r}(n), t) = \int f(\mathbf{r}(n'), p(n'), t) \prod_{n' \neq n} d^3\mathbf{r}(n') . \prod_{n'} d^3\mathbf{p}(n'), \tag{6.9}$$

so that we have

$$\rho(\mathbf{R}, t) = \sum_n q(n) g_n(\mathbf{R}, t), \tag{6.10a}$$

$$\mathbf{J}(\mathbf{R}, t) = \sum_n q(n)\mathbf{r}(n) g_n(\mathbf{R}, t). \tag{6.10b}$$

If f is a continuous function of its argument, then ρ and $\mathbf{J}$ are continuous functions.

Suppose that each particle of the system is known to be near a position $\mathbf{R}(n)$, for example, we might have

$$f = \prod_n \left(\frac{1}{2\pi\lambda^2}\right)^{\frac{3}{2}} \exp\left\{-\tfrac{1}{2}\left|\frac{\mathbf{R}(n)-\mathbf{r}(n)}{\lambda}\right|^2\right\} F(\mathbf{p}(n), t)$$

so that

$$g_n(R, t) = \left(\frac{1}{2\pi\lambda^2}\right)^{\frac{3}{2}} \exp\left\{-\tfrac{1}{2}\left|\frac{\mathbf{R}(n)-\mathbf{R}}{\lambda}\right|^2\right\},$$

then in this case both $\rho(\mathbf{R}, t)$ and $\mathbf{J}(\mathbf{R}, t)$ will depend explicitly on the $\mathbf{R}(n)$, i.e. the mean particle coordinates, thus

$$\rho(\mathbf{R}, t) = \sum_n q(n) \left(\frac{1}{2\pi\lambda^2}\right)^{\frac{3}{2}} \exp\left\{-\tfrac{1}{2}\left|\frac{\mathbf{R}(n)-\mathbf{R}}{\lambda}\right|^2\right\}. \tag{6.11}$$

If we were to express $\rho(\mathbf{R}, t)$ as a Fourier integral, e.g.

$$\rho(\mathbf{R}, t) = \int_{-\infty}^{\infty} \rho(\mathbf{k}, t) \exp(2\pi i\mathbf{k} \,.\, \mathbf{R})\, d^3\mathbf{k},$$

and the mean particle spacing were of the order of a_0, then information about the detailed position of the particles would be contained in Fourier components for which k is of the order of $1/a_0$. For any regular arrangement of the $\mathbf{R}(n)$, however, the Fourier components of (6.11) decrease as $\exp(-2\pi^2 k^2 \lambda^2)$ and so, if λ is larger than a_0 all details of the exact particle positions are eliminated from ρ. In this case ρ and $\mathbf{J}$ can be treated as continuous fluid variables without invoking the truncation procedure. Unfortunately there are few physical situations of this type, in which a particle's mean position is given by a regular law, and at the same time the width λ of the coordinate distribution is large compared with the particle spacing.

Fluctuations

Let us now consider a one dimensional example in which the sole macroscopic information available is that we have N particles distributed at random over an interval $-L/2 < x < L/2$. If the particles all have the same charge q, the charge density $\rho(x, v)$ for a particular configuration v can be expanded in a Fourier series as

$$\rho(x, v) = \sum_{m=-\infty}^{\infty} c_k(v) \exp\left(\frac{2\pi imx}{L}\right).$$

The expectation value of the coefficients in the ensemble is zero except for

$$\langle c_0 \rangle = \frac{Nq}{L},$$

but we also have

$$\langle c_m c_{m'}^* \rangle = \delta_{mm'} \frac{Nq^2}{L^2}, \qquad m \neq 0$$

and

$$\langle c_0 c_0^* \rangle = N^2 q^2 / L^2.$$

Thus, in the ensemble, since

$$\langle \rho \rangle = \frac{Nq}{L}$$

and

$$\langle \rho^2 \rangle = \sum_m \langle c_m c_m^* \rangle = \frac{N^2 q^2}{L^2} + N \frac{q^2}{L^2} \sum_{m=-\infty}^{\infty} 1,$$

we see that

$$\langle \Delta\rho^2 \rangle = \langle \rho^2 \rangle - \langle \rho \rangle^2 = 2N \frac{q^2}{L^2} \sum_{m=1}^{\infty} 1,$$

and ρ displays statistical fluctuations, which diverge if we do not cut off the sum at some value of m. If we let $\gamma = m/L$, the contribution to these fluctuations from Fourier components in a range $\mathrm{d}\gamma$ about γ and about $-\gamma$ is

$$\mathrm{d}\langle \Delta\rho^2 \rangle = 2 \frac{Nq^2}{L} \mathrm{d}\gamma = 2q \langle \rho \rangle \mathrm{d}\gamma.$$

Thus this ensemble, with a completely random distribution function, leads to fluctuations in ρ associated with both low and high spatial frequencies. Even if we truncate ρ at a high frequency k_0 associated with the nature of the macroscopic problem we are still left with the real low frequency fluctuations. Obviously these considerations are directly relevant to shot noise fluctuations in electron beams and to fluctuations in plasmas. They also have an indirect relevance to phenomena such as Rayleigh scattering.

Summary

It should be clear from the above discussion that, although an ensemble average preserves the form of the field equations it is not by itself likely to lead to any great simplification of the equations or to justify treating electricity as a continuous fluid. We may contrast this situation with hydrodynamics where, on the whole, the ensemble average is adequate.

The importance of knowing that the ensemble average preserves the form of the equations is simply that, in most macroscopic systems, we are dealing with ensembles and so are forced to regard all macroscopic quantities as ensemble averages. It is clearly necessary to know that forming an ensemble average does not by itself modify the field equations even if, in a statistical ensemble, the variables themselves may display statistical fluctuations. The reason why, in so many macroscopic contexts, we can use the continuum equations is, however, that the context itself is compatible with truncation of the spatial Fourier spectra at adequately low values of k. It does not usually depend on the existence of a statistical ensemble.

Truncation and forming an ensemble average are distinct and independent processes. We can obviously either average a truncated variable or truncate an averaged variable. The order is, however, of some significance. In the example above, of a random distribution of charges, it makes no difference whether we first truncate ρ and form $\langle[\rho]\rangle$ or first average ρ and form $[\langle\rho\rangle]$, but the effect on ρ^2 is quite different for $\langle\rho^2\rangle$ contains effects from oscillatory terms and so both $\langle\rho^2\rangle$ and $[\langle\rho^2\rangle]$ diverge (unless we give the charges themselves a finite size). On the other hand $\langle[\rho]^2\rangle$ is finite and since, as we have insisted in the last chapter, for any macroscopic variable $[\rho^2]$ has no meaning distinct from $[\rho]^2$, the sequence truncation followed by averaging is clearly the correct one. We can relate this to physical notions more clearly by considering a function $I(t)$ which varies with time. If $I(t)$ is recorded by an apparatus which responds only to frequencies below ν_0 the output of the apparatus will register $[I(t)]$. If $I(t)$ is the result of a statistical process the mean output of the apparatus will be $\langle[I(t)]\rangle$ and the mean square output $\langle[I(t)]^2\rangle$, not $[\langle I^2(t)\rangle]$.

Because, as we have remarked, ensemble averages play such a central role in physics it is not enough to know that a classical ensemble leaves the field equations unaltered: we also need the quantum mechanical result. We shall not treat this here; the reader is referred to the paper by Crowther and ter Haar (1971) quoted earlier and to Schram (1960).

In a macroscopic problem we first decide, in view of the resolution required, on a spatial frequency k_0 at which to truncate the variables. This tells us whether a description in terms of individual particles or in terms of a continuous fluid is applicable. We then perform the appropriate ensemble or quantum averages indicated by the laws of statistical mechanics and if necessary calculate the fluctuations.

Problems

1. Explain why in discussing shot noise in a diode, with a mean current of 1 mA, at radio frequencies we may express the result as a fluctuation in a macroscopic current, whereas in dealing with the output of a counter, with a resolving time of 1 μsec, observing fluctuations in the β emission of 1 μCurie of ^{60}Co this would be inappropriate.

2. Sound waves, of frequency 10^4 Hz, propagate in helium gas at s.t.p. The pressure is reduced. At what pressure would you expect a treatment of propagation in terms of continuum variables to become inappropriate?

CHAPTER 7

Macroscopic Equations of Motion

The Nature of a Macroscopic Equation of Motion

In the theory of electron tubes we treat the electron beam as a continuous medium both in discussing it as a source of fields and also in discussing its motion under the influence of electric and magnetic fields. Thus, for example, we define a local macroscopic velocity $\mathbf{v}$ and consider the equation of motion

$$\frac{d\mathbf{v}}{dt} = \frac{\partial \mathbf{v}}{\partial t} + (\mathbf{v} \cdot \nabla)\mathbf{v} = \frac{q}{m}(\mathbf{E} + \mathbf{v} \wedge \mathbf{B}). \tag{7.1}$$

Again, in the classical theory of conduction in metals we write equations such as

$$\frac{d\mathbf{v}}{dt} + \frac{\mathbf{v}}{\tau} = \frac{q}{m}\mathbf{E} \tag{7.2}$$

and include a term $\mathbf{v}/\tau$ which describes approximately the effects of electronic "collisions" with the lattice. Clearly the nature of the macroscopic velocity in both these equations requires further investigation, and this is one of the main topics of this chapter. The collision term in (7.2), however, raises a further question since, although we introduce it in an entirely phenomenological way, it must, in reality, be only a disguise for the effect of electrostatic or magnetic fields, due to the lattice, acting on the electrons. Thus it appears that the conventional equations used to describe the flow of charged particles rely on a number of tacit assumptions.

In an empty region of space the microscopic electric and magnetic field components satisfy Laplace's equation, or the wave equation, depending on whether we are dealing with static or time-dependent fields. From this it follows that the highest spatial frequencies which occur in these fields are determined, either by the distance to the nearest charged particles, or by the frequency of the oscillatory fields. It is reasonable to assume, though strictly speaking this requires a quantum electrodynamic discussion, that in an empty region the equation of motion of an isolated charged particle can be adequately described by the macroscopic fields, obtained either by truncation or ensemble averaging applied to the microscopic fields. The situation is, however, quite different for a particle moving in a region containing other charges. Consider for example a beam of 10 MeV protons traversing the region between the poles of an electromagnet and then encountering a thin plastic foil. Outside the foil the trajectory can be obtained simply from the force $q\mathbf{v} \wedge \mathbf{B}$, but the trajectories of the protons in the foil will, in macroscopic terms, be quite unpredictable. Some protons will traverse the foil undeviated and others will lose energy or even experience large angle deflections. The trajectories are clearly determined by the effects of microscopic components of the fields which are effective in near encounters between charged particles. These forces are predominantly electromagnetic but are clearly not tractable within the context of a macroscopic theory. Thus, in this problem, we make a clear distinction between the macroscopic effects of the applied magnetic field and essentially microscopic processes, which we regard as outside the scope of the electromagnetic problem. As far as the macroscopic treatment goes we are not concerned about the interactions responsible for the scattering processes. They happen to be electromagnetic but they might as well be due to nuclear forces or gravity. This is not to say that we cannot perform calculations about the microscopic processes using the electromagnetic equations, but simply that these calculations have then to be incorporated in an ensemble description, and that our knowledge of the details of the ensemble is so sketchy that we can do no more, in the way of obtaining macroscopic results, than obtain a statistical scattering formula. Of course, if in this calculation we first truncate the field equations using a cut-off k_0 of, say, 10^7 cm^{-1}, the scattering processes will disappear from the theoretical description. This merely shows that such a choice of the cut-off is inappropriate in a complete discussion of this type of problem. Consider, however, a slightly more complex example of a dilute neutral plasma of electrons and heavy positive ions. If we inject a moderately energetic electron into this plasma its macroscopic trajectory will be a straight line, although, if we set up macroscopic oscillations in the plasma, we may perhaps find other macroscopic trajectories. It is, however, possible that occasionally the electron will be appreciably deviated by a near encounter with a single electron or ion of the plasma. Thus again a complete discussion of this problem would require a rather high value of the cut-off k_0, certainly much higher than the inverse of the mean particle spacing in the plasma. This would then mean that we could not use continuum equations to describe the plasma and would probably eliminate the possibility of any useful discussion. If, however, we agreed to regard close encounters and scattering as a separate group of phenomena, whose effects were eventually to be included in a phenomenological form, we might well be able to discuss the more interesting macroscopic properties of the plasma in simple continuum language.

In a dilute plasma such as the electron beam of a microwave tube the effects of collisions are essentially a small statistical perturbation on a problem which could otherwise be treated by continuum methods. At the other extreme, in a metal, the effects of macroscopic fields become small perturbations on the essentially statistical and random motions of the colliding particles. In this case the interaction, on a microscopic scale, between the electrons and the ionic cores of the lattice is so overwhelming that, rather than attempt to treat it as part of the electromagnetic problem, we incorporate its effect into the dynamics of the electrons. Thus, to a first approximation we give the electrons an effective mass and introduce a frictional term into the equation of motion.

The treatment of the response of bound atomic electrons presents a further example of this dichotomy in which we implicitly separate microscopic and macroscopic features in a single problem. Thus if we have atoms in an applied electric field we do not usually include the atomic coulomb field in the electromagnetic problem. This is regarded as part of the internal dynamics of the atom and quite apart from any macroscopic discussion. As far as macroscopic electromagnetism goes, the electromagnetic nature of the forces within an atom is irrelevant, for all we are concerned atoms could equally well be held together with glue and elastic bands. Indeed many theories of atomic polarizability are almost equivalent to this hypothesis, i.e. we treat the atoms as harmonic oscillators. Since the actual potential in which the electrons move varies as $-1/r$, rather than $\frac{1}{2}\alpha r^2$, which does not even satisfy Laplace's equation, this is clearly not a model in which much attention is paid to the origin of forces within atoms.

In general, in macroscopic electromagnetism, we try as far as possible to separate off the intrinsically microscopic aspects of the problem and treat them separately as scattering processes, effective potentials or effective masses. The macroscopic theory then incorporates these features of the problem in a purely phenomenological way.

A somewhat different situation, in which the microscopic aspects of a system obtrude in a rather inconvenient way in a macroscopic calculation, occurs when the particles are the constituents of a material medium with dielectric or magnetic properties. Here we are concerned with the force acting on individual electrons in individual atoms of the medium when a field is applied to the medium as a whole. In the electric case the actual force acting on an electron is not simply $e\mathbf{E}$, where $\mathbf{E}$ is the macroscopic field in the medium, for the field $\mathbf{E}$ is smoothed out, either by truncation or ensemble averaging, over regions of atomic dimensions, while the electron is at a special position identifiable as such on a microscopic or atomic scale. Thus, for example, if in a regular, one dimensional lattice the microscopic field $\gamma = \mathbf{E}+\boldsymbol{\varepsilon}\cos 2\pi x/a_0$ where a_0 is the period of the lattice and the electrons are all located at points $x = na_0$ the effective field is $\mathbf{E}+\boldsymbol{\varepsilon}$ and not just $\mathbf{E}$. The calculation of the effective local field in a dielectric is one of the most difficult and important topics in the theory of dielectric behaviour. It combines in a complex way the microscopic and macroscopic aspects of the structure of a medium, two aspects which we normally aim to keep completely separate. We shall be discussing this problem at several points later in the book.

Truncated Equations

We now return to the problem of justifying the use of what are essentially hydrodynamic equations in discussing the motion of particles in electron beams, plasmas, and metals. In the discussion which ensues we shall regard the fields as already macroscopic fields and so will only have to consider the effects of truncation or ensemble averaging on the particles. The two processes can be treated together for, in this context, truncation is mathematically identical with forming a special type of ensemble average. Thus, we may compare the truncated charge density

$$[\rho(\mathbf{R}, t)] = \int \sum_n q(n)\delta(\mathbf{r}(n)-\mathbf{R}+\mathbf{s})f(\mathbf{s})\mathrm{d}^3\mathbf{s}, \tag{7.3a}$$

with the ensemble average

$$\langle\rho(\mathbf{R}, t)\rangle = \int \sum_n q(n)\delta(\mathbf{r}(n)-\mathbf{R})f(\mathbf{r}(n), \mathbf{p}(n), t)\prod_n \mathrm{d}^3\mathbf{r}(n)\prod_n \mathrm{d}^3\mathbf{p}(n). \tag{7.3b}$$

Although in (7.3a) the $\mathbf{r}(n)$ are to be thought of as exact, we could also interpret (7.3a) as an ensemble average in which particles in one system were located at $\mathbf{r}(n)+\mathbf{s}$. The fact that $f(\mathbf{s})$ can be negative prevents us from regarding it as a probability density but does not affect any mathematical manipulations. We can, therefore, discuss the particles in the language of ensembles even if eventually we decide to replace the ensemble average by truncation.

One continuum equation that we would certainly hope to be able to use is

$$\mathbf{J} = \rho\mathbf{v} \tag{7.4}$$

where $\mathbf{v}$ is a macroscopic transport velocity. This will only be possible when no more than one species of mobile carrier is present. In general we shall have several species (k), i.e. ions and electrons, present and we shall then have to define charge and current densities as well as transport velocities for each species. Thus

$$\mathbf{J} = \sum_k \rho_k \mathbf{v}_k = \sum_k \mathbf{J}_k \tag{7.5a}$$

$$\rho = \sum_k \rho_k. \tag{7.5b}$$

In many cases, e.g. in a plasma, one species will be much lighter than the others and then it may sometimes be enough to write

$$\mathbf{J} = \rho_1 \mathbf{v}_1 \tag{7.6}$$

while retaining (7.5b). This is the usual case in vacuum tubes. Thus in a microwave tube ρ is very nearly zero in the beam due to the simultaneous presence of electrons and ions, but $\mathbf{J}$ is large and almost solely due to the electrons.

For a single species, of charge q and mass m, we have

$$\mathbf{v}(\mathbf{R}, t) = \frac{\mathbf{J}(\mathbf{R}, t)}{\rho(\mathbf{R}, t)} = \frac{\langle \sum_n q\dot{\mathbf{r}}(n)\delta(\mathbf{r}(n)-\mathbf{R})\rangle}{\langle \sum_n q\delta(\mathbf{r}(n)-\mathbf{R})\rangle},$$

so that

$$\mathbf{v}(\mathbf{R}, t) = \frac{\langle \sum_n \dot{\mathbf{r}}(n)\delta(\mathbf{r}(n)-\mathbf{R})\rangle}{\langle \sum_n \delta(\mathbf{r}(n)-\mathbf{R})\rangle}. \tag{7.7}$$

The macroscopic momentum density is

$$\mathbf{P}(\mathbf{R}, t) = \langle \sum_n \mathbf{p}(n)\delta(\mathbf{r}(n)-\mathbf{R})\rangle \tag{7.8}$$

and, for non-relativistic particles,

$$\mathbf{P} = \frac{m}{q}\mathbf{J} = \frac{m}{q}\rho\mathbf{v}. \tag{7.9}$$

The hydrodynamic equation of motion of a fluid is obtained by considering momentum balance in an infinitesimal volume. The rate at which the momentum in the volume increases, together with the rate of efflux of momentum through the surface of the volume, is equal to the sum of the body force acting in the volume and the force acting on the surface of the volume. In Cartesian tensor notation, if $v_i(R, t)$ is the ith component of the local fluid velocity, X_k the body force and σ_{ik} the local stress tensor, balance of the kth component of momentum yields

$$\frac{\partial P_k}{\partial t} + \frac{\partial}{\partial R_i}(v_i P_k) = X_k + \frac{\partial}{\partial R_i}\sigma_{ik}, \tag{7.10}$$

where summation over $i = 1, 2, 3$ or $i = x, y, z$ is implied.

The other important equation is the equation of continuity, or charge conservation,

$$\frac{\partial \rho}{\partial t} + \nabla . \mathbf{J} = 0. \tag{7.11}$$

This equation can be obtained from the microscopic equations as follows. From our earlier result, if D/Dt denotes a total derivative in *phase* space (not the hydrodynamic derivative in real space),

$$\frac{\partial \rho}{\partial t} = \left\langle \frac{D}{Dt} \sum_n q(n)\delta(\mathbf{r}(n)-\mathbf{R}) \right\rangle,$$

but

$$\frac{D}{Dt} = \frac{\partial}{\partial t}+\sum_n (\dot{\mathbf{r}}(n)\cdot\nabla_{r(n)}+\dot{\mathbf{p}}(n)\cdot\nabla_{p(n)})$$

and so

$$\frac{\partial\rho}{\partial t} = \langle\sum_n q(n)\dot{\mathbf{r}}(n)\cdot\nabla_{r(n)}\delta(\mathbf{r}(n)-\mathbf{R})\rangle$$

or

$$\frac{\partial\rho}{\partial t} = -\langle\sum_n q(n)\dot{\mathbf{r}}(n)\cdot\nabla_R\delta(\mathbf{r}(n)-\mathbf{R})\rangle$$

which can be rearranged as

$$\frac{\partial\rho}{\partial t} = -\nabla_R\cdot\langle\sum_n q(n)\dot{\mathbf{r}}(n)\delta(\mathbf{r}(n)-\mathbf{R}) = -\nabla_R\cdot\mathbf{J}.$$

In the equation of motion, (7.10), we have

$$\frac{\partial P_k}{\partial t} = \sum_n\left\langle\frac{DP_k}{Dt}\right\rangle = \sum_n\langle\{\dot{\mathbf{r}}(n)\cdot\nabla_{r(n)}+\dot{\mathbf{p}}(n)\cdot\nabla_{p(n)}\}p_k(n)\delta(\mathbf{r}(n)-\mathbf{R})\rangle \tag{7.12}$$

since the microscopic momentum density $\sum_n \mathbf{p}(n)\delta(\mathbf{r}(n)-\mathbf{R})$ does not depend explicitly on t. Equation (7.12) yields, after some rather obvious manipulation,

$$\frac{\partial P_k}{\partial t}+\frac{\partial}{\partial R_i}\sum_n\langle\dot{r}_i(n)p_k(n)\delta(\mathbf{r}(n)-\mathbf{R})\rangle = \sum_n\langle\dot{p}_k(n)\delta(\mathbf{r}(n)-\mathbf{R})\rangle. \tag{7.13}$$

If we now introduce

$$\pi = \frac{q}{\rho}\mathbf{P} \tag{7.14}$$

as the average momentum per particle, we can express the ensemble average on the left hand side of (7.12) as

$$\sum_n\langle\dot{r}_i(n)p_k(n)\delta(\mathbf{r}(n)-\mathbf{R})\rangle = \sum_n\langle(\dot{r}_i(n)-v_i)(p_k(n)-\pi_k)\delta(\mathbf{r}(n)-\mathbf{R})\rangle+v_iP_k. \tag{7.15}$$

Equation (7.13) can now be written as

$$\frac{\partial P_k}{\partial t}+\frac{\partial}{\partial R_i}(v_iP_k) = \sum_n\langle\dot{p}_k(n)\delta(\mathbf{r}(n)-\mathbf{R})\rangle-\frac{\partial}{\partial R_i}\sum_n\langle(\dot{r}_i(n)-v_i)(p_k(n)-\pi_k)\delta(\mathbf{r}(n)-\mathbf{R})\rangle. \tag{7.16}$$

The last term on the right has the form of a gradient of a stress tensor. In the kinetic theory of gases this tensor is known as the kinetic stress tensor. It is used, for example, in calculating the viscosity of a perfect gas. The term describes momentum transport due to the macroscopically imperceptible spread of particle velocities $\dot{r}_i(n)$ and momenta $p_k(n)$ about their mean values v_i and π_k. We denote this term by

$$\sigma_{ik}^K = -\sum_n\langle(\dot{r}_i(n)-v_i)(p_k(n)-\pi_k)\delta(\mathbf{r}(n)-\mathbf{R})\rangle, \tag{7.17}$$

so that we can write (7.16) as

$$\frac{\partial P_k}{\partial t}+\frac{\partial}{\partial R_i}(v_iP_k) = \frac{\partial}{\partial R_i}\sigma_{ik}^K+\sum_n\langle\dot{p}_k(n)\delta(\mathbf{r}(n)-\mathbf{R})\rangle. \tag{7.18}$$

Macroscopic Equations of Motion

Comparing equation (7.18) with equation (7.10) we see that the process of taking an ensemble average has led to a satisfactory account of all the terms except the body force X_k. This was, of course, to be expected, since we know that hydrodynamics gives a very satisfactory account of the mechanical properties of fluids. In dealing with the body force $X_k = \sum_n \dot{p}_k(n)\delta(\mathbf{r}(n)-\mathbf{R})$ we shall be particularly concerned with those forces of electrical or magnetic origin which act on the individual charged particles. Neglecting, for brevity, magnetic forces, the equation of motion of the nth particle is

$$\dot{p}_k(n) = qe_k(\mathbf{r}(n), t, \mathbf{r}(n'), \mathbf{p}(n')) \tag{7.19}$$

where $\mathbf{r}(n)$ is the field point and n' is a generic label for all the particles other than n. The notation indicates that $\mathbf{e}$ at $\mathbf{r}(n)$ depends on the configuration of all the remaining particles. We now express the microscopic field $\mathbf{e}$ in terms of the macroscopic field $\mathbf{E}$, and a force field $(1/q)\mathbf{g}$ which represents the effect of local fluctuations from $\mathbf{E}$, i.e., just those short range forces which are responsible for scattering. It is indeed not essential that $\mathbf{g}$ should be purely electromagnetic in origin.

In the kinetic theory of gases, where we are not confused by knowing the nature of the forces incorporated in $\mathbf{g}$, we should write it as

$$\mathbf{g} = -\sum_{n' \neq n} \nabla_{r(n)} V_{n',n}, \tag{7.20}$$

in terms of an empirical or *ad hoc* interatomic potential $V_{n'n}$. Thus, with

$$\mathbf{e} = \mathbf{E}(\mathbf{r}(n), t) + \frac{1}{q}\,\mathbf{g}(\mathbf{r}(n), t, \mathbf{r}(n'), \mathbf{p}(n')), \tag{7.21}$$

we have

$$\frac{\partial P_k}{\partial t} + \frac{\partial}{\partial R_i}(v_i P_k) = E_k \sum_n \langle q\delta(\mathbf{r}(n)-\mathbf{R})\rangle + \frac{\partial}{\partial R_i}\sigma^K_{ik} - \sum\sum_{n' \neq n} \left\langle \frac{\partial}{\partial r_k(n)} V_{n'n}\delta(\mathbf{r}(n)-\mathbf{R})\right\rangle. \tag{7.22}$$

The last term on the right can, after some rather tedious and not very obvious manipulation, be rearranged as the gradient of a stress tensor $-\sigma^V_{ik}$, which for example describes viscous forces due to inter-particle interactions not explicitly contained in $\mathbf{E}$. We now have

$$\frac{\partial P_k}{\partial t} + \frac{\partial}{\partial R_i}(v_i P_k) = E_k \rho + \frac{\partial}{\partial R_i}(\sigma^V_{ik} + \sigma^K_{ik}), \tag{7.23}$$

where we have used

$$\rho = \sum_n \langle q\delta(\mathbf{r}(n)-\mathbf{R})\rangle.$$

The evaluation of the stress tensors σ^K_{ik} and σ^V_{ik} is essentially a problem in kinetic theory. The reader will find an account in the paper by Irving and Kirkwood (1950) on which this section has largely been based.

If we were dealing solely with truncation, rather than with the physical process of ensemble averaging, there would be no dispersion in the particle velocities or momenta and so σ^K_{ik} would be zero. The interaction stress tensor σ^V_{ik} would not, however, vanish. The effect of this stress on the macroscopic motion is proportional to the density. Thus, in kinetic theory it is important in liquids but can usually be neglected in gases. In the electromagnetic case we can neglect it in tenuous plasmas, where the electron density can be as low as 10^{10} m^{-3}, but not necessarily in metals where the density is nearer to 10^{31} m^{-3}.

If we identify $\rho\mathbf{E}$ as the body force, equation (7.23) is obviously equivalent to the hydrodynamic equation of motion (7.10). It is also easy to see that, if we include a macroscopic magnetic field, this will first appear in equation (7.19) and lead to an additional term $\mathbf{J} \wedge \mathbf{B}$ in (7.23).

If we introduce in (7.23) the average particle momentum π_k so that $qP_k = \rho\pi_k$ the result, omitting the stress terms, is

$$\frac{\partial}{\partial t}(\pi_k\rho)+\frac{\partial}{\partial R_i}(v_i\pi_k\rho) = q\rho(E_k+(\mathbf{v} \wedge \mathbf{B})_k),$$

which gives

$$\rho\left(\frac{\partial\pi_k}{\partial t}+v_i\frac{\partial\pi_k}{\partial R_i}\right)+\pi_k\left(\frac{\partial\rho}{\partial t}+\frac{\partial}{\partial R_i}(\rho v_i)\right) = q\rho(E_k+(\mathbf{v} \wedge \mathbf{B})_k).$$

By using the continuity equation to eliminate the second bracket on the left hand side we obtain

$$\frac{\partial\pi_k}{\partial t}+v_i\frac{\partial}{\partial R_i}\pi_k = q(E_k+(\mathbf{v} \wedge \mathbf{B})_k). \tag{7.24}$$

If we now define the total time derivative (*not* in phase space) as

$$\frac{d}{dt} = \frac{\partial}{dt}+\mathbf{v}\,.\,\boldsymbol{\nabla} \tag{7.25}$$

we obtain

$$\frac{d\boldsymbol{\pi}}{dt} = q(\mathbf{E}+\mathbf{v} \wedge \mathbf{B}). \tag{7.26}$$

For non-relativistic particles where $\boldsymbol{\pi} = m\mathbf{v}$ this reduces to the familiar equation

$$\frac{d\mathbf{v}}{dt} = \frac{q}{m}(\mathbf{E}+\mathbf{v} \wedge \mathbf{B}) \tag{7.27}$$

used in most discussions of electron-beam dynamics. We must remember, however, that this only applies to a system containing a single species of mobile charge.

If we turn back to equation (7.23) we see that we have succeeded in writing the equation of motion in a form in which the macroscopic body force $\rho\mathbf{E}$ is separated from the force $(\partial/\partial R_i)(\sigma_{ik}^K+\sigma_{ik}^V)$ which involves details of the microscopic structure of the medium. In many problems this term can be omitted. If the charged particles move against a more or less fixed background, e.g. if we consider electrons in a plasma moving past massive ions, or electrons in a metal moving past fixed ionic cores in a lattice, there will be an extra term in the equation of motion (7.19) due to the interaction with the background; if we write this term as $(\partial p_k(n)/\partial t)_{\text{int}}$ then, in the final equation of motion, we shall have, omitting the tensor terms,

$$\frac{\partial P_k}{\partial t}+\frac{\partial}{\partial R_i}(v_iP_k) = \rho E_k+\sum_n\left\langle\left(\frac{\partial p_k(n)}{\partial t}\right)_{\text{int}}\delta(\mathbf{r}(n)-\mathbf{R})\right\rangle. \tag{7.28}$$

If the collisions are frequent, and the medium is fine grained on a macroscopic scale, this last term can often be expressed in terms of a mean collision time τ so that we have

$$\frac{\partial P_k}{\partial t}+\frac{\partial}{\partial R_i}(v_iP_k)+P_k/\tau = \rho E_k \tag{7.29}$$

which, for non-relativistic particles, leads to the equation

$$\frac{d\mathbf{v}}{dt}+\mathbf{v}/\tau = \frac{q}{m}\mathbf{E}. \tag{7.30}$$

This is the familiar result used in the elementary theory of electrical conduction in metals.

There is no general reason why the interaction force should be equivalent to a frictional term: we might equally have a term proportional to acceleration, giving a change in the effective mass.

Whereas, both truncation and ensemble averaging applied to the field equations lead to new field equations of recognizably the same form, their application to the equations of motion leads to a familiar but essentially new set of equations. Although an equation such as (7.30) or (7.26) looks as though it is an immediately obvious consequence of the microscopic equation of motion

$$\dot{\mathbf{p}}(n) = q\mathbf{e}+q\dot{\mathbf{r}}(n) \wedge \mathbf{b},$$

the occurrence of the total derivative should serve to remind us that its physical content is entirely different.

Most macroscopic equations of motion result from an ensemble average together with a truncation of the microscopic equations. Usually we are not interested in the fluctuations which result from the use of a statistical ensemble but in, for example, the theory of noise in electron tubes their effects are significant.

Problem

In an electron beam, electrons cross a plane normal to the beam at a mean rate N_0. The kinetic energy of the electrons in the direction of the beam is given by the Boltzmann distribution, thus the mean rate at which electrons of energy between E and $E+\mathrm{d}E$ cross the plane is

$$\mathrm{d}N = N_0 \exp\left(-\frac{E}{kT}\right)\frac{\mathrm{d}E}{kT}, \qquad E \geqslant 0.$$

The rate at which electrons in each energy class cross the plane is, however, randomly and independently distributed about $\mathrm{d}N$ so that, in a time τ, the number will fluctuate about $\tau\,\mathrm{d}N$ with a Poisson distribution and a mean square fluctuation equal to $\tau\,\mathrm{d}N$. Show that the mean square fluctuation in the time average over τ of the macroscopic velocity v is given by

$$\langle(\overline{\delta v^{\tau}})^2\rangle = \frac{4-\pi}{2\tau}\frac{kT}{mN_0}.$$

This leads to a power spectrum

$$w(v) = (4-\pi)\frac{kT}{mN_0}$$

(Rack, 1938).

CHAPTER 8

Continuous Media

Introduction

The problem of the relation between the macroscopic field equations and the microscopic particle field equations is usually associated with a discussion of dielectric and magnetic materials and we now turn to this topic. However, as we have seen in the last three chapters, this is only one circumstance, though perhaps the most important, in which we have to justify using continuum equations to describe systems we know to consist of sub-microscopic charged particles.

The density of particles in condensed matter is usually so high that in static problems, and wave problems at frequencies below about 10^{16} Hz (in the u.v.), we have no difficulty in choosing a value of the cut-off k_0 which allows us to treat the medium as continuous. The new feature to be dealt with in this chapter is the sub-division of the elementary charges in a medium into bound charges on the one hand and free, extrinsic, accessible, or mobile charges on the other. In a medium such as for example SiO_2, quartz, all the charge is bound charge, that is to say that neither the nuclei, nor the atomic electrons, can make more than sub-atomic excursions from their mean positions without destroying the structure of the medium. The situation is only slightly more complicated in an ionic crystal such as NaCl since here we have the possibility that the nuclei and electrons move independently as in electronic polarization or that a complete ion moves as a unit about its mean lattice position. All the charge is, however, still bound. In some solid and liquid media it is possible to introduce a very small amount of extrinsic charge. Thus we can inject a limited number of extra electrons or H^+ ions into a crystal. Although these charges are usually not appreciably mobile it is useful to group them with the free or mobile charge since they lead to a local non-neutrality of the medium whose effects persist over macroscopically significant distances. In electrolytes, metals, and semiconductors we have mobile charges which can make unlimited excursions within the medium. It is unusual for the ions of an electrolyte to be able to pass through the boundaries of the medium, but usually electrons can be regarded as completely mobile and free to leave the medium.

In a medium containing an appreciable density of mobile charge, and especially in metals, the electric (but not the magnetic) effects of the bound charges are usually inappreciable, but in insulators, although the actual charge displacements are minuscule, the high density of charges leads to appreciable effects.

The discussion that follows is largely based on the paper by Mazur and Nijboer (1953) and will be presented in the language of ensemble theory; however, as we have seen, truncation is formally equivalent to a special type of ensemble, in so far as it affects the description of the charged particles, and so our results apply equally to both procedures. In practice, and certainly in solids, the truncation procedure is the more appropriate, but since we can have the results of both procedures for no extra work there is no point in insisting on one procedure rather than the other. For simplicity we shall ignore all complications due to retardation and quantum mechanics. The work of de Groot and Vlieger (1965) and of Crowther and ter Haar (1971) shows that the results we obtain are of general validity and even extend to spin magnetism.

Although in what follows it will be convenient to talk in terms of atoms and to ascribe all positive charges to nuclei and all negative charges to electrons, the treatment is in fact quite general. When we describe a unit as an atom we could equally well describe it as the contents of a unit cell, a distinct molecular or ionic group such as CBr_4 or PO_4^{3-}, or indeed as a single particle such as an electron or a positive hydrogen ion. All that we imply by the term "atom" is an identifiable structural unit with a definite point in the unit which can be taken as a local origin.

We have shown how both the processes of ensemble averaging and truncating the spatial Fourier spectrum yield the macroscopic field equations

$$\begin{aligned} \nabla \cdot \mathbf{B} &= 0, \qquad \nabla \wedge \mathbf{E}+\dot{\mathbf{B}} = 0, \\ \nabla \cdot \mathbf{E} &= \rho, \qquad \nabla \wedge \mathbf{B}-\dot{\mathbf{E}} = \mathbf{J}. \end{aligned} \tag{8.1}$$

(Again we are setting $\mu_0 = \varepsilon_0 = 1$ in order to keep the equations uncluttered.) If we have a material body as part of a system it is always possible to express the total charge and current densities as

$$\rho = \rho_f - \nabla \cdot \mathbf{P} \tag{8.2a}$$

$$\mathbf{J} = J_f + \dot{\mathbf{P}} + \nabla \wedge \mathbf{M} \tag{8.2b}$$

where the new variables have the following properties:

1. The net charge on the body is $\int \rho_f \, d^3\mathbf{R}$,
2. $\mathbf{P}$ is zero outside the body,
3. The net current leaving the surface of the body is $\int \mathbf{J}_f \cdot d\mathbf{S}$,
4. $\mathbf{M}$ is zero outside the body.

The proof is considered in the Appendix on vector analysis and we shall not give it here.

Our problem, therefore, is not to justify equations (8.2a) and (8.2b) but to show that **P** is the electric dipole moment density and that **M** is the magnetic dipole moment density obtained by the same averaging process or processes which lead to (8.1). In other words we have to connect **P**, $\mathbf{J}_f$ and **M** with atomic properties.

The Effective Sources of the Fields

We begin, then, by considering a material medium as an assembly of atomic sub-units and, since we shall need to discuss both the coordinates of these units and the relative coordinates of the particles within a unit we need a slight change in notation. We retain **R** and t as general space–time coordinates, but let $\mathbf{R}(n)$ and $\mathbf{P}(n)$ be the position and momentum of the nth sub-unit as a whole. The coordinates of the eth particle of the nth unit relative to $\mathbf{R}(n)$ are specified by $\mathbf{r}(n, e)$ and its momentum by $\mathbf{p}(n, e)$, thus its coordinates in space are given by $\mathbf{R}(n)+\mathbf{r}(n, e)$. We place the nuclear charge $q(n)$ at $\mathbf{R}(n)$ and other charges, e.g., electrons, $q(n, e)$ at $\mathbf{r}(n, e)$. The total charge of the atom or unit is $q(n)+\sum_e q(n, e)$ which will be zero for neutral atoms, but if we have $Z(n)$ electrons in the nth atom there is no need to make $Z(n)$ equal to the nuclear charge number. Thus we can also consider ions, or even electrons with $Z = 0$, $q(n) = -e$.

The electric dipole moment of the nth atom is

$$\pi(n) = \sum_e q(n, e)\mathbf{r}(n, e) \tag{8.3a}$$

and it is not necessary for the atom to be neutral for this definition to be unique, since we have assumed that the local origin $\mathbf{R}(n)$ can be identified, e.g. as a nucleus or as the centre of gravity of a unit cell. The magnetic dipole moment is defined by

$$\boldsymbol{\mu}(n) = \tfrac{1}{2} \sum_e q(n, e)\mathbf{r}(n, e) \wedge \dot{\mathbf{r}}(n, e) \tag{8.3b}$$

but if we intend to include $\boldsymbol{\mu}(n)$ we must also include the electric quadrupole moment. In Cartesian tensor notation this is

$$q_{ij}(n) = \tfrac{1}{2} \sum_e q(n, e)r_i(n, e)r_j(n, e) \tag{8.3c}$$

or, in dyadic notation,

$$\underline{\underline{\mathbf{q}}}(n) = \tfrac{1}{2} \sum_e q(n, e)\mathbf{r}(n, e)\mathbf{r}(n, e). \tag{8.3d}$$

These moments are to be regarded as point entities located at the point $\mathbf{R}(n)$, and, if we assume that these first few moments adequately describe an atom, we must already have decided that the relevant field components vary slowly over atomic dimensions. Thus we have already implied that the problem will be truncated at $k_0 \ll a_0^{-1}$ where a_0 is a typical atomic dimension.

The microscopic moment densities are now obtained as

$$\sum_n \{\delta(\mathbf{R}(n) - \mathbf{R}) \sum_e q(n, e)\mathbf{r}(n, e)\}$$

etc., and if is the ensemble distribution function in a phase space whose volume element is $\mathrm{d}\Gamma$, the electric dipole moment has an ensemble average density

$$\mathbf{P}(\mathbf{R}, t) = \sum_{n=1}^{n} \int \delta(\mathbf{R}(n) - \mathbf{R}) \sum_e q(n, e)\mathbf{r}(n, e) f \, \mathrm{d}\Gamma \tag{8.4}$$

where N is the total number of atoms in the system.

The distribution function $f(\mathbf{R}(n), \mathbf{P}(n), \mathbf{r}(n, e), \mathbf{p}(n, e), t)$ operates in a phase space of $6\sum_{n=1}^{N}(Z(n)+1)$ dimensions, but since all the variables that we shall have to discuss contain delta functions, we can make a considerable simplification in our calculations by introducing a new distribution function $g_n(\mathbf{R}(n), \mathbf{r}(n, e), t)$ obtained by integrating f over all the momenta and all the coordinates except those which refer to the nth atom and its particles. Thus we have

$$g_n(\mathbf{R}(n), \mathbf{r}(n, e), t) = \int f(\mathbf{R}(n'), \mathbf{P}(n'), \mathbf{r}(n', e'), \mathbf{p}(n', e'), t) \prod_{n'} \mathrm{d}^3\mathbf{P}(n') \cdot \prod_{n'e'} \mathrm{d}^3\mathbf{p}(n', e')$$
$$\prod_{n' \neq n} \mathrm{d}^3\mathbf{R}(n') \cdot \prod_{\substack{n' \neq n \\ e'}} \mathrm{d}^3\mathbf{r}(n', e'). \tag{8.5}$$

In terms of g_n we then have

$$\mathbf{P}(R, t) = \sum_n \int \delta(\mathbf{R}(n) - \mathbf{R}) \sum_e q(n, e)\, \mathbf{r}(n, e)\, g_n\, (\mathbf{R}(n), \mathbf{r}(n, e), t)\, \mathrm{d}^3\mathbf{R}(n) \prod_e \mathrm{d}^3\mathbf{r}(n, e),$$

which further reduces to

$$\mathbf{P}(\mathbf{R}, t) = \sum_n \int \sum_e q(n, e)\mathbf{r}(n, e)g_n(\mathbf{R}, \mathbf{r}(n, e), t) \prod_e \mathrm{d}^3\mathbf{r}(n, e). \tag{8.6}$$

We now introduce the notation

$$^{ne}\langle \ldots\ldots \rangle \equiv \int \ldots\ldots \prod_e \mathrm{d}^3\mathbf{r}(n, e). \tag{8.7}$$

so that

$$\mathbf{P}(\mathbf{R}, t) = \sum_n {}^{ne}\langle \sum_e q(n, e)\mathbf{r}(n, e)g_n(\mathbf{R}, \mathbf{r}(n, e), t)\rangle. \tag{8.8a}$$

Similarly

$$\mathbf{M}(\mathbf{R}, t) = \sum_n {}^{ne}\langle \tfrac{1}{2} \sum_e q(n, e)\mathbf{r}(n, e) \wedge \hat{\dot{\mathbf{r}}}(n, e)g_n(\mathbf{R}, \mathbf{r}(n, e), t)\rangle. \tag{8.8b}$$

where $\hat{\dot{\mathbf{r}}}(n, e)$ is the result of averaging $\dot{\mathbf{r}}(n, e)$ over all the *other* particle variables, and

$$Q_{ij}(\mathbf{R}, t) = \sum_n {}^{ne}\langle \tfrac{1}{2} \sum_e q(n, e)r_i(n, e)r_j(n, e)g_n(\mathbf{R}, \mathbf{r}(n, e), t)\rangle. \tag{8.8c}$$

The total charge density $\rho(\mathbf{R}, t)$ is the ensemble average of

$$\sum_n q(n)\delta(\mathbf{R}(n)-\mathbf{R})+\sum_n\sum_e q(n, e)\delta(\mathbf{R}(n)+\mathbf{r}(n, e)-\mathbf{R})$$

and we can express this in terms of g_n as

$$\rho(\mathbf{R}, t) = \sum_n {}^{ne}\langle q(n)g_n(\mathbf{R}, \mathbf{r}(n, e), t)\rangle + \sum_n\sum_e {}^{ne}\langle q(n, e)g_n(\mathbf{R}-\mathbf{r}(n, e), \mathbf{r}(n, e'), t)\rangle, \tag{8.9}$$

where the prime on e in the last term merely emphasizes that e' runs over all the electrons in the nth atom in the functional form of g_n.

We now assume that the distribution function varies slowly with its first argument $\mathbf{R}(n)$, over distances of atomic size. This may or may not be true for an actual statistical ensemble but, as we have already remarked, it must be true for the truncation function if we are to use only a few lower atomic multipole moments to describe the atoms. We can then expand g_n in the second term in (8.9) as a Taylor series about the point $\mathbf{R}$. If we let g_n with no explicit argument represent $g_n(\mathbf{R}, \mathbf{r}(n, e'), t)$ we have

$$g_n(\mathbf{R}-\mathbf{r}(n, e), \mathbf{r}(n, e'), t) = g_n - r_i(n, e)\frac{\partial g_n}{\partial R_i} + \tfrac{1}{2}r_i(n, e)r_j(n, e)\frac{\partial^2 g_n}{\partial R_i \partial R_j} + \ldots. \tag{8.10a}$$

We now have

$$\rho(\mathbf{R}, t) = \sum_n {}^{ne}\langle (q(n)+\sum_e q(n, e))g_n\rangle - \nabla_R \cdot \sum_n {}^{ne}\langle q(n, e)\mathbf{r}(n, e)g_n\rangle$$

$$+\frac{\partial^2}{\partial R_i \partial R_j}\sum_n {}^{ne}\langle \tfrac{1}{2} \sum_e q(n, e)r_i(n, e)r_j(n, e)g_n\rangle. \tag{8.11}$$

We identify the first term easily enough as the expectation value of the net charge density. It vanishes for neutral atoms. If we call this term $\rho_f(R, t)$ and note that the second and third terms are $-\nabla \cdot \mathbf{P}$ and $(\partial^2/\partial R_i \partial R_j)Q_{ij}$ we obtain the desired result,

$$\rho(\mathbf{R}, t) = \rho_f(\mathbf{R}, t) - \nabla \cdot \mathbf{P} + \frac{\partial^2}{\partial R_i \partial R_j} Q_{ij} \tag{8.12}$$

where ρ_f is the "true" or "free" charge density. $\mathbf{P}$ the dipole moment density and Q_{ij} the quadrupole moment density, expressed by (8.8a, b, c, d) in terms of the atomic moments. Thus we have

$$\rho_f = \langle \sum_n (q(n)+\sum_e q(n, e))\delta(\mathbf{R}(n)-\mathbf{R})\rangle, \tag{8.13a}$$

$$\mathbf{P} = \langle \sum_n \pi(n)\delta(\mathbf{R}(n)-\mathbf{R})\rangle, \tag{8.13b}$$

$$\mathbf{M} = \langle \sum_n \mu(n)\delta(\mathbf{R}(n)-\mathbf{R})\rangle, \tag{8.13c}$$

$$Q_{ij} = \langle \sum_n q_{ij}(n)\delta(\mathbf{R}(n)-\mathbf{R})\rangle. \tag{8.13d}$$

The total current density $\mathbf{J}(\mathbf{R}, t)$ is the expectation value of

$$\sum_n q(n)\dot{\mathbf{R}}(n)\delta(\mathbf{R}(n)-\mathbf{R})+\sum_n\sum_e q(n, e)(\dot{\mathbf{R}}(n)+\dot{\mathbf{r}}(n, e))\delta(\mathbf{R}(n)+\mathbf{r}(n, e)-\mathbf{R})$$

and its analysis is altogether more tedious. First of all we shall have to consider the values $\hat{\dot{\mathbf{R}}}(n)$ and $\hat{\dot{\mathbf{r}}}(n, e)$ obtained by averaging $\dot{\mathbf{R}}(n)$ and $\dot{\mathbf{r}}(n, e)$ over all the momenta and the co-ordinates of the other atoms and their particles. Thus we have

$$\hat{\dot{\mathbf{R}}}(n) = \frac{1}{g_n}\int \dot{\mathbf{R}}(n) f(\mathbf{R}(n'), \mathbf{P}(n'), \mathbf{r}(n', e), \mathbf{p}(n', e), t) \cdot \prod_{n'} d^3\mathbf{P}(n') \prod_{n'e} d^3\mathbf{p}(n', e) \prod_{n\neq n} d^3\mathbf{R}(n') \cdot \prod_{\substack{n\neq n\\ e}} d^3\mathbf{r}(n', e), \tag{8.14}$$

with a similar expression for $\hat{\dot{\mathbf{r}}}(n, e)$. Notice that the expectation value of $\dot{\mathbf{R}}(n)$ is

$$\langle \dot{\mathbf{R}}(n)\rangle = {}^{ne}\langle g_n \hat{\dot{\mathbf{R}}}(n)\rangle. \tag{8.15}$$

We now have for $\mathbf{J}(\mathbf{R}, t)$ the expression

$$\mathbf{J}(\mathbf{R}, t) = \sum_n {}^{ne}\langle q(n)\hat{\dot{\mathbf{R}}}(n) g_n(\mathbf{R}, \mathbf{r}(n, e), t)\rangle + \sum_n\sum_e {}^{ne}\langle q(n, e)(\hat{\dot{\mathbf{R}}}(n)+\hat{\dot{\mathbf{r}}}(n, e)) g_n(\mathbf{R}-\mathbf{r}(n, e), \mathbf{r}(n, e'), t)\rangle. \tag{8.16}$$

Again we expand g_n in the last term about the point $\mathbf{R}$ and this gives the rather forbidding expression

$$\begin{aligned}\mathbf{J}(\mathbf{R}, t) = {} & \sum_n {}^{ne}\langle (q(n)+\sum_e q(n, e))\hat{\dot{\mathbf{R}}}(n) g_n\rangle + \sum_n {}^{ne}\langle \sum_e q(n, e)\hat{\dot{\mathbf{r}}}(n, e) g_n\rangle \\ & -\sum_n {}^{ne}\langle q(n, e)\hat{\dot{\mathbf{R}}}(n)(\mathbf{r}(n, e)\cdot\nabla_R g_n)\rangle \\ & -\sum_n {}^{ne}\langle \sum_e q(n, e)\hat{\dot{\mathbf{r}}}(n, e)(\mathbf{r}(n, e)\cdot\nabla_R g_n)\rangle \\ & +\tfrac{1}{2}\sum_n {}^{ne}\left\langle \sum_e q(n, e)\hat{\dot{\mathbf{R}}}(n)\left(r_i(n, e) r_j(n, e)\frac{\partial^2}{\partial R_i \partial R_j} g_n\right)\right\rangle.\end{aligned} \tag{8.17}$$

The third and fifth terms in this expression correspond to the effects of moving polarized atoms and, although they are of some interest, we shall omit them for simplicity. The reader who is interested in these terms should refer to the paper by Mazur and Nijboer (1953). The first term is easily recognizable as the convection current $\mathbf{J}_f$ due to the motion of the units as wholes. The second term is clearly connected with $\dot{\mathbf{P}}$ and the fourth term with $\mathbf{M}$ and $\dot{Q}_{ij}$ but the connection is not entirely straightforward.

Continuous Media

We remember (equation (6.6)), that

$$\frac{\partial}{\partial t}\mathbf{P} = \frac{\partial}{\partial t}\langle\mathbf{P}\rangle = \left\langle\frac{\mathrm{d}}{\mathrm{d}t}\mathbf{P}\right\rangle$$

where d/dt is the total derivative in phase space. Now the microscopic expression from which **P** is derived is

$$\sum_n\sum_e q(n, e)\mathbf{r}(n, e)\delta(\mathbf{R}(n)-\mathbf{R})$$

which does not depend explicitly on t so that its total time derivative is

$$\sum_n\sum_e \{q(n, e)\dot{\mathbf{r}}(n, e)\delta(\mathbf{R}(n)-\mathbf{R})+q(n, e)\mathbf{r}(n, e)(\dot{\mathbf{R}}(n)\cdot\nabla_{R(n)}\delta(\mathbf{R}(n)-\mathbf{R}))\}.$$

The last term involves the motion of atomic units and so we shall omit it. If in the remaining term we form an ensemble average by first integrating over all the phase space variables except $\mathbf{r}(n, e)$, $\mathbf{R}(n)$ we obtain

$$\frac{\partial}{\partial t}\mathbf{P}(\mathbf{R}, t) = \sum_n {}^{ne}\langle\sum_e q(n, e)\hat{\mathbf{r}}(n, e)g_n\rangle$$

which coincides with the second term in (8.17). At this stage we have

$$\mathbf{J}(\mathbf{R}, t) = \mathbf{J}_f(\mathbf{R}, t)+\dot{\mathbf{P}}(\mathbf{R}, t)-\sum_n {}^{ne}\langle\sum_e q(n, e)\hat{\mathbf{r}}(n, e)(\mathbf{r}(n, e)\cdot\nabla_R g_n)\rangle$$

which may be written, in a judicious mixture of vector and tensor notation, as

$$\mathbf{J}(\mathbf{R}, t) = \mathbf{J}_f(\mathbf{R}, t)+\dot{\mathbf{P}}(\mathbf{R}, t)-\sum_n {}^{ne}\left\langle\sum_e q(n, e)\hat{\mathbf{r}}_i(n, e)r_j(n, e)\frac{\partial g_n}{\partial R_j}\right\rangle. \tag{8.18}$$

The curl of the vector **M** is

$$(\nabla\wedge\mathbf{M})_i = \tfrac{1}{2}\sum_n {}^{ne}\left\langle\sum_e q(n, e)\varepsilon_{ijk}\frac{\partial}{\partial R_l}(\varepsilon_{klm}r_l(n, e)\hat{\mathbf{r}}_m(n, e)g_n)\right\rangle \tag{8.19}$$

and, since the only function of **R** is g_n, we obtain, using the relation (see Appendix)

$$\sum_k \varepsilon_{ijk}\varepsilon_{klm} = \delta_{il}\delta_{jm}-\delta_{im}\delta_{jl},$$

$$(\nabla\wedge\mathbf{M})_i = \tfrac{1}{2}\sum_n {}^{ne}\left\langle\sum_e q(n, e)(r_i(n, e)\hat{r}_j(n, e)-r_j(n, e)\hat{r}_i(n, e))\frac{\partial g_n}{\partial R_j}\right\rangle.$$

The time derivative of Q_{ij}, if we again neglect atomic motion, is

$$\dot{Q}_{ij} = \tfrac{1}{2}\sum_n {}^{ne}\langle\sum_e q(n, e)(r_i(n, e)\hat{r}_j(n, e)+r_j(n, e)\hat{r}_i(n, e))g_n\rangle$$

and so the last term in equation (8.18) can be expressed as

$$(\nabla\wedge\mathbf{M})_i-\frac{\partial}{\partial t}\frac{\partial}{\partial R_j}Q_{ij},$$

which yields the total current density as

$$\mathbf{J} = \mathbf{J}_f+\frac{\partial}{\partial t}\left(\mathbf{P}-\frac{\partial}{\partial R_j}Q_{ij}\right)+\nabla\wedge\mathbf{M}, \tag{8.20}$$

where, in mixing vector and tensor notation, we imply that the ith component is to be considered. We have thus expressed $\mathbf{J}$ in terms of the atomic properties $\mathbf{J}_f$, $\mathbf{P}$, Q_{ij} and $\mathbf{M}$.

Atomic Motion

The terms that we have omitted are, in general, complicated but, if the motion of the atoms can be separated from the internal motion of the particles of the atoms, they can be expressed (see Mazur and Nijboer, 1953) as an extra term

$$\mathbf{J}^* = \nabla \wedge \left(\left(\mathbf{P}-\frac{\partial Q_{ij}}{\partial R_j}\right) \wedge \mathbf{v}\right) \tag{8.21}$$

where $\mathbf{v}$ is the local macroscopic velocity of the atoms. They are therefore equivalent to an additional magnetization

$$\mathbf{M}^* = \left(\mathbf{P}-\frac{\partial}{\partial R_j} Q_{ij}\right) \wedge \mathbf{v}. \tag{8.22}$$

This is usually small but may be significant if the atoms have large permanent electric moments but no permanent magnetic moments.

We notice in (8.20) that the terms in $\mathbf{M}$ and $\dot{Q}_{ij}$ arose from the same term in $\mathbf{J}$. This confirms our view that at high frequencies one will always have to consider the quadrupole term along with $\mathbf{M}$.

The Vectors D and H

If we now define the vectors

$$\mathbf{D} = \mathbf{E}+\mathbf{P}-\frac{\partial Q_{ij}}{\partial R_j} \tag{8.23a}$$

$$\mathbf{H} = \mathbf{B}-\mathbf{M}, \tag{8.23b}$$

the two inhomogeneous macroscopic field equations

$$\nabla \cdot \mathbf{E} = \rho_f - \nabla \cdot \left(\mathbf{P}-\frac{\partial Q_{ij}}{\partial R_j}\right) \tag{8.24a}$$

and

$$\nabla \wedge \mathbf{B}-\dot{\mathbf{E}} = \mathbf{J}_f+\frac{\partial}{\partial t}\left(\mathbf{P}-\frac{\partial Q_{ij}}{\partial R_j}\right)+\nabla \wedge \mathbf{M} \tag{8.24b}$$

take the familiar forms

$$\nabla \cdot \mathbf{D} = \rho_f$$

and

$$\nabla \wedge \mathbf{H}-\dot{\mathbf{D}} = \mathbf{J}_f.$$

We may perhaps remark that some authors define $\mathbf{D}$ as $\mathbf{E}+\mathbf{P}$ and omit the quadrupole term. This is usually adequate but can lead to difficulties in dealing with effects such as optical activity (Pershan, 1963).

We have now arrived at the canonical form of the macroscopic field equations in the presence of matter

$$\begin{aligned} \nabla \cdot \mathbf{B} &= 0, \qquad \nabla \wedge \mathbf{E}+\dot{\mathbf{B}} = 0, \\ \nabla \cdot \mathbf{D} &= \rho_f, \qquad \nabla \wedge \mathbf{H}-\dot{\mathbf{D}} = \mathbf{J}_f, \end{aligned} \tag{8.25}$$

in which the vectors **D** and **H** are related to **E** and **B** by equations (8.23), which involve quantities **P**, Q_{ij} and **M** which, in turn, can be expressed in terms of the atomic structure of the medium. In mks or S.I. units we have of course to replace (8.23) by

$$\mathbf{D} = \varepsilon_0\mathbf{E}+\mathbf{P}-\frac{\partial Q_{ij}}{\partial R_j}$$

and (8.26)

$$\mathbf{H} = \frac{1}{\mu_0}\mathbf{B}-\mathbf{M}.$$

The solid-state physicist is apt to regard equation (8.23) or (8.26) as expressing the most important properties of the vectors **D** and **H**. The electrical engineer, following Maxwell, is, however, more likely to regard the equations

$$\nabla \cdot \mathbf{D} = \rho_f$$

$$\nabla \wedge \mathbf{H} = \mathbf{J}_f+\dot{\mathbf{D}}$$

as more significant since they show that **D** is a flux vector whose sources are the true or free charges and **H** is a vector whose vortices are the true and displacement currents. The relations between these vectors and the vectors **E** and **B** which appear in the force equations are then to be treated as empirical rules. We shall pursue this point of view in the next chapter.

Constitutive Relations

Clearly Maxwell's equations in the form (8.25) are useless unless we have further relations between the vectors involved. In practical electrostatics the vector usually specified most directly by the conditions of the problem is **E**, and it is therefore usual to express the subsidiary vectors **P** or **D** in terms of **E**. On the other hand in magnetism the most readily specified vector is **H**, for, if the sources of a static field are permanent magnets of given magnetization $\mathbf{M}_0$, we have $\nabla \cdot \mathbf{H} = -\nabla \cdot \mathbf{M}_0$ and $\nabla \wedge \mathbf{H} = 0$, while if the sources are free or true currents we have $\nabla \wedge \mathbf{H} = \mathbf{J}_f$. For this reason it is usual to express **B** or **M** in terms of **H** rather than vice versa. These subsidiary relations are known as constitutive relations and they express in compact form the important electromagnetic properties of a medium. There is no reason why the relations should always be linear but the simplest relations are of course

$$\mathbf{P} = \varepsilon_0\chi_e\mathbf{E},$$

$$\mathbf{D} = \varepsilon_0(1+\chi_e)\mathbf{E} = \varepsilon\varepsilon_0\mathbf{E},$$

$$\mathbf{M} = \chi_m\mathbf{H},$$

$$\mathbf{B} = \mu_0(1+\chi_m)\mathbf{H} = \mu\mu_0\mathbf{H},$$

to which we might add Ohm's law

$$\mathbf{J}_f = \sigma\mathbf{E}.$$

Much more general relations than these are possible; for example, we have, in the Faraday effect,

$$P_i = \varepsilon_0\chi_e E_i+v_{ijk}E_jB_k$$

where v_{ijk} is a third rank tensor. We shall be discussing constitutive relations in several of the later chapters.

Polarization and Magnetization as Source Terms

It is sometimes convenient to treat **P** and **M** as source terms in the field equations. This is perhaps most useful in dealing with permanent magnets and in non-linear optics, but the method is quite general and can be set up in a number of different ways. In time-dependent problems the simplest approach is to use the vector and scalar potentials **A** and ϕ in the Lorentz gauge. We then obtain the inhomogeneous wave equations

$$\nabla^2\mathbf{A}-\mu_0\varepsilon_0\ddot{\mathbf{A}} = -\mu_0(\mathbf{J}_f+\dot{\mathbf{P}}+\nabla\wedge\mathbf{M})$$

$$\nabla^2\phi-\mu_0\varepsilon_0\ddot{\phi} = -\frac{1}{\varepsilon_0}(\rho_f-\nabla\cdot\mathbf{P})$$

and the fields are $\mathbf{B} = \nabla\wedge\mathbf{A}$, $\mathbf{E} = -\dot{\mathbf{A}}-\nabla\phi$. If the only term which need be considered is **P**, the retarded vector potential is

$$\mathbf{A}(\mathbf{r}_1, t_1) = \int\frac{\mu_0\dot{\mathbf{P}}(\mathbf{r}_2, t_1-(r_{12}/c))}{4\pi r_{12}}\,\mathrm{d}^3\mathbf{r}_2$$

and this can be used to give a compact treatment of radiation from an oscillating polarization.

If we have a static system in which there are no free charges or currents, the magnetic and electric equations separate and we have the two independent groups of equations:

$$\nabla\cdot\mathbf{B} = 0, \qquad \nabla\wedge\mathbf{B} = \mu_0\nabla\wedge\mathbf{M}, \tag{8.27a}$$

or

$$\nabla\wedge\mathbf{H} = 0, \qquad \nabla\cdot\mathbf{H} = -\nabla\cdot\mathbf{M}, \tag{8.27b}$$

and

$$\nabla\cdot\mathbf{D} = 0, \qquad \nabla\wedge\mathbf{D} = \nabla\wedge\mathbf{P}, \tag{8.28a}$$

or

$$\nabla\wedge\mathbf{E} = 0, \qquad \nabla\cdot\mathbf{E} = -\frac{1}{\varepsilon_0}\nabla\cdot\mathbf{P}. \tag{8.28b}$$

These equations can be integrated directly to give

$$\mathbf{B}_1 = \nabla_1\wedge\mathbf{A}_m = \nabla_1\wedge\int\frac{\mu_0\nabla_2\wedge\mathbf{M}_2\,\mathrm{d}^3\mathbf{r}_2}{4\pi r_{12}} \tag{8.29a}$$

$$\mathbf{H}_1 = -\nabla_1\phi_m = \nabla_1\int\frac{\nabla_2\cdot\mathbf{M}_2\,\mathrm{d}^3\mathbf{r}_2}{4\pi r_{12}} \tag{8.29b}$$

$$\mathbf{D}_1 = \nabla_1\wedge\mathbf{A}_e = \nabla_1\wedge\int\frac{\nabla_2\wedge\mathbf{P}_2\,\mathrm{d}^3\mathbf{r}_2}{4\pi r_{12}} \tag{8.30a}$$

$$\mathbf{E}_1 = -\nabla_1\phi_e = \nabla_1\int\frac{\nabla_2\cdot\mathbf{P}_2\,\mathrm{d}^3\mathbf{r}_2}{4\pi\varepsilon_0 r_{12}}, \tag{8.30b}$$

where $\mathbf{r}_1$ is the field point and $\mathbf{r}_2$ the location of the volume element. The quantities **A** and ϕ are the appropriate vector and scalar potentials. For somewhat obscure reasons it is unusual to make any use of (8.30a) in discussing electrostatics whereas (8.29a) and (8.29b) are mentioned with equal emphasis in discussion of magnetostatics.

If we have an isolated polarized or magnetized body in vacuum, **P** and **M** are discontinuous at the surface and thus there are singularities in $\nabla \cdot \mathbf{P}$ and $\nabla \cdot \mathbf{M}$ associated with the normal components P_n and M_n at the surface. The singularities in $\nabla \wedge \mathbf{P}$ and $\nabla \wedge \mathbf{M}$ are associated with the tangential components. In this case it is convenient to divide the volume of integration into two regions, one entirely within the body and whose boundary lies just inside the physical surface and the other which includes the surface layer and can be expressed as a surface integral. For the scalar potentials we then obtain the expressions

$$\phi_m = -\int_{\text{body}} \frac{\nabla_2 \cdot \mathbf{M}_2 \, d^3\mathbf{r}_2}{4\pi r_{12}} + \int_{\text{surface}} \frac{\mathbf{M}_2 \cdot d\mathbf{S}_2}{4\pi r_{12}}, \tag{8.31a}$$

$$\phi_e = -\int_{\text{body}} \frac{\nabla_2 \cdot \mathbf{P}_2 \, d^3\mathbf{r}_2}{4\pi\varepsilon_0 r_{12}} + \int_{\text{surface}} \frac{\mathbf{P}_2 \cdot d\mathbf{S}_2}{4\pi\varepsilon_0 r_{12}} \tag{8.31b}$$

while for the magnetic vector potential we obtain

$$\mathbf{A}_m = \int_{\text{body}} \frac{\mu_0 \nabla_2 \wedge \mathbf{M}_2}{4\pi r_{12}} d^3\mathbf{r}_2 - \int_{\text{surface}} \frac{\mu_0 \mathbf{M}_2 \wedge d\mathbf{S}_2}{4\pi r_{12}}. \tag{8.32}$$

In (8.31a) and (8.31b) the surface terms can be interpreted in terms of an equivalent surface density of either "magnetic poles" or charge

$$\sigma_m = M_n, \tag{8.33a}$$

$$\sigma_e = P_n. \tag{8.33b}$$

The reader will no doubt recognize (8.33a) as the basic concept used in an elementary discussion of permanent magnets.

The surface term in (8.32) can be interpreted in terms of an equivalent current in the surface perpendicular to the direction of **M** at the surface. If **n** is the unit outward normal we have a linear surface current density $\mathbf{I} = \mathbf{M} \wedge \mathbf{n}$. This result is of little practical use except in systems of very simple geometry in which **M** is tangential at the surface. Thus, if we have a long cylinder magnetized parallel to its axis, the "currents" will "flow" round the surface of the cylinder. The field **B** due to such a cylinder is then seen to be identical with the field of a long solenoid of the same shape.

Truncation and Averaging

The formula (8.31b) is often interpreted as meaning that the body has an effective volume charge $\rho_b = -\nabla \cdot \mathbf{P}$ together with the surface charge $\sigma = P_n$ given by (8.33b). We now consider how these concepts are related to our notions about the macroscopic variables as ensemble averages or truncations of the microscopic quantities. In Fig. 8.1 we show a familiar diagram illustrating conditions near the surface of a uniformly polarized dielectric. For simplicity we assume that the dipoles, of magnitude $q\delta$, are located on a cubic lattice with spacing a. At first sight it appears that the surface charge is $\sigma = q/a^2$ which is independent of

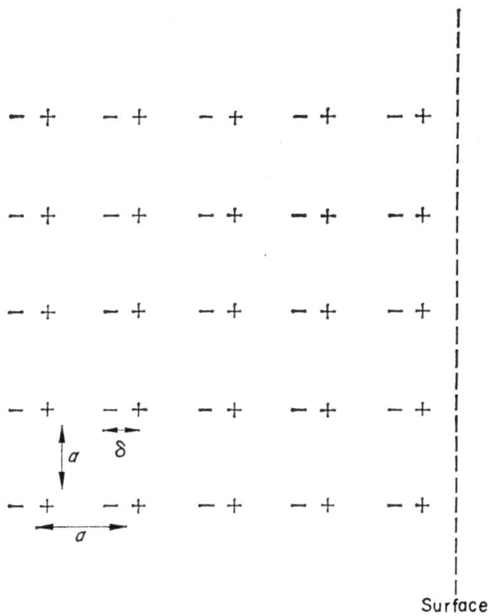

FIG. 8.1. Charges in a polarized dielectric.

the polarization **P**. This is of course fallacious. In any case, in real solids, the charge nearest the surface is always electronic and negative. If, however, we imagine a rectangular slab of area A and thickness l placed randomly on this diagram with l normal to the surface and one face outside the surface, the net charge covered by this slab is zero if the face in the medium lies between dipoles, and $+(qA/a^2)$ if it lies within dipoles. We see that the expectation value of the charge for a random position of the slab is

$$\frac{qA}{a^2}\frac{\delta}{a} = A\frac{q\delta}{a^3} = AP_n.$$

Thus the effective surface charge density is apparently P_n. Placing the slab randomly on the figure is in fact one way of illustrating the process of truncation. If $\rho(\mathbf{R})$ is the exact charge density and V the volume of the slab the expected charge in the slab is

$$\langle\rho\rangle = \int_V \rho(\mathbf{R}-\mathbf{s}) f(\mathbf{s})\, d^3\mathbf{s},$$

where $f(\mathbf{s})$ describes the probability that some fixed point in the slab falls at **s**. Although this gives us a pictorial representation of truncation it is obviously only rather approximate since we have some difficulty in imagining the effect of a distribution $f(\mathbf{s})$ which is negative for some values of **s**.

In Fig. 8.2 we show a linear array of dipoles increasing in strength towards the right. If we place a line element of length l on this diagram the net charge in l will be zero if the left hand end at x and the right hand end at $l+x$ *either* both lie between dipoles *or* both intercept dipoles. The charge will be $+q$ if only the left hand end intercepts a dipole. This has a probability

$$\left(\frac{\delta}{\delta}\right)_x\left(1-\frac{\delta}{a}\right)_{x+l}.$$

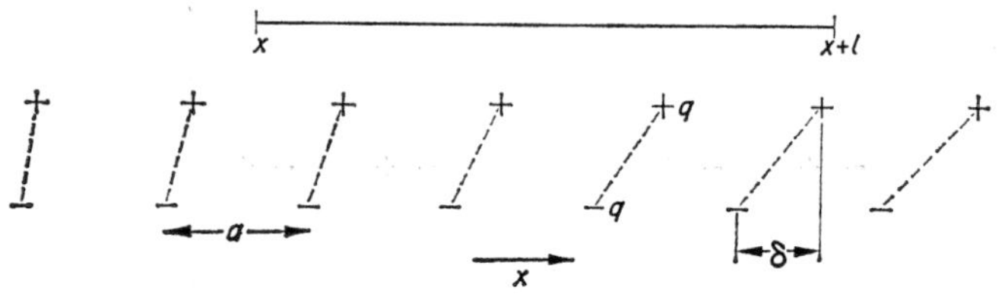

FIG. 8.2. A linear array of dipoles increasing in strength towards the right.

The charge is $-q$ if the left hand end lies between dipoles and the right hand end intercepts a dipole. This has a probability

$$\left(1-\frac{\delta}{a}\right)_x\left(\frac{\delta}{a}\right)_{x+l}$$

and so the expected charge is

$$q\left\{\left(\frac{\delta}{a}\right)_x\left(1-\frac{\delta}{a}\right)_{x+l}-\left(1-\frac{\delta}{a}\right)_x\left(\frac{\delta}{a}\right)_{x+l}\right\}$$

which gives

$$-\left\{\left(\frac{q\delta}{a}\right)_{x+l}-\left(\frac{q\delta}{a}\right)_x\right\}.$$

If we express this in terms of the linear polarization density $P = q\delta/a$ we see that the linear charge density is

$$\rho_b = -\frac{\partial}{\partial x}\left(\frac{q\delta}{a}\right) = -\frac{\partial P}{\partial x}.$$

By this means we obtain a pictorial representation of the relation $\rho_b = -\nabla . \mathbf{P}$. Notice that the representations obtained from Figs. 8.1 and 8.2 assume first of all that $\mathbf{P}$ has been defined and secondly involve the use of randomly positioned volume elements to obtain σ or ρ_b. It is not possible to draw on these figures a single volume element in which both $\mathbf{P}$ and σ or ρ_b simultaneously have well-defined values other than zero.

Summary

We may summarize our conclusions as follows. In material media in which we can identify definite structural units and when all fields of interest to us vary slowly over the dimensions of these units, we can describe the electromagnetic properties of the units in terms of their charge and a limited number of multipole moments. If, in addition, either we have a statistical ensemble which is coarse in coordinate space, or we are prepared to truncate all the variables at some wavelength larger than the "grain" of the medium, we can go on to define macroscopic charge, current, and multipole moment densities,

$$\rho_f,\ \mathbf{J}_f,\ \mathbf{P},\ Q_{ij},\ \mathbf{M}$$

and we can use these to express the field equations in the form

$$\nabla . \mathbf{B} = 0, \qquad \nabla \wedge \mathbf{E}+\dot{\mathbf{B}} = 0,$$

$$\varepsilon_0 \nabla . \mathbf{E} = \rho_f - \nabla . \left(\mathbf{P}-\frac{\partial Q_{ij}}{\partial R_j}\right),$$

$$\frac{1}{\mu_0}\nabla\wedge\mathbf{B}-\varepsilon_0\dot{\mathbf{E}} = \mathbf{J}_f+\dot{\mathbf{P}}-\frac{\partial\dot{Q}_{ij}}{\partial R_j}+\nabla\wedge\mathbf{M}.$$

We then define the auxiliary vectors **D** and **H** so that

$$\nabla\,.\,\mathbf{D} = \rho_f$$

$$\nabla\wedge\mathbf{H} = \dot{\mathbf{D}}+\mathbf{J}$$

which implies that

$$\mathbf{D} = \varepsilon_0\mathbf{E}+\mathbf{P}-\frac{\partial Q_{ij}}{\partial R_j}+\text{etc.},$$

$$\mathbf{H} = \frac{1}{\mu_0}\mathbf{B}-\mathbf{M}+\text{etc.}$$

In the last two equations we could, if we wished, add terms corresponding to higher order multipole moments. We have therefore shown that, in this intrinsically macroscopic context, the microscopic equations lead to the canonical Maxwell equations and that, in the Maxwell equations, the densities **P**, Q_{ij}, **M** do in fact correspond to averages of the atomic quantities. In practice, except perhaps in gases and liquids the statistical ensemble is rarely coarse enough to lead to useful definitions of **P**, Q_{ij} and **M** and we are therefore ultimately relying on the truncation procedure to justify these results. We may contrast truncation and the formation of an ensemble average, for in truncating a problem we reject information, potentially available to us, about structure on an atomic scale whereas in an ensemble we acknowledge that some potentially useful information is not available to us.

It should by now be apparent that all attempts to deduce the macroscopic equations from the microscopic equations in a general way, by appealing to statistical, time, or quantum averages, are inherently fallacious. The averaging process is not, in general, inherent in the physical system but only in the use we propose to make of the equations. If we are content to ignore all spatial Fourier components of wavelength λ_0 or less, and λ_0 is large compared with the grain of the medium, then we can always, even if the atoms happen to be at rest on exact lattice sites, pass from the microscopic to the macroscopic equations.

We conclude by noting that the process which leads to the definition of the macroscopic fields **E** and **B** and the polarization densities **P** and **M**, and therefore also to the definition of **D** and **H**, inherently excludes any possibility of our discussing the value of any of these quantities at points specified with microscopic precision. It is therefore entirely meaningless to enquire whether **E** or **D**, on the one hand, or **B** or **H**, on the other, is the "true field" either acting on or within an atom. Questions of this nature which are closely connected with the problem of the Lorentz local-field correction therefore lie outside the realm of purely macroscopic electromagnetism. This will not stop us discussing them in later chapters.

Problems

1. In an entirely hypothetical system the particles are at rest at the exact positions $\mathbf{R}(n, o)$, and $\mathbf{r}(n, e, o)$ relative to $\mathbf{R}(n, o)$. Construct a distribution function for an ensemble in which each configuration of the ensemble is obtained from the original system by a rigid translation **s** of all the particles, the probability of a translation **s** in $d^3\mathbf{s}$ being $\phi(\mathbf{s})\,d^3\mathbf{s}$. What is the form of the function $g_n(\mathbf{R}(n), \mathbf{r}(n, e), t)$ for this ensemble?

Continuous Media

2. If $\rho_b = -\nabla_R \cdot \mathbf{P}(\mathbf{R})$ show that, for a volume V bounded by a surface S

$$\int_V \mathbf{R}\rho_b \, dV = -\int_S \mathbf{R}\,\mathbf{P} \cdot d\mathbf{S} + \int_V \mathbf{P} \, dV.$$

Investigate the result of applying this relation to a dielectric body when (a) the surface S is taken outside the body and (b) the surface S is taken just inside the physical surface of the body. In case (b) consider the total dipole moment of the body when $\mathbf{P}$ is uniform within the body.

3. A long solenoid and a long uniformly magnetized cylinder produce the same magnetic field at points outside the solenoid or cylinder. How are the fields **B** and **H** within the solenoid or cylinder related?

4. Small permanent magnets with parallel magnetic moments **m** are embedded on a cubic lattice of spacing a in a linear homogeneous, isotropic medium of relative magnetic permeability μ in the form of a sphere. What are the values of the mean fields **H** and **B** in the medium?

5. A spherical sample of n type silicon of radius a contains $N = 10^{15}$ mobile electrons per cm^3. Calculate the induced dipole moment of the sphere when it is placed in a weak uniform electric field **E** due to fixed distant point charges. Indicate qualitatively how the moment varies with **E** in very strong fields.

6. The current density in a body can be expressed as $\mathbf{J} = \nabla \wedge \mathbf{I}$ where $\mathbf{I} = 0$ outside the body. Show that no net current leaves the body.

7. A crystal has units of equal and parallel electric dipole moments **p** placed on a cubic lattice. Make a sketch of this situation and draw on the diagram a volume, part of whose surface lies outside the crystal, chosen in such a way that it has a well-defined dipole moment independent of the choice of an origin. What is the net charge in this volume?

CHAPTER 9

The Macroscopic Fields

Introduction

In the last four chapters we have been concerned with the relation between the microscopic and macroscopic fields. In this chapter we devote our attention to the properties of the macroscopic fields.

In an exclusively macroscopic approach to electromagnetism we begin by considering charges and currents in vacuum and arrive at a definition of the fields **E** and **B** in terms of the body force $\rho\mathbf{E}+\mathbf{J} \wedge \mathbf{B}$ in a region where the charge density is ρ and the current density is **J**. The fields satisfy the pair of homogeneous equations

$$\nabla \cdot \mathbf{B} = 0 \quad (9.1a) \quad \text{and} \quad \nabla \wedge \mathbf{E}+\dot{\mathbf{B}} = 0 \quad (9.1b)$$

and a pair of inhomogeneous equations

$$\nabla \cdot \varepsilon_0\mathbf{E} = \rho \quad (9.1c) \quad \text{and} \quad \nabla \wedge \mathbf{B}-\mu_0\varepsilon_0\dot{\mathbf{E}} = \mu_0\mathbf{J} \quad (9.1d)$$

which relate the fields to charge and current. In the presence of matter, i.e. dielectric and magnetic media, these equations are found to be inadequate if we insist on including in ρ and **J** only those charges and currents which are directly accessible to measurement, i.e. the mobile charges and currents discussed in the last chapter. In order to maintain agreement with experiment we find that we have to modify the two inhomogeneous equations, and to see how this is done, we now consider two simple examples.

If the space between two parallel conducting plates forming a capacitor is filled with a medium such as paper or mica we can still attach a meaning to the potential difference ϕ between the plates by considering the work required to take unit charge from one plate to the other, by a path outside the plates in vacuum. There is therefore some meaning in stating that the field between the plates is $E = \phi/d$ where d is their separation. However, if the area of the plates is A, equation (9.1c) yields the charge q on the plates as $q = \varepsilon_0 A\phi/d$ whereas experimentally the charge is somewhat larger. To accommodate this result we assume that equation (9.1c) is to be replaced by $\nabla \cdot (\varepsilon_0\mathbf{E}+\mathbf{P}) = \rho$ where **P** is a vector which vanishes in vacuum and in some way represents a property of the medium between the plates. If we re-write this result as $\nabla \cdot (\varepsilon_0\mathbf{E}) = \rho-\nabla \cdot \mathbf{P}$, it appears that spatial variations in **P** generate electric fields in exactly the same way as charge density. Further, if we consider a finite body with zero net mobile charge, it is easy to show that it generates an electric field at a distance just as though it had a total dipole moment $\int_{\text{Vol}} \mathbf{P}\, \mathrm{d}V$. All this suggests that **P** introduced in this way represents the effects of bound, or immobile, atomic charge in the medium and that **P** is related to the atomic dipole moment density. We cannot, however, go on to identify **P** with this moment density and indeed, as we saw in the last chapter, if there are N atoms per unit volume each with a moment **p**, the vector **P** is not equal to $N\mathbf{p}$ but contains contributions from higher order multipole moment densities. We can, however, regard **P** as a quantity to be related to **E** by experiment and go on to define **D** as $\varepsilon_0\mathbf{E}+\mathbf{P}$ so that **D** satisfies $\nabla\cdot\mathbf{D}=\rho$. This is the usual definition of **D** and **P** but, of course, **P** in this relation contains the quadrupole and higher multipole order terms.

The magnetic field in a long solenoid with n turns per metre, carrying a current I, is $\mu_0 nI$ and, if the area of cross section is A, and the length l, the back e.m.f. when I changes can be derived from (9.1b) and is $\mu_0 n^2 lA\dot{I}$. If magnetic material is introduced into the solenoid we find a different e.m.f. and we can only retain equation (9.1b) if we modify equation (9.1d) which, for our present purposes, is $\nabla\wedge\mathbf{B}=\mu_0\mathbf{J}$. We make this modification by introducing a vector **M** which is zero in vacuum, and in some way represents the properties of the magnetic material, and write the new equation as $\nabla\wedge(\mathbf{B}-\mu_0\mathbf{M})=\mu_0\mathbf{J}$. In a medium in which $\mathbf{J}=0$ we have $\nabla\wedge\mathbf{B}=\mu_0\nabla\wedge\mathbf{M}$ and we can use this to show that the distant field due to a body in which **M** is non-zero is that of a magnetic dipole $\int_{\mathrm{Vol}}\mathbf{M}\,\mathrm{d}V$. Again it is clear that **M** introduced in this way represents the effect of bound atomic currents in the medium and is related to the magnetic dipole moment density. We cannot however go on to identify **M** with $N\mathbf{m}$ in a medium containing N atoms each of magnetic moment **m**, in unit volume. To do this we have to employ the methods used in the last chapter, and even so we can only show that **M** is equal to $N\mathbf{m}$ together with higher-order magnetic multipole terms which are, of course, usually utterly negligible. Also again this does not stop us giving **M** an empirical definition and going on to define **H** as $\mathbf{B}/\mu_0-\mathbf{M}$ so that **H** satisfies $\nabla\wedge\mathbf{H}=\mathbf{J}$. Finally of course we can show that, if mobile charge is to be conserved, $\varepsilon_0\dot{\mathbf{E}}$ in equation (9.1d) has to be replaced by $\varepsilon_0\dot{\mathbf{E}}+\dot{\mathbf{P}}=\dot{\mathbf{D}}$.

Although we have introduced the new vectors **P**, **D**, **M** and **H** by considering specific examples we can express the results more formally. In the presence of matter the fields **E** and **B** satisfy the equations

$$\nabla\cdot\mathbf{B}=0 \quad (9.2\text{a}) \qquad \nabla\wedge\mathbf{E}+\dot{\mathbf{B}}=0 \quad (9.2\text{b})$$

$$\nabla\cdot\varepsilon_0\mathbf{E}=\rho-\nabla\cdot\mathbf{P} \quad (9.2\text{c}) \qquad \nabla\wedge\mathbf{B}-\mu_0\varepsilon_0\dot{\mathbf{E}}=\mu_0(\mathbf{J}+\dot{\mathbf{P}}+\nabla\wedge\mathbf{M}), \quad (9.2\text{d})$$

where **P** and **M** are vectors associated with matter. If ρ, **J**, **P** and **M** are known, these equations can be used to calculate **E** and **B**, the macroscopic fields. We now have to consider the body force due to these fields. From the way we introduced the auxiliary vectors **P** and **M** it will be clear that, in parts of the system free of matter, the fields **E** and **B** obtained by solving the set of equations (9.2) are just the usual electric and magnetic fields and so, in these regions, the macroscopic body force is $\rho\mathbf{E}+\mathbf{J}\wedge\mathbf{B}$, and the force on a macroscopic body of charge q is $q\mathbf{E}$, while the force on a line element $\mathrm{d}\mathbf{l}$ of a circuit carrying a current I is $I\,\mathrm{d}\mathbf{l}\wedge\mathbf{B}$. We cannot, however, assert that these expressions also give the forces acting on macroscopic bodies immersed or embedded in a material medium. The value of these forces is a matter for calculation. There is no room in the theory for new definitions. In the next chapter we shall consider the nature of these forces but we may remark here that the concept of an electromagnetic force acting on a material body embedded in a material medium is not quite as straightforward as it appears to be. Thus if we embed a charged metal sphere in a solid dielectric medium held in a fixed position, the total force on the sphere must be zero. It certainly contains a purely electromagnetic term but it also contains a term arising from stress in the medium and, since this stress contains electromagnetic contributions, there is no clear separation between electromagnetic and mechanical terms in the force acting on the sphere. It is therefore pointless to attempt to define the macroscopic fields **E** and **B** inside a material medium in terms of forces acting on hypothetical material bodies in the medium. Rather we must take the view that **E** and **B** are the fields obtained as solutions of equations (9.2) and that these equations are so constructed that the body force in vacuum remains $\rho\mathbf{E}+\mathbf{J}\wedge\mathbf{B}$. This need not prevent us looking for special cases of heuristic value where the forces take a simple and familiar form. Thus, for example, in chapter 10 we shall show that the force on a charged macroscopic body in a *fluid* medium is $q\mathbf{E}$ and the force on an element $\mathrm{d}\mathbf{l}$ of wire carrying a current I in a *fluid* medium is $I\,\mathrm{d}\mathbf{l}\wedge\mathbf{B}$.

The Properties of E and B

Before we discuss the fields **D** and **H** it will be useful to establish two general properties of the fields **E** and **B** by writing (9.2a) and (9.2b) in integral form as

$$\oint B_n\,\mathrm{d}S = 0 \quad (9.3\mathrm{a}) \qquad \oint \mathbf{E}\,.\,\mathrm{d}l = -\int \dot{\mathbf{B}}_n\,\mathrm{d}S \quad (9.3\mathrm{b}).$$

From equation (9.3a) we discover that across an open surface element separating two regions (1) and (2), the normal components $B_n(1)$ and $B_n(2)$ in the two sides of the surface are equal. This is true whether it is the physical surface separating two media or merely a mathematical surface. From equation (9.3b) we find that, since **B** is non-singular, the tangential components $E_t(1)$ and $E_t(2)$ are continuous.

Equation (9.3a) also shows that the flux lines of **B** form closed loops, while (9.3b) shows that when $\dot{\mathbf{B}} = 0$ the line integral of **E** round any closed loop is zero.

The Properties of D and H

In integral form equation (9.2c) becomes

$$\oint D_n\,\mathrm{d}S = \int \rho\,\mathrm{d}V = Q, \tag{9.4}$$

where Q is the total charge inside the closed surface S and this expresses the property of **D** that its flux lines end on free charge (not the fictitious bound polarization charges). If we consider a thin laminar surface, of thickness δ, separating two regions (1) and (2), equation (9.4) yields the relation

$$D_n(2)-D_n(1) = \rho\delta \tag{9.5}$$

where the positive normal n is from (1) towards (2). If ρ is finite, then as $\delta \to 0$ we have

$$D_n(2) = D_n(1). \tag{9.6}$$

However, the numerical values of **D** which occur in practical problems are rarely greater than 10^{-1} coulomb m^{-2} while the density of mobile charge in metals is of the order of 10^{10} coulomb m^{-3}, so that, in many cases we can have an extreme discontinuity in D_n over distances of the order of less than 10^{-10} m or 1 Å. Thus, in dealing with metal surfaces it is often appropriate to replace (9.5) by

$$D_n(2)-D_n(1) = \sigma \tag{9.7}$$

where σ is a free (mobile) surface charge density. Great care must be exercised in applying (9.7) to semiconductor surfaces, since in semiconductors the mobile charge density is much less and the depth of the depletion layer at the surface can, as in a field effect transistor, assume macroscopic dimensions.

The Macroscopic Fields

The remaining Maxwell equation in integral form is

$$\oint \mathbf{H} \, . \, d\mathbf{l} = \int (J_n + \dot{D}_n) \, dS \tag{9.8}$$

and tells us that the line integral of **H** around any closed loop is equal to the sum of the linked free and displacement currents. From (9.8) we obtain, for a laminar layer of thickness δ separating two regions,

$$H_t(2) - H_t(1) = (J_\tau + \dot{D}_\tau)\delta \tag{9.9}$$

where the directions **t**, **n**, and $\boldsymbol{\tau}$ are related by $\boldsymbol{\tau} = \mathbf{n} \wedge \mathbf{t}$. Since $\dot{\mathbf{D}}$ is non-singular this leads as $\delta \to 0$ to

$$H_t(2) = H_t(1) \tag{9.10}$$

if **J** is also non-singular. In static problems **J** can only be non-zero in superconductors, but then it is possible for a substantial discontinuity to occur in a macroscopically small layer at the surface and so we might find it useful in some problems to introduce a surface current I_τ amp m^{-1} so that

$$H_t(2) - H_t(1) = I_\tau. \tag{9.11}$$

At high frequencies the currents in the surface of a metal are, as we shall see, confined to a layer—the skin depth—which, though not macroscopically infinitesimal, may yet be small compared to other dimensions of the problem. In this case we have $H_t(2) = 0$ well into the metal, and it is useful to replace the exact equation (9.10) by the approximate equation

$$I_\tau = -H_t(1) \tag{9.12}$$

with the understanding that we exclude the region near the surface of the metal from explicit consideration.

It is easiest to form a working notion of the properties of the fields by considering static problems and so we now turn from generalities to a discussion of particular electrostatic and magnetostatic problems.

Electrostatic Fields

In electrostatics we have the following relations:

$$\nabla \wedge \mathbf{E} = 0 \quad (9.13a) \qquad \nabla \, . \, \mathbf{E} = \frac{1}{\varepsilon_0}(\rho - \nabla \, . \, \mathbf{P}) = \frac{1}{\varepsilon_0}(\rho + \rho_b) \quad (9.13b)$$

$$\nabla \wedge \mathbf{D} = -\nabla \wedge \mathbf{P} \quad (9.13c) \qquad \nabla \, . \, \mathbf{D} = \rho \quad (9.13d)$$

Thus, whereas the line integral of **E** around a closed loop is zero, that of **D** is not necessarily zero. The lines of **D** end on free charge but the lines of **E** end on both free charge and changes in the polarization vector **P**. We illustrate these relations in the absence of free charge ρ by showing, in Fig. 9.1, the three vectors **E**, **P**, and **D** associated with a cylinder of an insulating material with a permanent polarization **P**.

To illustrate the effects of free charge we consider an archaic device, the electrophorus. This consists of a flat disc of permanently polarized dielectric resting on an earthed metal plate and, in addition, there is a movable metal plate which can be placed on top of the disc and either left isolated or earthed. The sequence of operations is shown in Figs. 9.2a, b, c, d, e).

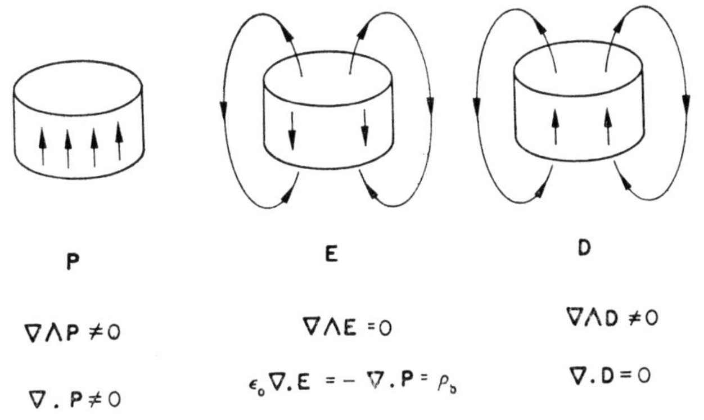

FIG. 9.1. Electric vectors associated with a polarized cylinder.

In Fig. 9.2a the metal plate is at a distance, in 9.2b it is in place and isolated. If it has unit area the charge on the lower surface is $-D$ and the charge on the upper surface $+D$. The plate is at a positive potential. In Fig. 9.2c the plate is earthed and so $E = 0$. The charge on the lower surface is $-D = -P$ and there is no charge on the upper surface. In Fig. 9.2d the earth connection is removed. The net charge $-P$ on the plate, thought of as a source of **E**, just cancels the effect of the polarization charge P on the surface of the dielectric and so **E** remains zero. In Fig. 9.2e the plate, with its true charge $q = -P$, is removed. The state of the dielectric reverts to 9.2a but the potential of the plate becomes more and more negative as it is withdrawn from the dielectric. The work done in withdrawing the plate is converted to electrostatic energy. The charge $q = -P$ can eventually be transferred to an external electrode at a negative potential.

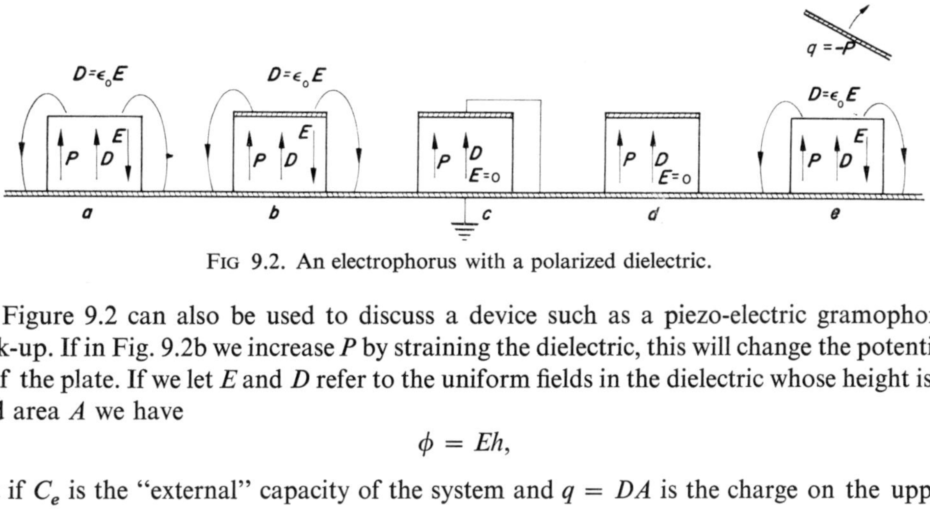

FIG 9.2. An electrophorus with a polarized dielectric.

Figure 9.2 can also be used to discuss a device such as a piezo-electric gramophone pick-up. If in Fig. 9.2b we increase P by straining the dielectric, this will change the potential ϕ of the plate. If we let E and D refer to the uniform fields in the dielectric whose height is h and area A we have

$$\phi = Eh,$$

but if C_e is the "external" capacity of the system and $q = DA$ is the charge on the upper surface of the plate, we also have

$$\phi C_e = q = DA = (P - \varepsilon_0)A,$$

so that

$$E = \frac{P}{\varepsilon_0}\frac{\varepsilon_0 A/h}{C_e + (\varepsilon_0 A/h)} = \frac{P}{\varepsilon_0}\frac{C_0}{C_0 + C_e},$$

where $C_0 = \varepsilon_0 A/h$ is the direct "internal" capacity of the plate to earth in vacuum (we neglect edge effects). Thus we obtain

$$\delta\phi = \frac{h}{\varepsilon_0}\frac{C_0}{C_e+C_0}\,\delta P.$$

In Fig. 9.2c it is obvious that the net charge which flows in the external lead when P changes is $\delta q = A\delta P$. If we assume that $C_e \to 0$ we have, for the open circuit voltage,

$$\delta\phi = \frac{h}{\varepsilon_0}\delta P$$

and, for the closed circuit charge, $\delta q = A\delta P = (\varepsilon_0 A/h)\delta\phi = C_0\delta\phi$ and so, as a circuit element the equivalent circuit is as shown in Fig. 9.3.

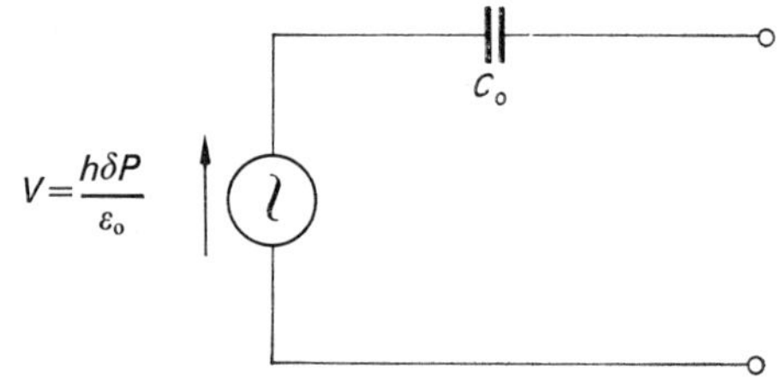

FIG. 9.3. Equivalent circuit of a piezo-electric crystal.

Note that this calculation neglects the normal linear dielectric response of the medium between the electrodes. If this is included the voltage is reduced by $1/\varepsilon$ and the capacitance C_0 increased by ε, where ε is the relative dielectric constant of the medium.

A.C. Circuits

The equation $\nabla \wedge \mathbf{E} = 0$ is the basis of one of the Kirchhoff circuit laws which states that the algebraic sum of the e.m.f.s (electromotive forces) acting in any closed loop of a circuit is zero. This is usually introduced by considering a circuit such as that shown in Fig. 9.4, but we soon extend it to a.c. circuits. If we have a piezo-electric crystal being strained by an oscillatory stress then we can regard this as a purely electrostatic a.c. source with the equivalent circuit shown in Fig. 9.3. There would be no objection to analysing the circuit shown in Fig. 9.5 in terms of the equation $\nabla \wedge \mathbf{E} = 0$. Suppose, however, that we have the circuit shown in Fig. 9.6 containing an inductor; the line integral of E taken round this circuit is no longer zero since the circuit links a changing flux. The line integral of E taken from A to B through the conductor is zero but the line integral taken by any other path is non-zero. The

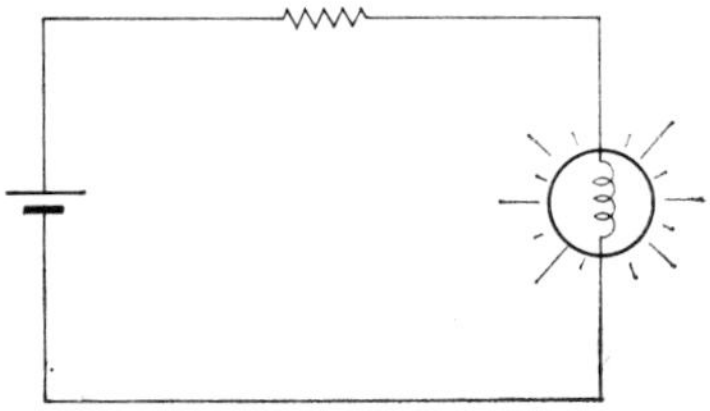

FIG. 9.4. A simple circuit.

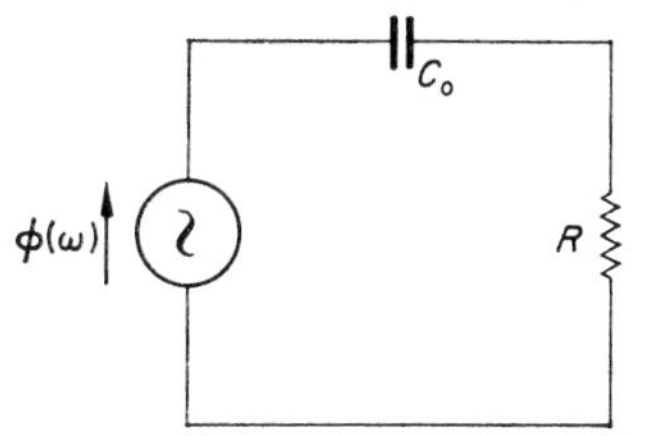

FIG. 9.5. A simple a.c. circuit.

difference between the two values is $\dot{\psi}$, the rate of change of the flux linkage. In a.c. circuit theory we replace this effect by attributing to the inductance a back e.m.f. $L\dot{I}$; for many purposes this is a useful simplification but it does not mean that there is an electrostatic potential difference $L\dot{I}$ across the inductance unless we exclude the interior of the inductance from the calculation. The reader will find an example of a system in which the too literal use of these notions is misleading in problems (7), (8) and (9) at the end of this chapter.

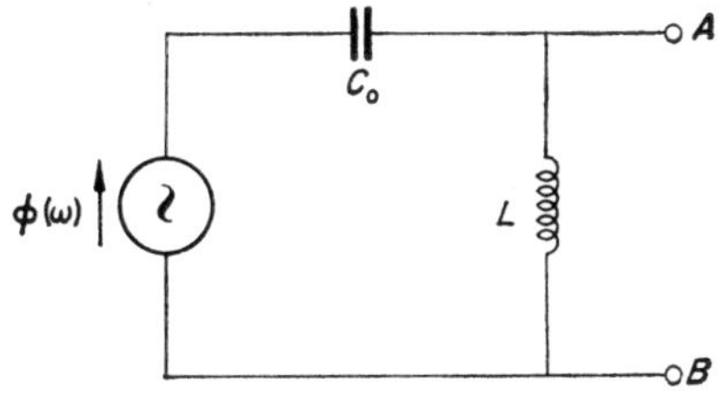

FIG. 9.6. A circuit with inductance.

In an infinite, homogeneous, isotropic, and linear dielectric medium $\mathbf{D} = \varepsilon\varepsilon_0\mathbf{E}$ and so the equations $\nabla \wedge \mathbf{E} = 0$ and $\nabla . \mathbf{D} = 0$ imply the equations $\nabla \wedge \mathbf{D} = 0$, $\nabla . \mathbf{E} = 0$, but one must beware of using these results out of context and especially of assuming that they apply at the boundary of a dielectric medium. The general relations, when $\rho = 0$, are $\nabla \wedge \mathbf{D} = \nabla \wedge \mathbf{P}$, $\varepsilon_0\nabla . \mathbf{E} = -\nabla . \mathbf{P}$ and there is no reason why either $\nabla \wedge \mathbf{P}$ or $\nabla . \mathbf{P}$ should be identically zero everywhere. Indeed both $\nabla \wedge \mathbf{P}$ and $\nabla . \mathbf{P}$ have singularities at the boundaries of dielectric media.

Magnetostatic Fields

The magnetostatic equations are

$$\nabla . \mathbf{B} = 0 \quad (9.14a) \qquad \nabla \wedge \mathbf{B} = \mu_0(\nabla \wedge \mathbf{M}+\mathbf{J}) \quad (9.14b)$$

$$\nabla . \mathbf{H} = -\nabla . \mathbf{M} \quad (9.14c) \qquad \nabla \wedge \mathbf{H} = \mathbf{J}. \quad (9.14d)$$

When $\mathbf{J} = 0$ these equations are similar to the electrostatic equations in the absence of free charge, i.e.

$$\nabla . \mathbf{D} = 0, \qquad \nabla \wedge \mathbf{D} = \nabla \wedge \mathbf{P},$$

$$\varepsilon_0\nabla . \mathbf{E} = -\nabla . \mathbf{P}, \qquad \nabla \wedge \mathbf{E} = 0,$$

and, just as in electrostatics, where we can derive **E** from a scalar potential, we can derive **H** from a magnetic scalar potential. This is much the simplest way of dealing with a system in which the fields are solely due to permanent magnets. To illustrate the basic properties of the fields in this case, we show in Fig. 9.7 the fields associated with a uniform, permanently magnetized cylinder.

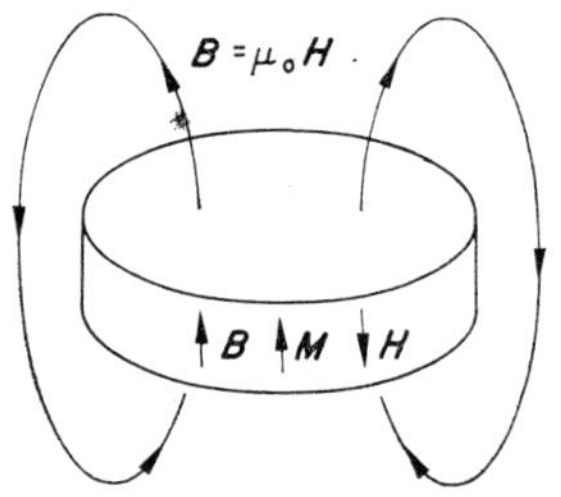

FIG. 9.7. A magnetized disc.

According to equation (9.14b) the field **B** due to a distribution of magnetization **M** is identical with the field due to a current distribution $\mathbf{J} = \nabla \wedge \mathbf{M}$. Conversely we can always find a distribution of magnetization **M** equivalent to a current distribution **J** and, if **J** is localized, the equivalent magnetization is also localized in the same region of space. Thus, at points outside the localized distribution, where $\mathbf{B} = \mu_0\mathbf{H}$ it is immaterial whether we consider the field to arise from current or magnetization. The situation is analogous to the use of induced e.m.f.s in electricity. The analogy is illustrated in Figs. 9.8 and 9.9. In both cases the situations within the boxes are quite different but the external effects are the same. If we pursue the analogy between **E** and **H** further, the box in Fig. 9.9 is the seat of a source of magnetomotive force and this concept leads to the usual formulation of the theory of magnetic circuits.

Because $\nabla \, . \, \mathbf{B} = 0$ is an invariable relation in magnetism there are no magnetic analogies to the effects of free charges in electrostatics. There are no magnetic equivalents to metal

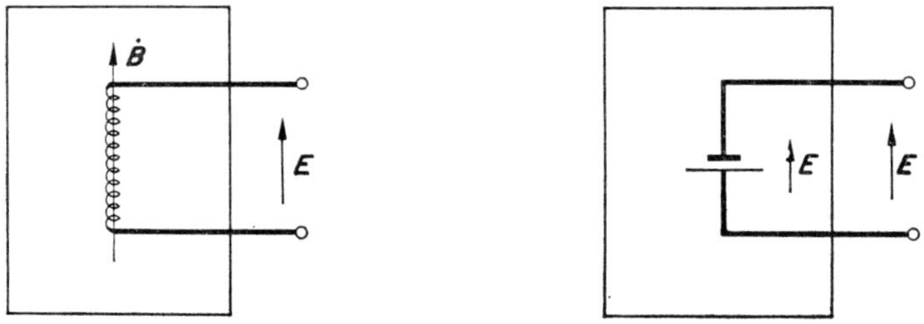

FIG. 9.8. External equivalence of a transformer winding and a source of e.m.f.

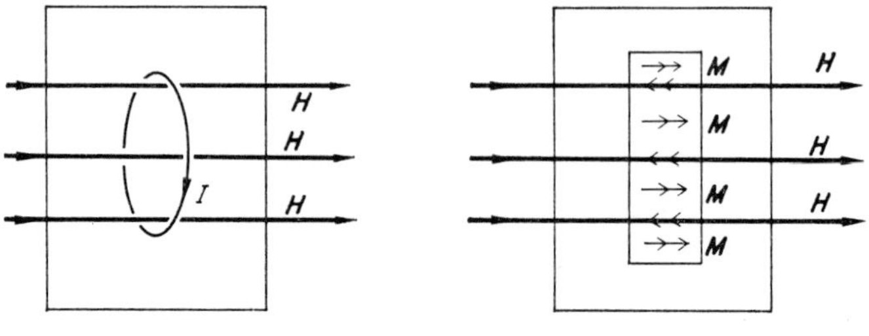

FIG. 9.9. External equivalence of a current loop and a permanent magnet.

surfaces on which lines of **D** terminate in surface charge. There is, however, one partial analogy to the behaviour of conductors in electrostatics, and this is provided by superconductors. A superconductor not only has zero resistance but also has zero magnetic field in its interior. Thus, just as a normal metal excludes electrostatic fields, a superconductor excludes static magnetic fields. The actual relations are however quite different. Thus in electrostatics the tangential electric field **E** is zero just outside the metal surface, and the normal component of **D** terminates on surface charge, whereas in magnetostatics the normal component of **B** is zero outside a superconducting surface and the tangential component of **H** terminates in surface currents (see problems 15 and 16).

In practical magnetism the vector **B** is the field whose flux lines are continuous, i.e. the most useful equation is $\nabla . \mathbf{B} = 0$. The vector **H** on the other hand is the field whose line integral is the linked current and whose flux lines end on discontinuities in the magnetization **M**. Since in most problems either the current distribution **J** or the magnetization **M** is prescribed we usually calculate **H** from the conditions of the problem. On the other hand **B** is the vector measured by fluxmeters, resonance experiments, or forces on currents. This makes magnetism very different from electrostatics where, on the whole, **E** is specified by electrode potentials and is generally obtained directly by measurement. Were it not for the fact that the charges on electrodes are related to **D** and not **E**, the vector **D** would have a very subsidiary role in electrostatics.

Shape Factors and Cavity Fields

The vectors **E** and **B** have an operational definition in terms of the forces they exert on charges and currents in vacuum, and it is sometimes difficult to see how these definitions relate to fields in the interior of solid media, especially as the force exerted on atomic particles, i.e. non-macroscopic objects, in the media are not necessarily directly related to the macroscopic fields. We shall be considering the problem of the fields acting on atoms in later chapters, but here we discuss some other aspects of the macroscopic fields. We shall be particularly concerned with the fields in cavities of different shapes cut out of uniformly polarized media, but before we consider this problem it will be useful to discuss the fields associated with uniformly polarized bodies of a definite shape.

The general problem of a body of arbitrary shape is quite inscrutable, and the only simple results are those for ellipsoidal bodies, i.e. bodies whose surface is described by an equation of the form

$$\frac{x^2}{a^2}+\frac{y^2}{b^2}+\frac{z^2}{c^2} = 1.$$

A uniform polarization **P** in any direction in the ellipsoid leads to a uniform contribution to the electric field in the ellipsoid. The problem was first discussed by Maxwell and there is an excellent account by Stoner (1949). If the polarization **P** is parallel to a principal axis of the ellipsoid the field is parallel and opposed to **P**; for the principal axis i we have $\varepsilon_0 E_i = -n_i P_i$ and the coefficient n_i is known as either a shape factor, a depolarization factor or a demagnetization factor. The sum of the principal shape factors is unity, i.e. $n_1+n_2+n_3 = 1$. For a sphere we have $n_1 = n_2 = n_3 = \frac{1}{3}$ and for an ellipsoid of revolution about the 3(z) axis $n_1 = n_2 = \frac{1}{2}(1-n_3)$. For a long ellipsoid of revolution, which approximates a long cylinder, we have $n_3 = 0$ and $n_1 = n_2 = \frac{1}{2}$ while for a flat ellipsoid of revolution or a thin circular disc $n_3 = 1$ and $n_1 = n_2 = 0$.

The Macroscopic Fields

If we have an ellipsoidal cavity in a medium in which there is a uniform polarization **P** the contribution $-n_i P_i$ to the field in the cavity is absent. If, therefore, E_i is the field in the medium in the absence of the cavity, the field in the cavity is

$$\varepsilon_0 E_i^{\text{Cav}} = \varepsilon_0 E_i + n_i P_i. \tag{9.15a}$$

The corresponding magnetic result is

$$H_i^{\text{Cav}} = H_i + n_i M_i. \tag{9.15b}$$

Thus, if we cut a long needle-shaped cavity parallel to the fields, for which $n_i = 0$, the cavity field is E or $\mu_0 H$ whereas for a thin disc, normal to the fields with $n_i = 1$, the cavity fields are $E+(1/\varepsilon_0)P = (1/\varepsilon_0)D$ or $\mu_0(H+M) = B$.

These results are often used to explain the significance of the vectors **E** and **D** or **H** and **B** in material media but it is not clear to the author exactly what experiments are envisaged which might be described in terms of the cavity fields.

Dielectric Relaxation

In a homogeneous conducting medium, characterized by a definite dielectric constant ε and conductivity σ, we have the two equations

$$\nabla . (\varepsilon\varepsilon_0 \mathbf{E}) = \rho$$

$$\nabla . \mathbf{J} + \dot{\rho} = \nabla . (\sigma \mathbf{E}) + \dot{\rho} = 0.$$

Within the medium, but not at the surface (where σ and ε change discontinuously), we can write these equations as

$$\varepsilon\varepsilon_0 \nabla . \mathbf{E} = \rho$$

$$\sigma \nabla . \mathbf{E} = -\dot{\rho}$$

and obtain

$$\dot{\rho} + \frac{\sigma}{\varepsilon\varepsilon_0}\rho = 0, \tag{9.16}$$

so that any local disturbance of the charge density collapses with a characteristic time $\tau_d = \varepsilon\varepsilon_0/\sigma$, known as the dielectric relaxation time. Values of τ_d vary from 10^{-18} seconds in metals to 10 weeks in good insulators. In semiconductors the range is roughly from 10^{-6} to 10^{-13} seconds. Thus, except at very high frequencies, there can be no space charge in the bulk of a metal and all charges must reside near the surface. We may remark, however, that the assumption that a medium can be characterized by a definite conductivity σ and dielectric constant ε is only fulfilled at frequencies low compared with the three parameters ω_p, $1/\tau_c$ and ω_0, where $\omega_p = (Ne^2/\varepsilon_0 m)^{\frac{1}{2}}$ is the plasma frequency associated with the electronic density N, τ_c is the carrier collision time and ω_0 is any atomic resonance frequency. Thus the result (9.16) is unlikely to be valid at extreme frequencies. We may note that a simple theory of electronic conduction leads to the re ation

$$\omega_p^2 \tau_c \tau_d = 1$$

so that the range of validity of (9.16) will be greatest when τ_c is small and ω_p large.

Skin Depth

Within a homogeneous medium characterized by constant relative permeability μ, dielectric constant ε and conductivity σ the field equations can be written in terms of **E** and **H** as

$$\nabla \cdot \mathbf{H} = 0, \qquad \nabla \cdot \mathbf{E} = 0,$$

$$\nabla \wedge \mathbf{E} = -\mu\mu_0 \dot{\mathbf{H}}, \qquad \nabla \wedge \mathbf{H} = \varepsilon\varepsilon_0 \dot{\mathbf{E}} + \sigma \mathbf{E},$$

and from these we obtain the wave equations

$$\nabla^2 \mathbf{E} = \mu\mu_0 \varepsilon\varepsilon_0 \ddot{\mathbf{E}} + \mu\mu_0 \sigma \dot{\mathbf{E}},$$

$$\nabla^2 \mathbf{H} = \mu\mu_0 \varepsilon\varepsilon_0 \ddot{\mathbf{H}} + \mu\mu_0 \sigma \dot{\mathbf{H}}.$$

For fields of a definite frequency, whose time dependence we may take as exp $(j\omega t)$, we obtain

$$\nabla^2 \mathbf{E} = j\omega\mu\mu_0\sigma \left(1 + \frac{j\omega\varepsilon\varepsilon_0}{\sigma}\right) \mathbf{E}$$

or

$$\nabla^2 \mathbf{E} = j\omega\mu\mu_0\sigma(1 + j\omega\tau_d)\mathbf{E}.$$

If $\omega\tau_d$ is small, which is the case in metals at optical and lower frequencies, the plane wave solutions of these equations are of the form exp $[j\{\omega t - (z/\delta)\} - (z/\delta)]$ where the skin depth δ is given by

$$\delta = \left(\frac{2}{\omega\mu\mu_0\sigma}\right)^{\frac{1}{2}} = \frac{\lambda}{2\pi n}(2\omega\tau_d)^{\frac{1}{2}} \tag{9.17}$$

and λ is the free space wavelength corresponding to ω while $n = (\mu\varepsilon)^{\frac{1}{2}}$. Thus, in many problems where the dimensions of the system are related to λ, the skin depth δ is small, though not necessarily infinitesimal, on a macroscopic scale. If we have fields in vacuum or an insulator near a metal surface the fields penetrate into the metal only to a depth of a few times δ. If H_t is the tangential component of **H** just outside the metal, then since $H_t = 0$ in the interior of the metal there must be a current $I = H_t$ flowing in unit length of the surface. In microwave problems, especially, it is convenient to treat this as a surface current.

Time-dependent Fields

The time-dependent fields in vacuum satisfy the equations

$$\mathbf{B} = \mu_0 \mathbf{H}, \qquad \mathbf{D} = \varepsilon_0 \mathbf{E},$$

$$\nabla \cdot \mathbf{B} = 0, \qquad \nabla \cdot \mathbf{D} = \rho,$$

$$\nabla \wedge \mathbf{H} = \mathbf{J} + \dot{\mathbf{D}}, \qquad \nabla \wedge \mathbf{E} = -\dot{\mathbf{B}},$$

and the terms $\dot{\mathbf{D}}$ and $\dot{\mathbf{B}}$ in the curl equations modify our notions about the fields **H** and **E** derived from static examples. To illustrate this we show, in Fig. 9.10, the fields in an axially symmetric mode of a cylindrical resonant cavity.

We see that there is no meaningful sense in which we can ascribe a potential difference to the points XX' in the second figure. In the third figure lines of D end on surface charge, and in the fourth figure the line integral of **H** is related to $\dot{\mathbf{D}}$. Outside the cavity **H** is zero because the wall currents **J** just cancel the effect of the displacement current $\dot{\mathbf{D}}$.

In many problems in optics and microwave theory we have to deal with systems in which, although there are conductors and dielectric media, no magnetic media are present. It is then rather more convenient to express the field equations in terms of **E** and **H** rather than **B**. The equations are now

$$\nabla . \mathbf{H} = 0, \quad (9.18a) \qquad \nabla . \varepsilon\varepsilon_0\mathbf{E} = \rho, \quad (9.18b)$$

$$\nabla \wedge \mathbf{H} = \partial/\partial t\,(\varepsilon\varepsilon_0\mathbf{E})+\mathbf{J}, \quad (9.18c) \qquad \nabla \wedge \mathbf{E} = -\mu_0\dot{\mathbf{H}}. \quad (9.18d)$$

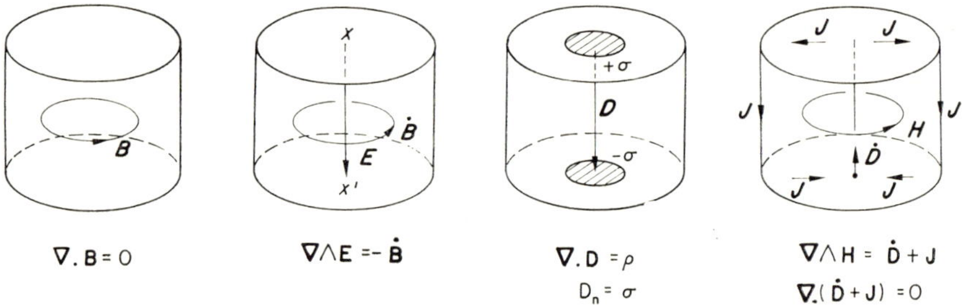

FIG. 9.10. Fields in a resonant cavity.

If, in addition, all the quantities vary with time as exp $(j\omega t)$ we have

$$j\omega\varepsilon\varepsilon_0\mathbf{E} = \nabla \wedge \mathbf{H}-\mathbf{J}, \tag{9.18e}$$

$$j\omega\mu_0\mathbf{H} = -\nabla \wedge \mathbf{E}, \tag{9.18f}$$

and these two equations imply the equations (9.18a) and (9.18b), which are therefore redundant. In particular, at a surface of discontinuity the boundary conditions on the tangential components of **E** and **H** are, by themselves, sufficient. This leads to a considerable simplification in many problems.

The use of equations (9.18) is widespread and there is an unfortunate tendency in some books to write (9.18b) as $\nabla . \mathbf{E} = \rho/\varepsilon\varepsilon_0$ which is only correct in an infinite homogeneous medium. The reader should beware of taking results derived for particular situations as general truths. For example, in some cases ε is a function of t and the passage from (9.18c) to (9.18e) is not permissible.

The canonical form of the macroscopic equations is

$$\begin{aligned} \nabla . \mathbf{B} &= 0, & \nabla \wedge \mathbf{E}+\dot{\mathbf{B}} &= 0, \\ \nabla . \mathbf{D} &= \rho, & \nabla \wedge \mathbf{H}-\dot{\mathbf{D}} &= \mathbf{J}, \end{aligned} \tag{9.19}$$

with

$$\mathbf{D} = \varepsilon_0\mathbf{E}+\mathbf{P}, \qquad \mathbf{H} = \frac{1}{\mu_0}\mathbf{B}-\mathbf{M},$$

so that we also have

$$\begin{aligned} \nabla . \mathbf{B} &= 0, & \nabla \wedge \mathbf{E}+\dot{\mathbf{B}} &= 0, \\ \varepsilon_0\nabla . \mathbf{E} &= \rho-\nabla . \mathbf{P}, & \frac{1}{\mu_0}\nabla \wedge \mathbf{B}-\varepsilon_0\dot{\mathbf{E}} &= \mathbf{J}+\dot{\mathbf{P}}+\nabla \wedge \mathbf{M}. \end{aligned} \tag{9.20}$$

We could, of course, write the field equations in terms of other pairs of vectors of which the only one of any interest is

$$\begin{aligned} \nabla \cdot \mathbf{H} &= \nabla \cdot \mathbf{M}, & \nabla \wedge \mathbf{E} + \mu_0 \dot{\mathbf{H}} &= -\mu_0 \nabla \wedge \mathbf{M}, \\ \varepsilon_0 \nabla \cdot \mathbf{E} &= \rho - \nabla \cdot \mathbf{P}, & \nabla \wedge \mathbf{H} - \varepsilon_0 \dot{\mathbf{E}} &= \mathbf{J} + \dot{\mathbf{P}}. \end{aligned} \tag{9.21}$$

Unless, however, $\mathbf{M} \equiv 0$ this contains no homogeneous equations and we would, for example, find it difficult to express **E** and **H** in terms of potentials.

Problems

1. Show that B_n and E_t are continuous across any surface.

2. A closed circular loop of wire of resistance R and radius r rotates at an angular rate ω about a diameter normal to a steady uniform magnetic field **B**. What is the current in the wire? Is the integral of the electric field taken round the wire loop zero?

3. An anchor ring is formed from a permanent magnet material with a fixed magnetization **M** in a direction round the circumference of the ring. What are the values of **B** and **H** in the ring?

4. A short dipole is placed at the origin of a polar coordinate system. Show that, on a sphere of radius r surrounding the dipole, the normal component of **D** must take both positive and negative signs.

5. In a semiconductor *pn* junction diode the bulk *p* material is at a negative potential with respect to the bulk *n* material and the fields in both materials are zero except near and in the depletion layer. Show that the net charge in the depletion layer is zero.

6. One plate of a parallel plate capacitor consists of a thick slab of metal, the other plate consists of a thin layer of *n* type silicon 10^{-6} m thick containing 10^{21} mobile electrons per m^3. The dielectric is a thin film of thickness 10^{-7} m and relative dielectric constant 10, equal to that of the silicon. Contact to the silicon is made by a thick plate on the side away from the dielectric. This contact is earthed. Sketch the form of the charge–voltage relation for this capacitor when the metal plate of the condenser is negative with respect to earth.

7. An electron beam from a cathode at zero potential passes through two grids at fixed potentials V_0 and V_0+v and impinges on a collector at a potential V_0. Show that the kinetic energy of the electrons arriving at the collector is independent of v.

8. In the system of problem (7) the voltage v is an alternating voltage applied between the grids by connecting them to the secondary winding of a transformer whose primary is excited by an a.c. generator. Show that the kinetic energy of the electrons at the collector is still V_0.

9. In a klystron the grids form part of a closed cavity resonator excited by an external source. Explain how, in view of the results of problems 7 and 8, the electrons leaving the second grid are velocity modulated.

The Macroscopic Fields

10. The hysteresis loop of a hard magnetic material can be approximated by

$$\left(\frac{B}{\mu_0 H_r}\right)^2 + \left(\frac{H}{H_c}\right)^2 = 1$$

where H_r and H_c are constants. Calculate the fields B and H and the magnetic moment for an isolated sphere of radius a.

11. A small hole parallel to the direction of magnetization is bored through a block of material whose coercive force is H_c. Show that the field in the hole cannot exceed $\mu_0 H_c$.

12. Show that the external fields B and H due to a circular current loop I of radius r are equivalent to the external fields of a thin disc of material of radius r, thickness t and magnetization M where $Mt = I$. Show in addition that the fields **B** due to the two systems are everywhere the same.

13. Show that, in a system free from true currents, the volume integral of $\mathbf{H}\cdot\mathbf{B}$ is zero (hint, express **B** as $\nabla \wedge \mathbf{A}$). Use this result to show that in an isolated permanent magnet $\mathbf{H}\cdot\mathbf{B} \leqslant 0$. If the equality sign holds what is the stray external field?

14. An anchor ring of radius 10 cm is formed from a hard magnetic material whose coercive force is 10^4 A m^{-1} and remanent flux 1 T. A field H of 3×10^4 A m^{-1} is required to magnetize the material fully. The ring is wound with 100 turns of wire. What current is needed to magnetize the ring? With the ring magnetized a saw-cut 1 mm thick is made through the ring. What is the field B in the resulting air gap?

15. A metal sphere of radius a is placed in a uniform electric field **E** due to fixed distant charges. Calculate the induced dipole moment of the sphere and determine the distribution of surface charge.

16. A superconducting sphere of radius a is placed in a uniform magnetic field $\mathbf{B}_0$ due to fixed distant currents. Calculate the induced dipole moment of the sphere and the surface current distribution.

17. A spherical sample of a material with a fixed permanent magnetization M has a small hole cut through it, on a diameter perpendicular to M, and a non-magnetic wire carrying a current I is inserted through the hole. What is the force per unit length acting on the wire?

18. A sphere of dielectric constant ε is placed in a uniform external field E_0. What is the field in a spherical cavity at the centre?

19. A flat disc of ferromagnetic material, in which the magnetization **M** and the angular momentum density **J** are related by $\mathbf{M} = \gamma\mathbf{J}$, is placed in an applied field parallel to a diameter of the disc. Show that ferromagnetic resonance occurs at a frequency $\gamma(\mu_0 HB)^{\frac{1}{2}}$ where H and B are the fields in the disc (reference Kittel, *Introduction to Solid State Physics*, Wiley, 1971).

20. The tangential component of **H** just outside a metal surface has the r.m.s. value H_0. If the metal has a conductivity σ and a skin depth δ at the relevant frequency show that the mean power dissipation per unit area of the surface is $H_0^2/\sigma\delta$.

21. What is the skin depth in copper (for which $\sigma = 6\times10^7$ mho m^{-1}) at 50 Hz?

22. A long solenoid of radius b and n turns per metre carries a current $I \cos \omega t$. What is the electric field at radius $r < b$?

23. A plane parallel condenser consists of two discs of radius b separated by a dielectric of thickness d and relative dielectric constant ε. The lower plate is grounded and the potential of the upper plate increases slowly at a rate of $\dot{V}$ volts s^{-1}. What is the value of H at a radius $r < b$ between the plates?

24. Two small toy balloons are connected by a pipe of length l fitted with a tap. Initially the volume of one balloon is v and that of the other $V > v$. The tap is opened and the two volumes slowly equalize. Obtain an expression for the velocity of the air at a distance from the system while the two volumes are changing.

25. The electrostatic capacity between two thick circular plates of radius r placed a distance equal to their radius apart is approximately $3\pi r\varepsilon_0$. An electromagnet with circular pole faces of radius r a distance r apart produces a field of $\frac{1}{4}$ tesla at the centre of the gap. What is the flux in the yoke of the magnet?

26. When two electrodes are totally immersed in a liquid of conductivity $\sigma = 1 \text{ ohm}^{-1} \text{ m}^{-1}$ the resistance between the electrodes is 1 ohm. What is the capacitance of the electrode system *in vacuo*?

CHAPTER 10

Energy, Power and Stress

Electrostatic Energy

Any attempt to define an energy function for the e.m. field in a system containing point charges leads to a divergent expression. This is true in both classical and quantum electrodynamics although the nature and the origin of the divergence are different in the two cases. The most that we can ask of any theory of the field energy is that it should give finite results for the difference in energy between two configurations of the same system. The formulation of a general theory with this property which is at the same time Lorentz invariant is a formidable task. The reader will find it discussed at length by Rohrlich (1965).

Fortunately, in macroscopic electromagnetism, we are not faced with this problem. The divergent terms in the energy arise from high frequency components of the fields and the process of truncating the spectrum automatically excludes just these components. Thus any expression we form from the truncated macroscopic fields will be convergent and, although it may not represent the true total field energy, will adequately describe energy differences between macroscopically different configurations.

We begin by considering the work $\delta\mathfrak{W}$ that has to be done against electrostatic forces to change the configuration of a set of N charges $q(n)$ from $\mathbf{r}(n)$ to $\mathbf{r}(n)+\delta\mathbf{r}(n)$. This is not yet a macroscopic calculation and we can take the charges to be points. It is not difficult to see that this work is equal to the change in the value of the finite expression

$$\mathfrak{U} = \sum_n \sum_{m<n} \frac{q(n)q(m)}{4\pi\varepsilon_0|\mathbf{r}(n)-\mathbf{r}(m)|} = \tfrac{1}{2}\sum_{n\neq m}\sum \frac{q(n)q(m)}{4\pi\varepsilon_0|\mathbf{r}(n)-\mathbf{r}(m)|}\,. \tag{10.1}$$

The electric field $\mathbf{E}$ is also a function of the particle configuration and so we ought to be able to express $\mathfrak{U}$ as a function of $\mathbf{E}$. To explore this possibility we consider the function

$$\mathfrak{U}_T = \tfrac{1}{2}\int \varepsilon_0 E^2\,\mathrm{d}^3\mathbf{R}. \tag{10.2}$$

If we write $\mathbf{E} = -\nabla\phi(\mathbf{R})$ where

$$\phi(\mathbf{R}) = \sum_m \frac{q(m)}{4\pi\varepsilon_0|\mathbf{R}-\mathbf{r}(m)|} \tag{10.3}$$

with

$$\nabla^2\phi(\mathbf{R}) = -\frac{1}{\varepsilon_0}\sum_m q(m)\delta(\mathbf{r}(m)-\mathbf{R}), \tag{10.4}$$

we can express $\mathfrak{U}_T$ as

$$\mathfrak{U}_T = \tfrac{1}{2}\varepsilon_0\int(\nabla\phi)^2\,\mathrm{d}^3\mathbf{R} = -\tfrac{1}{2}\varepsilon_0\int\phi\nabla^2\phi\,\mathrm{d}^3\mathbf{R}+\tfrac{1}{2}\varepsilon_0\int_{S\to\infty}\phi\frac{\partial\phi}{\partial n}\,\mathrm{d}S, \tag{10.5}$$

in which the surface integral at infinity can be discarded. We now obtain

$$\mathfrak{U}_T = \tfrac{1}{2}\int\sum_n\sum_m \frac{q(n)q(m)}{4\pi\varepsilon_0|\mathbf{R}-\mathbf{r}(m)|}\,\delta(\mathbf{r}(n)-\mathbf{R})\,\mathrm{d}^3\mathbf{R}. \tag{10.6}$$

The terms with $n \neq m$ cause no trouble and, in fact, yield $\mathfrak{U}$ of equation (10.1) so that

$$\mathfrak{U}_T = \mathfrak{U} + \tfrac{1}{2}\sum_n \int \frac{q^2(n)\delta(\mathbf{r}(n)-\mathbf{R})}{4\pi\varepsilon_0|\mathbf{R}-\mathbf{r}(n)|}\, \mathrm{d}^3\mathbf{R}. \tag{10.7}$$

The remaining term is independent of the particle configuration and is a divergent self energy term $\mathfrak{U}_S$. Thus we have

$$\mathfrak{U}_T = \mathfrak{U} + \mathfrak{U}_S \tag{10.8}$$

but nevertheless

$$\delta\mathfrak{W} = \delta\mathfrak{U} = \delta\mathfrak{U}_T. \tag{10.9}$$

Now had we, at the outset, formed truncated variables, i.e., had we expressed the charge density as

$$\rho(\mathbf{R}) = \int_{-\infty}^{\infty} \sum_n q(n)\delta(\mathbf{r}(n)-\mathbf{R}+\mathbf{S})f(\mathbf{S})\, \mathrm{d}^3\mathbf{S}, \tag{10.10}$$

the function $\rho(R)$ would have become a non-singular continuous variable of $\mathbf{R}$. The truncated potential ϕ would then have satisfied

$$\nabla^2\phi = -\rho/\varepsilon_0 \tag{10.11}$$

so that, from equation (10.5), we would have obtained the finite expression

$$\mathfrak{U}_T = \int \tfrac{1}{2}\rho\phi\, \mathrm{d}^3\mathbf{R}, \tag{10.12}$$

which is exactly the expression for $\mathfrak{U}$ obtained by using truncated variables in equation (10.1). Thus, in a macroscopic context, the work done in changing a configuration can be expressed as the change in (10.2), i.e., we have

$$\delta\mathfrak{W} = \delta\mathfrak{U} = \int \varepsilon_0\mathbf{E}\cdot\delta\mathbf{E}\, \mathrm{d}^3\mathbf{R} \tag{10.13}$$

and this is the only result that we shall need.

We now turn to quasi-static electric phenomena, omitting magnetic effects. The result

$$\nabla\cdot(\mathbf{J}+\dot{\mathbf{D}}) = 0 \tag{10.14}$$

is a direct consequence of the field equations and we can use it in the integral

$$\int \mathbf{E}\cdot(\mathbf{J}+\dot{\mathbf{D}})\, \mathrm{d}^3\mathbf{R} = -\int (\mathbf{J}+\dot{\mathbf{D}})\cdot\nabla\phi\, \mathrm{d}^3\mathbf{R}$$

to obtain

$$\int \mathbf{E}\cdot(\mathbf{J}+\dot{\mathbf{D}})\, \mathrm{d}^3\mathbf{R} = -\oint_S \phi(J_n+\dot{D}_n)\, \mathrm{d}S. \tag{10.15}$$

If we take the surface far enough away in a region where $\mathbf{J} = 0$ and $\phi D_n \sim 1/R^3$, we can discard the surface integral. Thus, for an isolated slowly changing system,

$$\int (\mathbf{E}\cdot\mathbf{J}+\mathbf{E}\cdot\dot{\mathbf{D}})\, \mathrm{d}^3\mathbf{R} = 0. \tag{10.16}$$

Since $\mathbf{E}\cdot\mathbf{J}$ is the rate at which the electric field does work on the charges in unit volume the term $-\mathbf{E}\cdot\dot{\mathbf{D}}$ must be the rate at which the field energy is decreasing.

Instead of considering an isolated system we might consider a system totally enclosed in an equipotential metal surface, within which there is a set of metal electrodes each maintained at potentials ϕ_k by batteries within the system. We now obtain

$$\int (\mathbf{E}\cdot\mathbf{J}+\mathbf{E}\cdot\dot{\mathbf{D}})\, \mathrm{d}^3\mathbf{R} = -\sum_k \phi_k \int (J_n+\dot{D}_n)\, \mathrm{d}S_k.$$

But $-D_n$ is the surface charge on the electrodes and so

$$\int (\mathbf{E}\,.\,\mathbf{J}+\mathbf{E}\,.\,\dot{\mathbf{D}})\,\mathrm{d}^3\mathbf{R} = \sum_k I_k\phi_k \tag{10.17}$$

which is the rate at which energy is being supplied by the batteries.

Again, suppose that we have an isolated localized charge distribution in which we make a charge $\delta\rho(\mathbf{R})$. The work involved is

$$\int \phi\delta\rho\,\mathrm{d}^3\mathbf{R} = \int \phi\nabla\,.\,(\delta\mathbf{D})\,\mathrm{d}^3\mathbf{R} = -\int \nabla\phi\,.\,\delta\mathbf{D}\,\mathrm{d}^3\mathbf{R}+\int_S \phi\delta D_n\,\mathrm{d}S.$$

We discard the surface integral at a distance and obtain

$$\int \phi\delta\rho\,\mathrm{d}^3\mathbf{R} = \int \mathbf{E}\,.\,\delta\mathbf{D}\,\mathrm{d}^3\mathbf{R}. \tag{10.18}$$

Each of these results (10.16), (10.17), and (10.18) indicates that the general expression for the change in the electrostatic field energy is

$$\delta\mathfrak{U}_E = \int \mathbf{E}\,.\,\delta\mathbf{D}\,\mathrm{d}^3\mathbf{R}. \tag{10.19}$$

It is usual to regard

$$\delta U_E = \mathbf{E}\,.\,\delta\mathbf{D} \tag{10.20}$$

as a change in the field energy density. It is a moot point whether this has any physical significance or is merely a verbal convenience.

Magnetic Energy and Poynting's Vector

In order to demonstrate that the change in the magnetic field energy is

$$\delta\mathfrak{U}_M = \int \mathbf{H}\,.\,\delta\mathbf{B}\,\mathrm{d}^3\mathbf{R} \tag{10.21}$$

or that the change in the energy density is

$$\delta U_M = \mathbf{H}\,.\,\delta\mathbf{B} \tag{10.22}$$

we have to invoke both Ampère's law and Faraday's law of induction, or else to use the properties of idealized permanent magnets, which may not actually exist in the real world. It is, therefore, more sensible to obtain these results as part of a more general scheme.

The field equations

$$\nabla\wedge\mathbf{E} = -\dot{\mathbf{B}}, \qquad \nabla\wedge\mathbf{H} = \mathbf{J}+\dot{\mathbf{D}},$$

when multiplied by $\mathbf{H}\,.$ and $\mathbf{E}\,.$ and subtracted, yield

$$-\nabla\,.\,(\mathbf{E}\wedge\mathbf{H}) = \mathbf{E}\,.\,(\nabla\wedge\mathbf{H})-\mathbf{H}\,.\,(\nabla\wedge\mathbf{E}) = \mathbf{E}\,.\,\mathbf{J}+\mathbf{H}\,.\,\dot{\mathbf{B}}+\mathbf{E}\,.\,\dot{\mathbf{D}}. \tag{10.23}$$

This is exact, provided that $\mathbf{D}$ and $\mathbf{H}$ contain all relevant multipole order terms, i.e.,

$$\begin{aligned}\mathbf{D} &= \varepsilon_0\mathbf{E}+\mathbf{P}-\frac{\partial Q_{ij}}{\partial R_j}+\text{etc.}\\ \mathbf{H} &= \frac{1}{\mu_0}\mathbf{B}-\mathbf{M}+\text{etc.}\end{aligned} \tag{10.24}$$

If we integrate (10.23) over the whole of space we obtain

$$\int (\mathbf{E}\,.\,\mathbf{J}+\mathbf{E}\,.\,\dot{\mathbf{D}}+\mathbf{H}\,.\,\dot{\mathbf{B}})\,\mathrm{d}^3\mathbf{R} = -\int_S (\mathbf{E}\wedge\mathbf{H})_n\,\mathrm{d}S, \tag{10.25}$$

where S is a closed surface. If S is a sphere of radius R then all static fields at a distance fall off at least as fast as $1/R^2$ and oscillatory fields, of period τ, as fast as $1/R\tau$. Thus as $R \to \infty$ the right hand side of (10.25) is at most of order $1/\tau^2$. The left hand side, however, contains only single time derivatives and so is only of order $1/\tau$. For low frequency, or quasi-static, fields, therefore, the left hand side must be zero and, since we have already identified $\mathbf{E} \cdot \mathbf{J}$ and $\mathbf{E} \cdot \dot{\mathbf{D}}$, the term $\mathbf{H} \cdot \dot{\mathbf{B}}$ must be the rate of increase of magnetic field energy, which confirms (10.21) and (10.22).

If instead we take S to be a closed surface around a finite volume V, then the right hand side will not vanish and we must interpret the expression as the flux of energy through the surface S. The vector $\mathbf{E} \wedge \mathbf{H}$ is known as Poynting's vector and we shall generally write it as

$$\mathbf{N} = \mathbf{E} \wedge \mathbf{H}. \tag{10.26}$$

It is the energy flux vector of the e.m. field. In general, if T is the non-electromagnetic energy flux vector (heat, mechanical energy, etc.) and U_c the non-electromagnetic energy density, we have

$$\nabla \cdot (\mathbf{N}+\mathbf{T})+\frac{\partial}{\partial t}(U_c+U_M+U_E) = 0. \tag{10.27}$$

Note that $(\partial/\partial t)U_c$ includes the term $\mathbf{E} \cdot \mathbf{J}$.

The fundamental property of $\mathbf{N}$ is that, for a *closed* surface A, the outward flux of energy carried by the field is

$$\oint N_n \, \mathrm{d}A.$$

The flux across an *open* surface element $\mathrm{d}A$ is not necessarily $N_n \, \mathrm{d}A$. (See problem 4.)

Linear Media

In a linear, non-dissipative medium we can express $\mathbf{D}$ as

$$D_i = \varepsilon_{ij}\varepsilon_0 E_j \tag{10.28}$$

so that

$$\delta U_E = \varepsilon_{ij}\varepsilon_0 E_i \delta E_j. \tag{10.29}$$

If the final field is $(E_1, E_2, 0)$ we obtain

$$U_E = \tfrac{1}{2}\varepsilon_0\varepsilon_{11}E_1^2+\tfrac{1}{2}\varepsilon_0\varepsilon_{22}E_2^2+\varepsilon_0\varepsilon_{12}E_1E_2$$

by first increasing E_1, holding $E_2 = 0$ and then increasing E_2. If instead we first increase E_2 and then E_1 we obtain

$$U_E = \tfrac{1}{2}\varepsilon_0\varepsilon_{11}E_1^2+\tfrac{1}{2}\varepsilon_0\varepsilon_{22}E_2^2+\varepsilon_0\varepsilon_{21}E_1E_2.$$

These two expressions must be equal if the medium is non-dissipative and so $\varepsilon_{12} = \varepsilon_{21}$. Thus, in non-dissipative media, ε_{ij} and the corresponding magnetic quantity are symmetric tensors.

In general, in *linear, non-dissipative* media

$$U_E = \tfrac{1}{2}\varepsilon_0\varepsilon_{ij}E_iE_j = \tfrac{1}{2}\mathbf{E} \cdot \mathbf{D}, \tag{10.30}$$

$$U_M = \tfrac{1}{2}\mathbf{H} \cdot \mathbf{B}. \tag{10.31}$$

Linear and non-dissipative magnetic media, other than vacuum, are rare and so (10.31) must be treated with even more reserve than (10.30). We emphasize that (10.30) and (10.31) are much less generally valid than the fundamental relations

$$\delta U_E = \mathbf{E} \cdot \delta \mathbf{D},$$

$$\delta U_M = \mathbf{H} \cdot \delta \mathbf{B}.$$

It is unfortunate that the expressions (10.30) and (10.31) are treated so uncritically in many elementary texts.

In vacuum, the linear medium *par excellence*, we have, of course,

$$U_E = \tfrac{1}{2}\mathbf{E} \cdot \mathbf{D} = \tfrac{1}{2}\varepsilon_0 E^2 = \frac{1}{2\varepsilon_0} D^2,$$

$$U_M = \tfrac{1}{2}\mathbf{H} \cdot \mathbf{B} = \tfrac{1}{2}\mu_0 H^2 = \frac{1}{2\mu_0} B^2. \tag{10.32}$$

It is perhaps surprising that so fundamental a concept as the energy flux vector $\mathbf{N}$ should involve the auxiliary field $\mathbf{H}$ rather than $\mathbf{B}$. The reason for this is that, on a microscopic scale, the magnetization $\mathbf{M}$ is associated with circulating currents which transfer energy at a rate $-\mathbf{E} \wedge \mathbf{M}$ per unit area (Pershan, 1963).

In later sections we shall be discussing other aspects of the field energy in connection with thermodynamics and there we shall relate the energy to external influences such as impressed currents or charges. The connection is established as follows.

If an electrostatic field is due to charges q_k on electrodes at potentials ϕ_k, and there are no true or mobile charges in the system other than those on the electrodes, we have

$$\int \mathbf{E} \cdot \mathbf{D}\, d^3\mathbf{R} = \int -\mathbf{D} \cdot (\nabla\phi)\, d^3\mathbf{R} = -\int \nabla \cdot (\phi\mathbf{D})\, d^3\mathbf{R} + \int \phi \nabla \cdot \mathbf{D}\, d^3\mathbf{R}. \tag{10.33}$$

The last term is zero because $\nabla \cdot \mathbf{D} = 0$ and the remaining term gives a surface integral so that

$$\int \mathbf{E} \cdot \mathbf{D}\, d^3\mathbf{R} = -\int \phi D_n\, dS = -\sum_k \phi_k \int_{S_k} D_n\, dS_k \tag{10.34}$$

where the sum is over the equipotential electrodes. Since $-D_n$ is the surface charge on these electrodes we have

$$\int \mathbf{E} \cdot \mathbf{D}\, d^3\mathbf{R} = \sum_k \phi_k q_k,$$

and also

$$\int \mathbf{E} \cdot \delta\mathbf{D}\, d^3\mathbf{R} = \sum_k \phi_k \delta q_k, \tag{10.35}$$

$$\int \delta\mathbf{E} \cdot \mathbf{D}\, d^3\mathbf{R} = \sum_k \delta\phi_k q_k.$$

These results express quantities referring to fields in the interior of the system in terms of parameters accessible to experimental manipulation at the surface of the system.

In the magnetic case we express $\mathbf{B}$ as $\nabla \wedge \mathbf{A}$ and use the result

$$\int \mathbf{H} \cdot \mathbf{B}\, d^3\mathbf{R} = \int \mathbf{H} \cdot (\nabla \wedge \mathbf{A})\, d^3\mathbf{R} = \int \{(\nabla \wedge \mathbf{H}) \cdot \mathbf{A} - \nabla \cdot (\mathbf{H} \wedge \mathbf{A})\}\, d^3\mathbf{R}.$$

The divergence in the last term gives a surface integral at infinity, which we can discard for slowly varying fields. We then replace $\nabla \wedge \mathbf{H}$ by $\mathbf{J}$ and obtain

$$\int \mathbf{H} \cdot \mathbf{B}\, d^3\mathbf{R} = \int \mathbf{J} \cdot \mathbf{A}\, d^3\mathbf{R}.$$

If the fields are due to a current I in a conductor characterized by a line element $\mathbf{dl}$ we can write this as

$$\int \mathbf{H} \cdot \mathbf{B}\, d^3\mathbf{R} = I \oint \mathbf{A} \cdot \mathbf{dl}$$

so that

$$\int \mathbf{H} \cdot \dot{\mathbf{B}}\, d^3\mathbf{R} = I \oint \dot{\mathbf{A}} \cdot \mathbf{dl}.$$

In the absence of charge accumulation, i.e. with currents flowing in closed loops, we have $\mathbf{E} = -\dot{\mathbf{A}}$ and so

$$\int \mathbf{H} \cdot \dot{\mathbf{B}}\, d^3\mathbf{R} = -I \oint \mathbf{E} \cdot \mathbf{dl} = I\psi \tag{10.36a}$$

where ψ is the back e.m.f. in the circuit. We can also express this result in terms of the flux linking the circuit as

$$\int \mathbf{H} \cdot \dot{\mathbf{B}}\, d^3\mathbf{R} = I \int \dot{B}_n\, dS. \tag{10.36b}$$

In a system containing only permanent magnets $I = 0$ and

$$\int \mathbf{H} \cdot \delta\mathbf{B}\, d^3\mathbf{R} = 0. \tag{10.37}$$

If there is a single permanent magnet of given (controlled) magnetization $\mathbf{M}$, then we can express (10.37) as

$$\underset{\text{outside magnet}}{\int \mathbf{H} \cdot \delta\mathbf{B}\, d^3\mathbf{R}} + \underset{\text{inside magnet}}{\int \mu_0 \mathbf{H} \cdot \delta\mathbf{H}\, d^3\mathbf{R}} = -\underset{\text{inside magnet}}{\int \mu_0 \mathbf{H} \cdot \delta\mathbf{M}\, d^3\mathbf{R}} \tag{10.38}$$

and regard the left hand side of the equation as the change in field energy while the right hand side is the work done on the system by whatever external agencies change $\mathbf{M}$.

Stress, Force and Momentum

We now leave the topics of energy and power and address ourselves to a discussion of force and stress. No simple general formulation for the body force and stress in a macroscopic medium is known and the results that we shall derive, though of considerable complexity, are restricted to a rather small number of special cases. A demonstration that even in simple cases the results are complex is, in itself, worth while, for an erroneous belief that they are simple is obviously a potential source of confusion. In vacuum the force on a stationary macroscopic body carrying a charge q is $q\mathbf{E}$ and the force on a stationary element of a macroscopic circuit is $I\,\mathbf{dl} \wedge \mathbf{B}$. However, as we noted at the beginning of chapter 9, we are not entitled to assume that these expressions are valid for macroscopic bodies embedded in material media. One of our aims is to show how these forces may be calculated and under what circumstances they reduce to $q\mathbf{E}$ and $I\,\mathbf{dl} \wedge \mathbf{B}$.

We begin by considering a system of charged particles in vacuum. In this case we know, from first principles, that the macroscopic force acting on matter in unit volume is $\rho\mathbf{E}+\mathbf{J} \wedge \mathbf{B}$, this being that part of the force which does not arise from microscopically close collisions between the particles. The two homogeneous field equations $\nabla \cdot \mathbf{B} = 0$ and $\nabla \wedge \mathbf{E}+\dot{\mathbf{B}} = 0$ yield the identity

$$\varepsilon_0\mathbf{E}(\nabla \cdot \mathbf{E})+\frac{1}{\mu_0}(\nabla \wedge \mathbf{B}-\mu_0\varepsilon_0\dot{\mathbf{E}}) \wedge \mathbf{B}+\frac{\partial}{\partial t}(\varepsilon_0\mathbf{E} \wedge \mathbf{B})$$

$$= \varepsilon_0\{\mathbf{E}(\nabla \cdot \mathbf{E})-\mathbf{E} \wedge (\nabla \wedge \mathbf{E})\}+\frac{1}{\mu_0}\{\mathbf{B}(\nabla \cdot \mathbf{B})-\mathbf{B} \wedge (\nabla \wedge \mathbf{B})\}, \tag{10.39}$$

and the right hand side of this equation can be expressed as the divergence of a tensor T^0_{ij}

where

$$T^0_{ij} = \frac{\varepsilon_0}{2}\{E_iE_j+E_jE_i-E^2\delta_{ij}\}+\frac{1}{2\mu_0}\{B_iB_j+B_jB_i-B^2\delta_{ij}\}. \tag{10.40}$$

On the left hand side we have $\varepsilon_0\nabla \cdot \mathbf{E} = \rho$ and $\nabla \wedge \mathbf{B}-\mu_0\varepsilon_0\dot{\mathbf{E}} = \mu_0\mathbf{J}$ so that

$$\rho E_i+(\mathbf{J} \wedge \mathbf{B})_i+\frac{\partial}{\partial t}\varepsilon_0(\mathbf{E} \wedge \mathbf{B})_i = \frac{\partial T^0_{ij}}{\partial r_j}. \tag{10.41}$$

Because in this case we know that $\rho E_i+(\mathbf{J} \wedge \mathbf{B})_i$ is the ith component of the force on matter in unit volume, or the ith component of the rate of change of material momentum density, we can recognize this as the momentum conservation equation and identify the other two terms. The momentum density of the field is $\varepsilon_0\mathbf{E} \wedge \mathbf{B}$ which can be written as

$$\boldsymbol{\Pi} = \varepsilon_0\mathbf{E} \wedge \mathbf{B} = \mu_0\varepsilon_0\mathbf{E} \wedge \mathbf{H} = \mathbf{N}/c^2 \tag{10.42}$$

where $\mathbf{N}$ is Poynting's vector. This implies that a photon, of vector wave number $\mathbf{k}$ and energy $h\nu = hk/c$, carries a momentum $h\mathbf{k}$. The tensor T^0_{ij} is the stress tensor of the field, the components give the flux of the ith component of momentum in the j direction as $-T^0_{ij}$, and the rate of momentum flow into a system within a surface S with a positive outward normal is

$$F_i = \int_S T^0_{ij}\, \mathrm{d}S_j. \tag{10.43}$$

For static fields this is the force on matter within S. For time-dependent fields this is not in general true, because of the term $(\partial/\partial t)(\varepsilon_0\mathbf{E} \wedge \mathbf{B})$ in (10.41), but, for harmonic or sinusoidal fields, the time average of this expression is zero and then F_i correctly gives the time average force on matter within S. The total force exerted by radiation on an isolated body can therefore be calculated using (10.43) and a surface S in vacuum outside the body.

In matter we have $\varepsilon_0\nabla \cdot \mathbf{E} = \rho-\nabla \cdot \mathbf{P}$ and $\nabla \wedge \mathbf{B}-\mu_0\varepsilon_0\dot{\mathbf{E}} = \mu_0(\mathbf{J}+\dot{\mathbf{P}}+\nabla \wedge \mathbf{M})$ so that (10.41) remains an identity if the body force term is replaced by

$$\mathbf{f}' = (\rho-\nabla \cdot \mathbf{P})\mathbf{E}+(\mathbf{J}+\dot{\mathbf{P}}+\nabla \wedge \mathbf{M}) \wedge \mathbf{B}. \tag{10.44}$$

For static or steady harmonic fields the average force on an isolated body is still given by (10.43) and so $\int \mathbf{f}'\, \mathrm{d}V$ is an alternative expression for the force on the body. We cannot however deduce that $\mathbf{f}'$ is the correct internal body force. Indeed it is not. This may seem surprising since $-\nabla \cdot \mathbf{P}$ and $\dot{\mathbf{P}}+\nabla \wedge \mathbf{M}$ are the effective macroscopic charge and current densities due to bound atomic charge, but we must remember that all the macroscopic variables ρ, $\mathbf{J}$, $\mathbf{E}$ and $\mathbf{B}$, not only $\mathbf{P}$ and $\mathbf{M}$, are defined by smoothing or truncation, so that they correctly describe interactions between two systems when all the particles of one system are at a macroscopic distance from all the particles of the other system. We cannot use them to discuss forces acting across a surface within a continuous medium without further investigation. To obtain a microscopic interpretation of a macroscopic stress we should first have to calculate, with microscopic exactitude, the force acting across a plane located with microscopic precision in the medium, and then, regarding this force as a function of the position of the plane, truncate its spatial Fourier spectrum to obtain the macroscopic force. This procedure has not so far been attempted, although somewhat similar ideas are discussed by de Groot (1969). Fortunately, in certain circumstances, the desired result can be obtained by the following, purely macroscopic, argument generally attributed to Helmholtz.

If **f** is the body force in a medium and each material point of the medium is given a small impressed velocity **v**, which may be an arbitrary continuous function of position, the rate of change of the total energy of the system is

$$\frac{dU}{dt} = -\int \mathbf{f}\cdot\mathbf{v}\,dV. \tag{10.45}$$

If then we can also express dU/dt in terms of a function **X** of the electromagnetic variables, so that

$$\frac{dU}{dt} = -\int \mathbf{X}\cdot\mathbf{v}\,dV \tag{10.46}$$

we can, because **v** is arbitrary, identify **X** with the body force **f**.

We begin by considering an electrostatic system containing dielectric media and extrinsic charge of density ρ. This charge is anchored in matter so that, as each material point moves, it takes its charge with it.

The energy required to assemble this system against elastic and electrostatic forces contains two terms. Only the electrostatic term need concern us and, if the system is non-dissipative, we only have to consider

$$U^e = \int \{\int_0^D \mathbf{E}\cdot d\mathbf{D}\}\,dV = \int \{\mathbf{E}\cdot\mathbf{D} - \int_0^E \mathbf{D}\cdot d\mathbf{E}\}\,dV.$$

The time derivative of this expression is

$$\frac{dU^e}{dt} = \int\left\{\mathbf{E}\cdot\dot{\mathbf{D}} + \mathbf{D}\cdot\dot{\mathbf{E}} - \mathbf{D}\cdot\dot{\mathbf{E}} - \int_0^E \left(\frac{\partial \mathbf{D}}{\partial t}\right)_E \cdot d\mathbf{E}\right\}dV = \int\left\{\mathbf{E}\cdot\dot{\mathbf{D}} - \int_0^E \left(\frac{\partial \mathbf{D}}{\partial t}\right)_E \cdot d\mathbf{E}\right\}dV.$$

The last term is quite intractable unless there is a simple relation between **D** and **E**. We confine our attention to linear media in which $D_i = \varepsilon_0\varepsilon_{ij}D_j$ and ε_{ij}, although it may be a function of position, does not depend on **E**. Because the medium is non-dissipative $\varepsilon_{ij} = \varepsilon_{ji}$. We therefore have to consider

$$\frac{dU^e}{dt} = \int (\mathbf{E}\cdot\dot{\mathbf{D}} - \tfrac{1}{2}\varepsilon_0\dot{\varepsilon}_{ij}E_iE_j)\,dV. \tag{10.47}$$

If **E** is expressed as $-\nabla\phi$, then since $\nabla\cdot\dot{\mathbf{D}} = \dot{\rho} = -\nabla\cdot(\rho\mathbf{v})$ a series of partial integrations, in which surface terms at infinity are discarded, yields

$$\begin{aligned}\int \mathbf{E}\cdot\dot{\mathbf{D}}\,dV &= -\int \dot{\mathbf{D}}\cdot\nabla\phi\,dV = \int \phi\nabla\cdot\dot{\mathbf{D}}\,dV = -\int \phi\nabla\cdot(\rho\mathbf{v})\,dV = \int \rho\mathbf{v}\cdot\nabla\phi\,dV\\ &= -\int \rho\mathbf{v}\cdot\mathbf{E}\,dV.\end{aligned}$$

Thus one term in the body force is, as we might expect, $\rho\mathbf{E}$.

The partial derivative $\dot{\varepsilon}_{ij}$, at a fixed point in space, contains a contribution from the motion which brings new matter to the point, a contribution from the rotation of the medium and a contribution from the change in the properties of the medium as it is strained by the impressed motion. These terms are discussed in section 11 of the Appendix and if

$$\alpha_{ijkl} = \frac{d\varepsilon_{ij}}{dS_{kl}}$$

is the rate of change of ε_{ij} with strain S_{kl}, we have

$$\dot{\varepsilon}_{ij} = -v_k \frac{\partial \varepsilon_{ij}}{\partial r_k} + \tfrac{1}{2}\left\{\varepsilon_{kj}\left(\frac{\partial v_i}{\partial r_k} - \frac{\partial v_k}{\partial r_i}\right) + \varepsilon_{ik}\left(\frac{\partial v_j}{\partial r_k} - \frac{\partial v_k}{\partial r_j}\right)\right\} + \alpha_{ijkl}\frac{\partial v_k}{\partial r_l}. \tag{10.48}$$

When this is substituted in (10.49) and the terms are rearranged by partial integration and the relabelling of dummy tensor suffixes, the result is

$$\frac{\mathrm{d}U_e}{\mathrm{d}t} = \int -v_i\left\{\rho E_i - \frac{\varepsilon_0}{2} E_j E_k \frac{\partial \varepsilon_{jk}}{\partial r_i} + \tfrac{1}{2}\frac{\partial}{\partial r_j}(E_i D_j - E_j D_i) + \frac{\partial}{\partial x_j} T^{es}_{ij}\right\} \mathrm{d}V$$

where the electrostrictive tensor is

$$T^{es}_{ij} = -\tfrac{1}{2}\varepsilon_0 \alpha_{klij} E_k E_l. \tag{10.49}$$

The electrostatic body force in linear media is therefore

$$f^e_i = \rho E_i - \frac{\varepsilon_0}{2} E_j E_k \frac{\partial \varepsilon_{jk}}{\partial r_i} - \tfrac{1}{2}\frac{\partial}{\partial r_j}(E_i D_j - E_j D_i) + \frac{\partial}{\partial r_j} T^{es}_{ij}. \tag{10.50}$$

The interpretation of the first term is obvious. The second term gives forces at the boundary between two media of different dielectric properties, it tends to attract the more polarizable medium into the field. The third term, which is the ith component of $-\tfrac{1}{2}\mathbf{\nabla} \wedge (\mathbf{E} \wedge \mathbf{D})$, leads to torques and couples in anisotropic media and at surfaces. The final term gives a force which generally tends to attract dielectric media into regions of stronger field. It is usually comparable in magnitude with the other terms.

If, in (10.50), ρ is replaced by $\partial D_j/\partial x_j$, then using the symmetry of ε_{ij}, and

$$\frac{\partial E_i}{\partial x_j} = \frac{\partial E_j}{\partial x_i},$$

which follows from $\mathbf{\nabla} \wedge \mathbf{E} = 0$, we can rearrange the force so that it can be expressed as

$$f_i = \frac{\partial}{\partial r_j}(T^e_{ij} + T^{es}_{ij}) \tag{10.51}$$

where

$$T^e_{ij} = \tfrac{1}{2}(E_i D_j + E_j D_i - \mathbf{E} \cdot \mathbf{D}\delta_{ij}). \tag{10.52}$$

This quantity is the electrostatic stress tensor.

In vacuum the total stress tensor is just T^e_{ij}, but in a material medium it is not just $T^e_{ij} + T^{es}_{ij}$, for the medium can support stress. If the elastic stress tensor is Y_{ij} the body force is

$$f_i = \frac{\partial}{\partial r_j}(T^e_{ij} + T^{es}_{ij} + Y_{ij}) \tag{10.53}$$

and the total force acting on matter within a closed surface S is

$$F_i = \int (T^e_{ij} + T^{es}_{ij} + Y_{ij})\, \mathrm{d}S_j. \tag{10.54}$$

Now, if the system is in static equilibrium, the body force f_i must vanish except at points where external impressed forces are applied. Thus, if we wish to calculate the external force, $-F^e_i$, that must be applied to maintain a charged body embedded in a dielectric medium in static equilibrium, we can certainly assume that $f_i = 0$ in the medium. This means that the required force $F_i^e = -F_i$ can be obtained by evaluating (10.54) over any convenient surface entirely within the medium. Furthermore, in the medium we must have

$$\frac{\partial}{\partial r_j}(T_{ij}^e + T_{ij}^{es} + Y_{ij}) = 0. \tag{10.55}$$

This unfortunately, except in one special case, is not enough to determine Y_{ij} uniquely and so there is no unique value for F_i^e. Whatever force is applied to the body it will remain at rest, and Y_{ij} will adjust itself so that

$$-F_i^e - \int Y_{ij}\,\mathrm{d}S_j = \int (T_{ij}^e + T_{ij}^{es})\,\mathrm{d}S_j.$$

Thus, in general, the most that we can calculate is the force that has to be supported by the sum of elastic stress and externally applied forces. The one special case occurs when the medium is a fluid. Since this cannot support shear stress, but only exhibit a hydrostatic pressure p, the elastic stress tensor has the form $-p\delta_{ij}$. Equation (10.55) is then enough to determine Y_{ij} uniquely. The force required to restrain the body has a unique value.

In a homogeneous fluid medium, free of extrinsic charge, the body force given by (10.50) vanishes except for the term $(\partial/\partial r_j)T_{ij}^{es}$. In section 11 of the Appendix we show that, for a fluid of density τ, the electrostrictive stress tensor is $T_{ij}^{es} = \frac{1}{2}\varepsilon_0 E^2 \tau(\mathrm{d}\varepsilon/\mathrm{d}\tau)\delta_{ij}$. Thus the condition for static equilibrium in the fluid is simply

$$\frac{\partial}{\partial r_i}\left(\tfrac{1}{2}\varepsilon_0 E^2 \tau \frac{\mathrm{d}\varepsilon}{\mathrm{d}\tau} - p\right) = 0,$$

and the fluid will distort until the pressure distribution is

$$p = \tfrac{1}{2}\varepsilon_0 E^2 \tau \frac{\mathrm{d}\varepsilon}{\mathrm{d}\tau} + \text{constant}. \tag{10.56}$$

A constant term in a stress tensor produces no physical effects and so the effective stress tensor in the fluid is simply

$$T_{ij}^e = \varepsilon\varepsilon_0(E_i E_j - \tfrac{1}{2}E^2\delta_{ij}). \tag{10.57}$$

Consider then a macroscopic body carrying a charge q immersed in a fluid medium in which there is a uniform impressed field $\mathbf{E}^*$, parallel to the z axis, due to distant charges. To evaluate the force on the body we integrate the stress tensor over the surface of a sphere of large radius r centred on the body. Using polar coordinates the only appreciable field components on the surface of the sphere are

$$E_x = \frac{q \sin\theta \cos\phi}{4\pi\varepsilon\varepsilon_0 r^2}, \qquad E_y = \frac{q \sin\theta \sin\phi}{4\pi\varepsilon\varepsilon_0 r^2}$$

and

$$E_z = E^* + \frac{q\cos\theta}{4\pi\varepsilon\varepsilon_0 r^2},$$

while the Cartesian components of a surface element

$$r^2 \sin\theta\,\mathrm{d}\theta\,\mathrm{d}\phi$$

are

$$\mathrm{d}S_x = r^2 \sin^2\theta \cos\phi\,\mathrm{d}\theta\,\mathrm{d}\phi,$$
$$\mathrm{d}S_y = r^2 \sin^2\theta \sin\phi\,\mathrm{d}\theta\,\mathrm{d}\phi$$

and

$$\mathrm{d}S_z = r^2 \cos\theta \sin\theta\,\mathrm{d}\theta\,\mathrm{d}\phi.$$

When these fields are substituted in (10.57) and $\int T^e_{ij}\, dS_j$ is evaluated, the angular integrals eliminate all components of the force except

$$F_z = \int_0^{2\pi}\int_0^{\pi} \frac{qE^*}{4\pi}\{\sin^3\theta\cos^2\phi+\sin^3\theta\sin^2\phi+2\cos^2\theta\sin\theta-\cos^2\theta\sin\theta\}\,d\theta\,d\phi = qE^*. \tag{10.58}$$

In this case, where the special elastic properties of the medium impose a unique value of Y_{ij}, we obtain a unique force and it corresponds to our naïve expectations.

Next we consider an isotropic solid medium and the force

$$F_i = \int (T^e_{ij}+T^{es}_{ij})\, dS_j,$$

which has to be supported by elastic stress and applied forces. In section 11 of the Appendix the electrostrictive stress tensor for an isotropic solid is found to be

$$T^{es}_{ij} = \tfrac{1}{2}\varepsilon_0(a_1E_iE_j+a_2E^2\delta_{ij}).$$

If this is compared with equation (10.57), which leads to (10.58), we see that the additional force is

$$F'_z = \int_0^{2\pi}\int_0^{\pi} -\frac{qE^*}{8\pi\varepsilon}\{a_1(\sin^3\theta+2\cos^2\theta\sin\theta)+2a_2\cos^2\theta\sin\theta)\}\,d\theta\,d\phi = -qE^*\frac{2a_1+a_2}{3\varepsilon}.$$

Thus the total force in this case is

$$F = q\mathbf{E}^*\left(1-\frac{2a_1+a_2}{3\varepsilon}\right);$$

since, for simple dielectric media, a_1 is small and $a_2 \approx (\varepsilon-1)$ this is approximately $qE^*(\frac{4}{3}-\frac{1}{3}\varepsilon)$, which may be quite different from $q\mathbf{E}^*$. We see from these two examples that the notion that there is any simple general formula for the force on a charged body in a dielectric medium does not stand up to examination.

To calculate the magnetostatic forces and stress we consider the work that has to be done against magnetic forces,

$$U^m = \int \{\textstyle\int^{\mathbf{B}} \mathbf{H}\,.\,d\mathbf{B}\}\,dV,$$

to assemble a system of magnetic media in which currents of density $\mathbf{J}$ flow. If the media are linear this leads to

$$\frac{dU^m}{dt} = \int (\mathbf{H}\,.\,\dot{\mathbf{B}}-\tfrac{1}{2}\mu_0H_iH_j\dot{\mu}_{ij})\,dV. \tag{10.58}$$

The term in $\dot{\mu}_{ij}$ presents no new problems but the term $\mathbf{H}\,.\,\dot{\mathbf{B}}$ needs careful attention. In this case the flow of current is connected with a change in the thermal and chemical energy of the system. Batteries in the system are required to maintain the current flow and the flow generates ohmic heat in the media.

The total rate at which energy is stored in the media by charging batteries and ohmic heating is the volume integral of $\mathbf{E}\,.\,\mathbf{J}$, where $\mathbf{E}$ is the electric field relative to the moving medium. Since, in a coordinate system moving with the medium, $\nabla^m \wedge \mathbf{H} = \mathbf{J}$ we have

$$\int \mathbf{E}\,.\,\mathbf{J}\,dV = \int \mathbf{E}\,.\,(\nabla^m \wedge \mathbf{H})\,dV = \int \mathbf{H}\,.\,(\nabla^m \wedge \mathbf{E})\,dV$$

where, in the partial integration in the last step, we have, as usual, discarded a surface term. In section 11 of the Appendix it is shown that $\nabla^m \wedge \mathbf{E}$ is related to $\dot{\mathbf{B}}$ at a point fixed in space by $\nabla^m \wedge \mathbf{E} = -\dot{\mathbf{B}}+\nabla \wedge (\mathbf{v} \wedge \mathbf{B})$ and so we obtain

$$\int \mathbf{E}\,.\,\mathbf{J}\,\mathrm{d}V+\int \mathbf{H}\,.\,\dot{\mathbf{B}}\,\mathrm{d}V = \int \mathbf{H}\,.\,[\nabla \wedge (\mathbf{v} \wedge \mathbf{B})]\,\mathrm{d}V.$$

A partial integration converts this to

$$\int \mathbf{E}\,.\,\mathbf{J}\,\mathrm{d}V+\int \mathbf{H}\,.\,\dot{\mathbf{B}}\,\mathrm{d}V = \int (\mathbf{v} \wedge \mathbf{B})\,.\,(\nabla \wedge \mathbf{H})\,\mathrm{d}V = \int (\mathbf{v} \wedge \mathbf{B})\,.\,\mathbf{J}\,\mathrm{d}V = -\int \mathbf{v}\,.\,(\mathbf{J} \wedge \mathbf{B})\,\mathrm{d}V.$$

The left hand side of this expression is the sum of the required term in $\mathrm{d}U^m/\mathrm{d}t$ and the rate at which chemical and thermal energy is being stored in the system. The term on the right must be the required rate at which the forces, which drive the impressed motion $\mathbf{v}$, do work. Thus the body force is $\mathbf{J} \wedge \mathbf{B}$. The total magnetostatic body force is therefore

$$f_i^m = (\mathbf{J} \wedge \mathbf{B})_i - \frac{\mu_0}{2} H_j H_k \frac{\partial \mu_{jk}}{\partial r_i} - \tfrac{1}{2}\frac{\partial}{\partial r_i}(H_i B_j - H_j B_i) + \frac{\partial}{\partial r_j} T_{ij}^{ms} \tag{10.60}$$

in which the magnetostrictive stress tensor is

$$T_{ij}^{ms} = -\tfrac{1}{2}\mu_0 H_k H_l \beta_{klij}, \tag{10.61}$$

where

$$\beta_{klij} = \frac{\mathrm{d}\mu_{kl}}{\mathrm{d}S_{ij}} \tag{10.62}$$

gives the rate of change of permeability with strain. If $\mathbf{J}$ is replaced by $\nabla \wedge \mathbf{H}$ and the terms are rearranged the force can be expressed as

$$f_i^m = \frac{\partial}{\partial r_j}(T_{ij}^m + T_{ij}^{ms}) \tag{10.63}$$

where the magnetostatic stress tensor is

$$T_{ij}^m = \tfrac{1}{2}(H_i B_j + H_j B_i - \mathbf{H}\,.\,\mathbf{B}\delta_{ij}). \tag{10.64}$$

We now use this result to calculate the force per unit length acting on a round enamelled non-magnetic wire carrying a current I parallel to the z axis in a homogeneous fluid medium in which there is a uniform impressed field $\mathbf{B}^* = \mu\mu_0\mathbf{H}^*$ parallel to the x axis. Because the medium is a homogeneous fluid it will only be in equilibrium if the body force due to T_{ij}^{ms} is cancelled by a pressure gradient and so the effective stress tensor is simply

$$T_{ij}^m = \mu\mu_0(H_i H_j - \tfrac{1}{2}H^2\delta_{ij}).$$

To calculate the force per unit length we integrate this stress tensor over unit length of a large cylinder of radius r, coaxial with the wire. In terms of cylindrical polar coordinates (r, θ, z) the Cartesian components of the fields on this surface are

$$H_x = H^* - \frac{I}{2\pi r}\sin\theta \quad \text{and} \quad H_y = \frac{I\cos\theta}{2\pi r}$$

while the area elements are $\mathrm{d}S_x = r\cos\theta\,\mathrm{d}\theta$ and $\mathrm{d}S_y = r\sin\theta\,\mathrm{d}\theta$. It is easy to see that integration over θ eliminates all force components except

$$F_y = \int T_{yx}\,\mathrm{d}S_x + \int T_{yy}\,\mathrm{d}S_y = \int_0^{2\pi} \{H_y H_x \cos\theta + \tfrac{1}{2}(H_y^2 - H_x^2)\sin\theta\}\, r\,\mathrm{d}\theta$$

and that this gives $\mu\mu_0 H^* I = B^* I$. Again, for a fluid medium we obtain an expected and trivial result. To see that it is not so trivial we consider the behaviour of the body force.

Inside the wire the only term is $\mathbf{J} \wedge \mathbf{B}$ and, since $\mathbf{B}$ inside a cylindrical body of permeability 1 immersed in a medium of permeability μ with its axis normal to the applied field is $(2/\mu+1)B^*$, this gives a force $[2/(\mu+1)]IB^*$. In the fluid the body force due to magnetostriction is cancelled by the pressure gradient and the rest of the force is zero. However, f^m given by equation (10.60) contains a term $-(\mu_0/2)H^2(\partial\mu/\partial x_i)$ and μ changes discontinuously at the surface of the enamelled wire. The resulting force acting on the wire in the y direction is

$$F_y = \int_0^\infty \int_0^{2\pi} -\frac{\mu_0}{2} H^2 \frac{\partial \mu}{\partial r} \sin\theta \, r \, d\theta \, dr.$$

At the surface of the wire the radial component of H changes discontinuously from $[2\mu/(\mu+1)]H^* \cos\theta$ to $[2/(\mu+1)]H^* \cos\theta$, but the tangential component

$$H_\theta = \frac{I}{2\pi r} - \frac{2\mu}{\mu+1} H^* \sin\theta$$

is continuous. Since the only non-vanishing angular integral in F_y comes from the cross term

$$\frac{-4\mu I H^* \sin\theta}{2\pi r(\mu+1)}$$

in H_θ^2, we have

$$F_y = \int_0^{2\pi} \frac{\mu_0 \mu I H^* \sin^2\theta}{\pi(\mu+1)} (\mu-1) \, d\theta = \frac{\mu-1}{\mu+1} IB^*.$$

The total force $F_y = IB^*$ is therefore made up of two parts, a direct magnetic force $[2/(\mu+1)]IB^*$ acting on the conductor, and a force $[(\mu-1)/(\mu+1)]IB^*$ localized at the outer surface of the enamel and transmitted mechanically, through the enamel, to the conductor.

We need hardly add that the situation is much more complicated in solid media. The important practical problem of calculating the stress in an iron transformer yoke or armature wound with wire clearly has no simple elementary solution.

If we let f_i^e and f_i^m stand for the expressions on the right of equations (10.50) and (10.60) then, whether or not they represent forces, the equation

$$f_i^e + f_i^m + \frac{\partial}{\partial t}(\mathbf{D} \wedge \mathbf{B}) = \frac{\partial}{\partial r_j}(T_{ij}^{em} + T_{ij}^{es} + T_{ij}^{ms}), \tag{10.65}$$

with

$$T_{ij}^{em} = \tfrac{1}{2}\{E_i D_j + E_j D_i + H_i B_j + H_j B_i - \mathbf{E} \cdot \mathbf{D}\delta_{ij} - \mathbf{H} \cdot \mathbf{B}\delta_{ij}\}, \tag{10.66}$$

is an identity and can be derived directly from the field equations. It is valid whatever the relation between $\mathbf{D}$ and $\mathbf{E}$ or $\mathbf{B}$ and $\mathbf{H}$ and also whatever the form of the electrostrictive and magnetostrictive stress tensors T_{ij}^{es} and T_{ij}^{ms} since these appear on both sides of the equation. Equation (10.65) is clearly not the general momentum conservation equation for non-linear media, since the terms f_i^e and f_i^m only give the static body force correctly in linear media. We might, however, suspect that it is the conservation equation for time-dependent fields, in which case $\mathbf{D} \wedge \mathbf{B}$ is to be interpreted as the field momentum density, we shall return to this question later. First we consider the average force exerted on an isolated body by wave fields or radiation incident from vacuum. For this we need only the vacuum result (10.41). Since, for steady sinusoidal fields, the time average of an expression such as $(\partial/\partial t)(\varepsilon_0 \mathbf{E} \wedge \mathbf{B})$

vanishes identically we can equate the average force to the average value of

$$F_i = \int_S T^0_{ij}\, \mathrm{d}S_j, \tag{10.67}$$

evaluated over any closed surface in vacuum outside the body. We begin by considering the normal component of the force exerted on a large, but thin, plane slab of a perfectly absorbing medium, when a plane wave is incident at an angle θ to the normal to its surface. The surface of integration is indicated in Fig. 10.1. The fields are zero on the part S''' of the surface and, by a suitable choice of geometry, the contributions to (10.67) from S'' and S'''' can be made negligible. Thus we need only evaluate (10.67) over the plane surface S' just

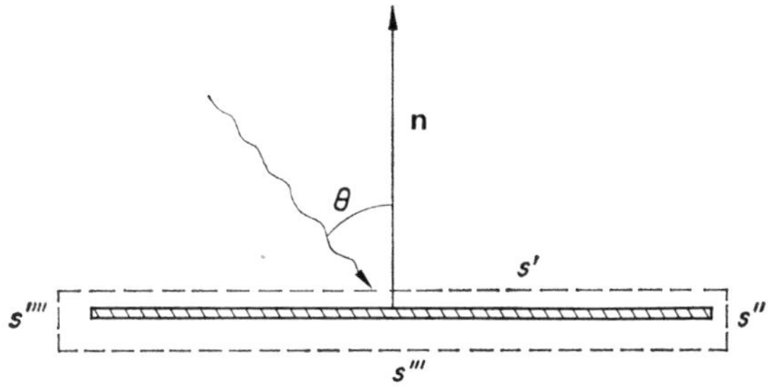

FIG. 10.1. Wave incident at an angle θ to the normal n on a plane surface.

above the medium. Clearly the pressure, or normal force per unit area, is just the component $-T^0_{nn}$ of the tensor. If the electric vector of the wave makes an angle ϕ with the plane of incidence the normal component of $\mathbf{E}$ is $E \sin\theta \cos\phi$, where E is the amplitude, and the normal component of $\mathbf{B}$ is $B \sin\theta \sin\phi$. Thus we have

$$T^0_{nn} = \varepsilon_0(E_n^2 - \tfrac{1}{2}E^2) + \frac{1}{\mu_0}(B_n^2 - \tfrac{1}{2}B^2) = \tfrac{1}{2}\varepsilon_0 E^2(2\sin^2\theta\cos^2\phi - 1) + \frac{1}{2\mu_0}B^2(2\sin^2\theta\sin^2\phi - 1).$$

But, for a plane wave $\tfrac{1}{2}\varepsilon_0 E^2 = (1/2\mu_0)B^2$ and so

$$p = -T_{nn} = \varepsilon_0 E^2(1 - \sin^2\theta) = \tfrac{1}{2}\left(\varepsilon_0 E^2 + \frac{1}{\mu_0}B^2\right)\cos^2\theta. \tag{10.68}$$

Since the average energy density u in front of the surface is the average of $\tfrac{1}{2}[\varepsilon_0 E^2 + (1/\mu_0)B^2]$, we can express the average pressure as

$$p = u\cos^2\theta. \tag{10.69}$$

If the plane waves are incident on a perfectly reflecting surface there is a standing wave pattern of fields in front of the surface and this appreciably complicates the calculation (see problems 20 and 21). For waves at normal incidence we certainly find $p = u$ but for other angles, although p, as it must be, is independent of the choice of the surface S', the time average energy density u varies periodically with the position of S', in a way which depends on the wave polarization. Nevertheless, if N_n^i and N_n^r are the components of Poynting's vector $\mathbf{E} \wedge \mathbf{H}$ for the incident and reflected waves, resolved along the outward normal $\mathbf{n}$, we find that we can express the pressure as

$$p = (\mu_0\varepsilon_0)^{\frac{1}{2}}(N_n^r - N_n^i) = \frac{1}{c}(N_n^r - N_n^i). \tag{10.70}$$

Since N_n^i is negative this is consistent with the notion that the waves transport momentum at a rate $(1/c)\mathbf{N}$ per unit area.

If the radiation is unpolarized, an average over all directions of polarization eliminates the spatial variation of u, and we recover the relation (10.69). For thermal radiation, incident equally from all directions, an average over θ yields

$$p = \tfrac{1}{3}u. \tag{10.71}$$

Thus for thermal radiation incident on either a perfect absorber or a perfect reflector, or indeed any medium which does not transmit radiation, the average pressure is one-third of the energy density outside the surface. This result is the basis for the thermodynamic treatment of black-body radiation.

If radiation falls on a perfect reflector from an isotropic, homogeneous and linear fluid medium of relative dielectric constant ε and permeability μ and, if the expression on the right of (10.65) is indeed the correct stress tensor, then the average pressure is the average value of

$$p = -T_{nn}^{em} = \varepsilon\varepsilon_0(\tfrac{1}{2}E^2 - E_n^2) + \frac{1}{\mu\mu_0}(\tfrac{1}{2}B^2 - B_n^2) = \varepsilon\varepsilon_0(\tfrac{1}{2}E^2 - E_n^2) + \mu\mu_0(\tfrac{1}{2}H^2 - H_n^2).$$

This has the effect of increasing $(\mu_0\varepsilon_0)^{\frac{1}{2}}$ in equation (10.70) to $n(\mu_0\varepsilon_0)^{\frac{1}{2}}$ where $n = (\mu\varepsilon)^{\frac{1}{2}}$ is the refractive index of the medium. Thus, for a given intensity of incident radiation the pressure should be increased by a factor n by the presence of the medium. This effect has been confirmed experimentally by Jones and Richards (1954) and indicates that, at least as far as steady-state conditions are concerned, T_{ij}^{em} given by (10.66) is the correct stress tensor for linear media. We already know that it is correct for static fields. This still does not mean that equation (10.65) is the general momentum conservation equation, and that $\mathbf{D} \wedge \mathbf{B}$ is the field momentum density. There are two possibilities. First of all we might add to both sides of (10.65) a term which vanished for steady-state conditions; however, this term must have the dimensions of energy density and also be symmetric in the tensor suffixed i and j, for conservation of angular momentum requires the stress tensor to have this symmetry. It is difficult to see how a new term with all these properties can be constructed from the available variables. The other possibility is that the term $(\partial/\partial t)(\mathbf{D} \wedge \mathbf{B})$ conceals a contribution to the body force which only shows up in transient, i.e. non-steady-state, conditions. Thus, for example, we might write equation (10.65) as

$$f_i^e + f_i^m + \frac{\partial}{\partial t}(\mathbf{P} \wedge \mathbf{B} + \mathbf{E} \wedge \mathbf{M})_i + \mu_0\varepsilon_0\frac{\partial}{\partial t}(\mathbf{E} \wedge \mathbf{H})_i = \frac{\partial}{\partial r_j}(T_{ij}^{em} + T_{ij}^{es} + T_{ij}^{ms}) \tag{10.72}$$

and identify only the term $\mu_0\varepsilon_0\mathbf{E} \wedge \mathbf{H}$ as the field momentum density and treat

$$\frac{\partial}{\partial t}(\mathbf{P} \wedge \mathbf{B} + \mathbf{E} \wedge \mathbf{M})$$

as a contribution to the body force. There are some theoretical grounds, based on considerations of relativistic invariance, for believing that, in so far as any meaning can be ascribed to the concept, the momentum density should be $\mu_0\varepsilon_0\mathbf{E} \wedge \mathbf{H} = \mathbf{N}/c^2$, but unfortunately there is no experimental evidence for the existence of the extra term in the body force. An experimental test could only be made using transient pulses of radiation.

Despite the uncertainty that attaches to the interpretation of equation (10.65) the notion of an electromagnetic stress tensor is of considerable use in practical calculations. In material media fortunately we are usually only concerned with static fields, where, at least for linear media, the results are unambiguous. In wave problems most practical questions can be answered in terms of the vacuum stress tensor whose general validity is unquestioned.

We conclude with a brief discussion of the general structure of the arguments which lead to conservation equations in macroscopic electromagnetism. The four field equations $\nabla \cdot \mathbf{B} = 0$, $\nabla \wedge \mathbf{E}+\dot{\mathbf{B}} = 0$, $\nabla \cdot \mathbf{D} = \rho$ and $\nabla \wedge \mathbf{H}-\dot{\mathbf{D}} = \mathbf{J}$, together with the law for the body force on particles in vacuum $\mathbf{f} = \rho\mathbf{E}+\mathbf{J} \wedge \mathbf{B}$, and the notion that $\mathbf{J}$ is the flux vector for charge, make up the basic laws. By manipulating these equations we derive a series of relations, of which the simplest is $\nabla \cdot \mathbf{J}+\dot{\rho} = 0$, which we recognize as possible conservation laws. In this case the law is charge conservation. A second example is the Poynting equation: $\nabla \cdot (\mathbf{E} \wedge \mathbf{H})+\mathbf{E} \cdot \dot{\mathbf{D}}+\mathbf{H} \cdot \dot{\mathbf{B}}+\mathbf{E} \cdot \mathbf{J} = 0$. In this case the basic laws allow us to identify $\mathbf{E} \cdot \mathbf{J}$ as the rate of transfer of energy to matter in unit volume by the action of the field on mobile carriers. This is enough to justify interpreting Poynting's equation as the energy conservation law, and allow us to make the usual identification of the other terms. When we come to construct a momentum conservation law there is no shortage of equations, of the right form, that can be constructed; the problem is to make a general, and unambiguous, identification of any single one of the terms that appear. Intuition is not enough, the identification must be justified by reference to the basic laws. In the most general time-dependent case this is only possible for a system of charges in vacuum. In material media with known constitutive relations between $\mathbf{E}$ and $\mathbf{D}$ or $\mathbf{B}$ and $\mathbf{H}$ it is also possible for static fields, but the expression for the body force is likely to be exceedingly complicated and the stress tensor even more so, unless the constitutive relation is linear. The vacuum momentum law is also valid on a microscopic scale and this can sometimes be used to derive general results (we have not done so here), and we might also expect to be able to derive the macroscopic momentum law directly from the microscopic equations. Thus, in general, the momentum law and the associated notions of force and stress present serious problems which are entirely absent from the simpler energy conservation law.

Problems

1. A charged particle q moves with velocity $\mathbf{v}$ in a field $\mathbf{E}$ due to two electrodes whose potential difference ϕ is maintained constant by a battery. Show that the current in the battery lead is $(q/\phi)\mathbf{v} \cdot \mathbf{E}$.

2. A *p–n* junction diode is formed from two high conductivity regions joined externally by a wire making ohmic contacts to the two regions. An electron traverses the diode from the n to the p region; show that current flows in the external lead only as the electron crosses the depletion layer.

3. By considering a long solenoid in a linear medium show that the energy stored in the solenoid can be expressed as $\int_V \frac{1}{2}\mathbf{H} \cdot \mathbf{B}\, \mathrm{d}V$ where V is the volume of the solenoid.

4. A charged parallel-plate condenser, with its plates normal to the z axis, is placed between the poles of a permanent magnet producing a field $B_x = \mu_0 H_x$. Show that, although $\mathbf{N} = \mathbf{E} \wedge \mathbf{H}$ is non-zero, the surface integral of N_n over any closed surface is zero.

5. A coaxial cable parallel to the z axis carries a direct current I and there is a potential difference V between the conductors. Verify that the power flow IV is equal to the integral of $(\mathbf{E} \wedge \mathbf{H})_z$ over the cross-section of the cable.

Energy, Power and Stress

6. Use equation (10.27) to discuss the operation of a car dynamo. What happens if the output leads are shorted within the dynamo?

7. Calculate the reaction, due to radiation pressure, on the surface of a black body at 3000 K in surroundings at room temperature. (Stefan's constant is 6×10^{-8} watt degree^{-4} m^{-2}.)

8. A plane wave is normally incident from vacuum on the surface of a medium with $\varepsilon = \mu = 1$ and $\sigma \ll \omega\varepsilon_0$. Show, by considering the force exerted on the currents in the medium, that the pressure acting on the medium is $(1/c)(\mathbf{E} \wedge \mathbf{H})$, where $\mathbf{E}$ and $\mathbf{H}$ are the r.m.s. fields at the surface of the medium and have a sinusoidal variation with time.

9. A transformer consists of a ring of iron with a primary winding on one side of the ring and a secondary winding on the opposite side as shown in the figure. Discuss the energy transfer from the primary to the secondary in terms of Poynting's vector.

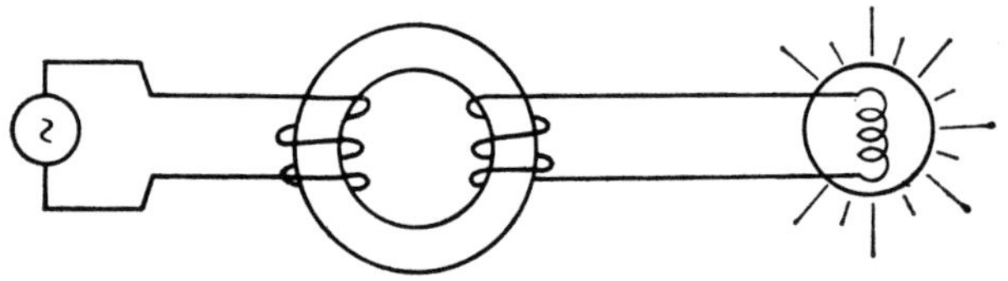

FIG. 10.2.

10. Two equal pendulums A and B carry equal charges $+q$. Initially pendulum A is stationary and B is swinging. Describe the transfer of energy from B to A in terms of Poynting's vector.

11. Verify equations (10.39) and (10.40).

12. If $\mathbf{E} = \sum_{n=0}^{\infty} \mathbf{E}_n \cos 2\pi n(t/T)$ and $\mathbf{B} = \sum_{n=0}^{\infty} \mathbf{B}_n \cos(2\pi n(t/T) + \phi_n)$ where $\mathbf{E}_n$, $\mathbf{B}_n$ and ϕ_n are constants show that the time average of $(\partial/\partial t)(\mathbf{E} \wedge \mathbf{B})$ is identically zero.

13. Use the electrostatic stress tensor to calculate the force acting between the plates of a parallel vacuum spaced capacitor of area A and plate separation d, charged to a potential ϕ.

14. Obtain an expression for the pressure in an infinite homogeneous fluid medium, in which $\varepsilon - 1$ is proportional to density, when a metal sphere of radius a and charge q is immersed in the medium.

15. A vertical U tube containing a fluid, of dielectric constant ε and density τ, is open at both ends to the atmosphere. One end of the tube is in a region where the electric field is $\mathbf{E}$, the other in a field free region. Calculate the difference in height between the columns in the two limbs of the tube.

16. A fluid of dielectric constant ε, in which there is a uniform field $\mathbf{E}^*$ due to distant charges, contains a spherical bubble of radius a. At the centre of the bubble there is, instantaneously, a speck of matter carrying a charge q. Calculate the force acting on the speck and on the bubble.

17. An iron wire, of permeability μ, carries a current I at right-angles to a uniform applied magnetic field $\mathbf{B}^*$. Find the force per unit length acting on the wire. Discuss the nature of the stresses in the wire at an instant in time when the wire, which has initially been held in position, is released.

18. Derive equation (10.65) from the field equations.

19. A plane wave in vacuum has the two field components $E_x = E_0 \sin(\omega t - \beta z)$ and $B_y = B_0 \sin(\omega t - \beta t)$ for $ct > z$ and is zero for $ct < z$. Find the component T^{em}_{Zz} of the electromagnetic stress tensor on the plane $z = 0$ as a function of time.

20. Plane waves with **E** in the plane of incidence fall at an angle θ to the normal on a plane reflecting surface. Find expressions for the fields in front of the plane in terms of the amplitude of **B** in the incident wave. Calculate the mean energy density on a plane distant l from the reflecting surface and the average value of the normal component of the stress tensor on this plane.

21. Repeat problem 20 with the roles of **E** and **B** interchanged.

CHAPTER 11

Plane Electromagnetic Waves

Introduction

The next three chapters discuss the constitutive relations which express, in compact form, the influence of the microscopic structure of a medium on its macroscopic electromagnetic behaviour. Since many of the most interesting manifestations of this behaviour are associated with waves this chapter provides a background for their discussion. An unrelieved diet of plane-wave theory is liable to convey a false impression of the nature of electromagnetic waves and, in the first two problems at the end of the chapter, the reader's attention is drawn to the existence of other types of wave. Additional reading will be found in the books by Ramo *et al.* (1965) and by Jackson (1962).

This is not a general discussion, even of plane waves, and we assume from the outset that all the quantities in the field equations vary with time as exp $(j\omega t)$. Thus the equations are

$$\nabla \cdot \mathbf{B} = 0, \quad (11.1a) \qquad \nabla \wedge \mathbf{E} = j\omega\mathbf{B}, \quad (11.1b)$$

$$\nabla \cdot \mathbf{D} = \rho, \quad (11.1c) \qquad \nabla \wedge \mathbf{H} = \mathbf{J}+j\omega\mathbf{D}. \quad (11.1d)$$

We note immediately that equation (11.1b) makes equation (11.1a) redundant. If E_t is continuous across a surface between two media then B_n is automatically continuous and need not be considered separately. Furthermore, if either H_t is continuous across a surface and **J** is finite or there is a surface current related to the discontinuity in H_t, the associated charge density, obtained from $j\omega\rho = -\nabla \cdot \mathbf{J}$, is sufficient to satisfy the boundary condition for D_n. Thus for harmonic fields the two boundary conditions for the tangential components of **E** and **H** are, by themselves, enough. There is, therefore, a considerable advantage in wave problems in working with equations expressed in terms of **E** and **H** as the basic vectors.

Plane Waves in Vacuum

In vacuum the equations will be

$$\nabla \cdot \mathbf{H} = 0, \quad (11.2a) \qquad \nabla \cdot \mathbf{E} = 0, \quad (11.2b)$$

$$\nabla \wedge \mathbf{H} = \varepsilon_0\dot{\mathbf{E}} = j\omega\varepsilon_0\mathbf{E}, \quad (11.2c) \qquad \nabla \wedge \mathbf{E} = -\mu_0\dot{\mathbf{H}} = -j\omega\mu_0\mathbf{H}, \quad (11.2d)$$

and, if we assume that we seek only plane-wave solutions propagating along the z axis, we can set $\partial/\partial x = \partial/\partial y = 0$ wherever they occur. The z components of the curl equations give $E_z = H_z = 0$ and so the waves are transverse. The remaining components give two uncoupled sets of equations,

$$\frac{\partial H_y}{\partial z} = -\varepsilon_0 \frac{\partial E_x}{\partial t}, \qquad (11.3a)$$

$$\frac{\partial E_x}{\partial z} = -\mu_0 \frac{\partial H_y}{\partial t}, \tag{11.3b}$$

and

$$\frac{\partial H_x}{\partial z} = \varepsilon_0 \frac{\partial E_y}{\partial t}, \tag{11.3c}$$

$$\frac{\partial E_y}{\partial z} = \mu_0 \frac{\partial H_x}{\partial t}, \tag{11.3d}$$

which correspond to two independent linear polarizations. If we let

$$\eta = \left(\frac{\mu_0}{\varepsilon_0}\right)^{\frac{1}{2}} \tag{11.4}$$

and define

$$X_+ = E_x + \eta H_y,$$

$$X_- = E_x - \eta H_y,$$

then equations (11.3a) and (11.3b), on addition, give

$$\frac{\partial X_+}{\partial z} = -(\mu_0 \varepsilon_0)^{\frac{1}{2}} \frac{\partial X_+}{\partial t},$$

and, on subtraction,

$$\frac{\partial X_-}{\partial z} = +(\mu_0 \varepsilon_0)^{\frac{1}{2}} \frac{\partial X_-}{\partial t}.$$

These are the equations for waves propagating with a velocity $c = (\mu_0 \varepsilon_0)^{-\frac{1}{2}}$ in the $+z$ and $-z$ directions respectively. There are similar equations for the orthogonally polarized waves described by (11.3c) and (11.3d). In a pure forward wave $E_x = +(\mu_0/\varepsilon_0)^{\frac{1}{2}} H_y$ and the z component of Poynting's vector is positive as we should expect. The separation into orthogonally, linearly polarized waves does not depend on the assumption of harmonic time dependence but, if the waves are harmonic, we can also decompose the basic plane waves into left and right circularly polarized waves. Thus, with $j = \sqrt{-1}$, the two waves described by

$$R_+ = (E_x + \eta H_y) + j(E_y - \eta H_x)$$

and

$$L_+ = (E_x + \eta H_y) - j(E_y - \eta H_x)$$

satisfy the wave equations

$$\frac{\partial R_+}{\partial z} = -j\omega(\mu_0 \varepsilon_0)^{\frac{1}{2}} R_+$$

and

$$\frac{\partial L_+}{\partial z} = -j\omega(\mu_0 \varepsilon_0)^{\frac{1}{2}} L_+$$

and represent waves propagating in the positive direction. Thus if $E_x = \eta H_y$, $E_y = -\eta H_x$ and $E_y = -jE_x$ we see that L_+ vanishes and only the wave R_+ exists. At $z = 0$ if $E_x = \cos \omega t$ then $E_y = \sin \omega t$ and the electric vector rotates clockwise about the positive z axis. We may also note that it rotates anticlockwise about the z axis as we proceed along this axis at a fixed time. At a fixed instant in time the electric vector lies on a left-handed corkscrew.

Plane Electromagnetic Waves

A wave propagating along the positive z axis varies with t and z as $\exp j(\omega t - kz)$ where

$$k = \omega(\mu_0\varepsilon_0)^{\frac{1}{2}}, \tag{11.5}$$

and a wave propagating in an arbitrary direction parallel to a vector $\mathbf{k}$ can be described in terms of $\exp j(\omega t - \mathbf{k} \cdot \mathbf{r})$ where

$$k^2 = k_x^2 + k_y^2 + k_z^2 = \omega^2\mu_0\varepsilon_0. \tag{11.6}$$

Thus a typical plane wave can be written as

$$\mathbf{E}(\mathbf{r}, t) = \mathscr{R}[\mathbf{E} \exp j(\omega t - \mathbf{k} \cdot \mathbf{r})],$$
$$\mathbf{H}(\mathbf{r}, t) = \mathscr{R}[\mathbf{H} \exp j(\omega t - \mathbf{k} \cdot \mathbf{r})], \tag{11.7}$$

and the field equations yield the relations

$$\mathbf{k} \cdot \mathbf{E} = 0, \qquad \mathbf{k} \cdot \mathbf{H} = 0,$$

$$\mathbf{k} \wedge \mathbf{E} = \omega\mu_0\mathbf{H}, \tag{11.8}$$

and

$$\mathbf{k} \wedge \mathbf{H} = -\omega\varepsilon_0\mathbf{E}.$$

The time average of the Poynting vector is

$$\overline{\mathbf{N}} = \overline{\mathbf{E}(\mathbf{r}, t) \wedge \mathbf{H}(\mathbf{r}, t)} = \tfrac{1}{2}\mathscr{R}(\mathbf{E}^* \wedge \mathbf{H}) \tag{11.9}$$

and this gives

$$\overline{\mathbf{N}} = \frac{1}{2\omega\mu_0} E^* \wedge (\mathbf{k} \wedge \mathbf{E}) = \frac{\mathbf{E} \cdot \mathbf{E}^*}{2\omega\mu_0}\mathbf{k} \tag{11.10a}$$

or

$$\overline{\mathbf{N}} = \frac{\mathbf{H} \cdot \mathbf{H}^*}{2\omega\varepsilon_0}\mathbf{k}, \tag{11.10b}$$

or

$$\overline{\mathbf{N}} = \tfrac{1}{2}\frac{EH}{k}\mathbf{k}, \tag{11.10c}$$

where E, H, and k are the magnitudes of $\mathbf{E}$, $\mathbf{H}$, and $\mathbf{k}$. Note that these results apply only to a single Fourier component of the field.

Waves in Homogeneous Media

In a homogeneous insulating medium characterized by isotropic constitutive relations $\mathbf{D} = \varepsilon\varepsilon_0\mathbf{E}$ and $\mathbf{B} = \mu\mu_0\mathbf{H}$ we can move ε and μ back and forth past differential operators, i.e., replace $\nabla \cdot (\varepsilon\varepsilon_0\mathbf{E}) = 0$ by $\varepsilon\varepsilon_0\nabla \cdot \mathbf{E} = 0$, and clearly the only change in the wave propagation is the replacement of μ_0 by $\mu\mu_0$ and ε_0 by $\varepsilon\varepsilon_0$. Thus if the waves propagate as $\exp j(\omega t - \boldsymbol{\beta} \cdot \mathbf{r})$ we have

$$\beta^2 = \omega^2\mu\mu_0\varepsilon\varepsilon_0 = \mu\varepsilon k^2, \tag{11.11}$$

and the wave velocity is reduced from $c = (\mu_0\varepsilon_0)^{-\frac{1}{2}}$ to

$$c' = \frac{c}{(\mu\varepsilon)^{\frac{1}{2}}}. \tag{11.12}$$

Thus $(\mu\varepsilon)^{\frac{1}{2}}$ is the refractive index. In optics we always take $\mu = 1$, for reasons that will be discussed in later sections, and so the refractive index in optics is

$$n = \varepsilon^{\frac{1}{2}}. \tag{11.13}$$

The ratio between E_x and H_y in a forward wave, i.e. the wave impedance of the medium, is

$$Z = \left(\frac{\mu}{\varepsilon}\right)^{\frac{1}{2}}\left(\frac{\mu_0}{\varepsilon_0}\right)^{\frac{1}{2}} = \left(\frac{\mu}{\varepsilon}\right)^{\frac{1}{2}}\eta. \tag{11.14}$$

At optical frequencies

$$Z = \eta/n. \tag{11.15}$$

The situation is, however, quite different in either an anisotropic insulator or a conducting medium. We shall consider anisotropic dielectrics in more detail in the next chapter and so, here, we deal only with a medium in which $\mu = \varepsilon = 1$, but there is a finite electrical conductivity σ.

The field equations are now

$$\nabla \cdot \mathbf{E} = 0, \quad (11.16\text{a}) \qquad \nabla \cdot \mathbf{H} = 0, \quad (11.16\text{b})$$

$$\nabla \wedge \mathbf{E} = -j\mu_0\mathbf{H}, \quad (11.16\text{c}) \qquad \nabla \wedge \mathbf{H} = (j\omega\varepsilon_0 + \sigma)\mathbf{E} \quad (11.16\text{d})$$

and if the wave propagates as $\exp j(\omega t - \boldsymbol{\beta} \cdot \mathbf{r})$ we have

$$\beta^2 = \omega^2\mu_0\varepsilon_0 - j\omega\mu_0\sigma. \tag{11.17}$$

Thus β is in general complex and the wave is attenuated. The surfaces of constant phase and constant amplitude need no longer coincide, for if

$$\boldsymbol{\beta} = \boldsymbol{\gamma} - j\boldsymbol{\alpha} \tag{11.18}$$

the surfaces of constant phase satisfy $\boldsymbol{\gamma} \cdot \mathbf{r} =$ constant, and those of constant amplitude satisfy $\boldsymbol{\alpha} \cdot \mathbf{r} =$ constant. These surfaces need not be parallel. Furthermore, the equation $\nabla \cdot \mathbf{E} = 0$ now only implies that $\boldsymbol{\gamma} \cdot \mathbf{E} = j\boldsymbol{\alpha} \cdot \mathbf{E}$ and there is no reason why we should also have $\boldsymbol{\gamma} \cdot \mathbf{E} = 0$ or $\boldsymbol{\alpha} \cdot \mathbf{E} = 0$. We can therefore have waves with a component of $\mathbf{E}$ normal to the planes of constant phase. Thus propagation in a conducting medium (or indeed any dissipative medium) is altogether more complicated than in a non-dissipative insulator. For an excellent and very full discussion the reader is referred to Stratton (1941).

Although the general wave type is no longer simple we can, of course, also have simple plane waves. Thus, for example, the wave with only the two field components

$$E_x(z, t) = \mathscr{R}\{E_x \exp j(\omega t - \beta z)\},$$

$$H_y(z, t) = \mathscr{R}\{H_y \exp j(\omega t - \beta z)\},$$

is still a solution if β satisfies (11.17) and

$$\frac{E_x}{H_y} = \left\{\frac{\mu_0}{[\varepsilon_0 + (\sigma/j\omega)]}\right\}^{\frac{1}{2}}. \tag{11.19}$$

In metals the conduction current density $\sigma\mathbf{E}$ far exceeds the displacement current density $\omega\varepsilon_0\mathbf{E}$ at all frequencies below the ultra-violet, and so

$$\beta^2 \approx -j\omega\mu_0\sigma$$

and

$$\beta \approx \left(\frac{\omega\mu_0\sigma}{z}\right)^{\frac{1}{2}}(1-j).$$

The waves are attenuated with a characteristic length

$$\delta = \left(\frac{2}{\omega\mu_0\sigma}\right)^{\frac{1}{2}}, \tag{11.20}$$

known as the skin depth. The wave impedance is approximately

$$Z = \left(\frac{\mu_0}{\varepsilon_0}\right)^{\frac{1}{2}}\left(\frac{j\omega\varepsilon_0}{\sigma}\right)^{\frac{1}{2}} \ll \left(\frac{\mu_0}{\varepsilon_0}\right)^{\frac{1}{2}}. \tag{11.21}$$

In complex notation dielectric loss can be described by a complex dielectric constant

$$\varepsilon = \varepsilon' - j\varepsilon'' \tag{11.22}$$

and the phase angle δ between **D** and **E** is

$$\delta = \tan^{-1}\frac{\varepsilon''}{\varepsilon'}. \tag{11.23}$$

The propagation constant is given by

$$\beta^2 = \omega^2\mu_0\varepsilon_0(\varepsilon' - j\varepsilon'') \tag{11.24}$$

and, since usually $\tan\delta = \varepsilon''/\varepsilon'$ is small, we have

$$\beta \approx \omega(\mu_0\varepsilon'\varepsilon_0)^{\frac{1}{2}}(1-\tfrac{1}{2}j\tan\delta). \tag{11.25}$$

The intensity of a wave decreases with distance as

$$\exp\{-2\omega(\mu_0\varepsilon'\varepsilon_0)^{\frac{1}{2}}z(\tfrac{1}{2}\tan\delta)\} = \exp\{-\omega(\mu_0\varepsilon'\varepsilon_0)^{\frac{1}{2}}z\tan\delta\}$$

and so is reduced by a factor $\exp(-2\pi\tan\delta)$ in a distance equal to one wavelength in the medium.

The phase velocity of a wave is $v_\Phi = \omega/\beta$ and, if the wave is unattenuated, the group velocity is $v_g = \mathrm{d}\omega/\mathrm{d}\beta$. For plane waves in vacuum $v_g = v_\Phi = c$ but in a material medium where ε or μ may depend on ω the group and phase velocities need not be equal. In dissipative media the group velocity loses its significance and its indiscriminate use can lead to fallacious results. In most cases energy is convected with the group velocity in the sense that if u is the energy density the energy flux is uv_g but the reader should realize that this is a special property of some wave types, e.g. electromagnetic waves, and not a general result. Counter examples in other branches of physics are not hard to find.

Reflection and Refraction

We have seen that in a homogeneous medium the propagation constant β and the wave impedance Z which relates **E** to **H** can be expressed in terms of the frequency and the electromagnetic constants of the medium. At a surface between two different media both the tangential components E_t and H_t must be continuous. If **n** is the unit normal to the surface and β_1 and β_2 the two propagation constants this requires

$$\mathbf{n} \wedge (\mathbf{n} \wedge \boldsymbol{\beta}_1) = \mathbf{n} \wedge (\mathbf{n} \wedge \boldsymbol{\beta}_2). \tag{11.26}$$

Applied to the interface between two dielectrics this leads to Snell's law and the equality of the angles of incidence and reflection. In the incident wave we have $E_i = Z_1 H_i$ and in the reflected wave $E_r = -Z_1 H_r$. In the refracted or transmitted wave $E_\tau = Z_2 H_\tau$. If **E** is in the plane of incidence, θ the angle of incidence and ρ the angle of refraction we have the two relations

$$(E_i + E_r)\cos\theta = E_\tau \cos\rho$$

$$H_i + H_r = H_\tau.$$

We see that the reflection coefficient is

$$E_r/E_i = (Z_2 \cos\rho - Z_1 \cos\theta)/(Z_2 \cos\rho + Z_1 \cos\theta).$$

Since also Snell's law gives

$$n_2 \sin\theta = n_1 \sin\rho$$

or

$$Z_1 \sin\theta = Z_2 \sin\rho$$

we can express the reflection coefficient in terms of Z_1/Z_2 and θ alone. We leave it to the reader to show that, for this polarization of the waves there is no reflection at Brewster's angle given by $\theta = \tan^{-1} Z_1/Z_2 = \tan^{-1} n_2/n_1$. For our purposes it is more important to note that $\boldsymbol{\beta}$ and Z together are enough to determine the conditions at an interface between two media. For a fuller discussion of reflection problems the reader should consult the books by Jackson (1962), Stratton (1941), or Ramo *et al.* (1965).

Problems

1. Give examples of waves propagating in vacuum near a material structure with each of the following properties:

(a) The wave is transverse but not plane.

(b) The wave is neither transverse nor plane but the phase velocity is c.

(c) The phase velocity is greater than c.

(d) The phase velocity is less than c.

2. If **g** is a constant vector verify that, for large values of **r**, $\mathbf{H} = \mathbf{r} \wedge \mathbf{g}/r^2 \exp j\omega(t - r/c)$ satisfies the field equations in vacuum. What is the corresponding electric field?

3. In the plane $z = 0$ the electric field has a single component $E_x = E_0 \exp\{-(x^2/2\delta^2)\}$ $\cos\omega t$. Describe the approximate form of the field in vacuum for $z > 0$.

4. Unpolarized light of intensity I falls at Brewster's angle to the normal on a plane glass surface of refractive index n. What is the intensity of the reflected light?

5. In a conducting medium the propagation constant $\boldsymbol{\beta}$ has components $(\gamma_x, 0, \gamma_z - j\alpha_z)$. Show that there is a wave with only the components E_x, E_z and H_y and discuss the relation between the direction of Poynting's vector and the normal to the surfaces of constant phase.

6. The best optical glass is reputed to have an attenuation of 20 db per kilometre at 5×10^{14} *Hz*. If the refractive index is 1·5 what is $\tan\delta$?

7. A resonant circuit has a quality factor $Q = 100$. What is the change in Q if it is immersed in a medium for which dielectric loss is negligible but $\mu = 1 - j \times 10^{-7}$?

CHAPTER 12

Macroscopic Aspects of the Constitutive Relations

Introduction

The constitutive relations amongst the vectors **E**, **D**, **B**, and **H** express in a convenient form the effects of the microscopic atomic structure of a medium on its macroscopic electromagnetic properties. We shall consider the connection between the constitutive relations and structure in the next two chapters; our present topic is a discussion of the relations themselves and the phenomena with which particular types of relation are associated.

The simplest constitutive relations are

$$\mathbf{D} = \varepsilon\varepsilon_0\mathbf{E} = (1+\chi_e)\varepsilon_0\mathbf{E}, \tag{12.1a}$$

$$\mathbf{B} = \mu\mu_0\mathbf{H} = (1+\chi_m)\mu_0\mathbf{H}, \tag{12.1b}$$

$$\mathbf{P} = \chi_e\varepsilon_0\mathbf{E}, \tag{12.1c}$$

$$\mathbf{M} = \chi_m\mathbf{H}, \tag{12.1d}$$

where ε, μ, χ_m and χ_e are dimensionless constants. These relations are single valued, frequency independent, non-dissipative, isotropic, inactive, and purely electric or magnetic. In general the constitutive relations need have none of these properties and it is indeed rare for all these properties to occur in any real medium. In addition the constitutive relations are often dependent on pressure, strain, or temperature.

In ferroelectric and ferromagnetic media we usually observe hysteresis and the relations are not single valued. Almost all media display some degree of dispersion or frequency dependence and smaller or greater degrees of dissipation. Isotropic dielectric media are relatively common but many magnetic media are highly anisotropic. Dielectrics rarely display pronounced anisotropy but even a small anisotropy is readily observable as optical birefringence. An active medium may be interpreted in two quite unconnected ways as either a medium which generates energy in response to a field or more usually as a medium which rotates the plane of polarized light. Finally, in the Faraday effect we have a typical example of a mixed constitutive relation in which **P** depends on both **E** and **B**.

Media in which strain results in a dielectric polarization are known as piezo-electric. The effect, though complex, is of some technical and experimental importance. Most constitutive relations depend in some degree on temperature but in addition, in pyro-electric crystals a change in temperature is by itself sufficient to generate a dielectric polarization. We propose to discuss most but not all of these effects although in some cases we shall consider only the electric and in others only the magnetic phenomena. We shall also pay only brief attention to the complex phenomena of magnetic resonance which, from a macroscopic point of view, are, nevertheless, properly regarded as constitutive relations.

There is a profound difference between magnetism and electricity. All media display electronic polarizability and, in condensed media, this leads to values of χ_e exceeding unity at all frequencies below the ultra-violet. The corresponding magnetic effect diamagnetism is almost always of negligible magnitude. Dielectric polarization due to ionic motion in solids leads to values of χ_e which are usually greater than unity and may sometimes be very large. There is no corresponding magnetic phenomenon. On the other hand, dielectric effects due to the re-orientation of molecules with permanent dipole moments are confined to liquids and gases. Furthermore, single atoms cannot display a permanent electric dipole moment. The corresponding magnetic effect, paramagnetism, though small, is of frequent occurrence and has interesting consequences. In a few media cooperative interactions lead to ferromagnetism and values of χ_m as great as 10^4. Ferroelectricity is an altogether more complex phenomenon largely because the interactions which lead to ferroelectricity are comparable in strength to those which determine the structure of the medium.

There are other major differences between electricity and magnetism which arise from the different nature of the electric and magnetic vectors. The electric vectors are polar vectors but the magnetic vectors are axial vectors, intimately associated with rotation and angular momentum. This has consequence for both the dynamical behaviour and symmetry of the two types of relation. In practice, however, the two principal differences are that magnetism is a static or low frequency phenomenon whereas dielectric effects persist at optical frequencies and that, because there are free charges but no free magnetic poles, observable permanent dielectric polarization is uncommon while permanent magnetization is of everyday occurrence.

Causality

Whatever the microscopic origin of the constitutive relations they must be consistent with the laws of thermodynamics and with our notions of causality. Causality implies that the polarization of a medium at a time t can depend only on the fields at earlier times. Thus if we write

$$P_i(t) = \varepsilon_0 \int_{-\infty}^{\infty} f_{ij}(E_j(t_1), t, t_1)\, dt_1,$$

the function f_{ij} must vanish for $t_1 > t$. If we make the entirely reasonable assumption that the constitutive relation is the same on Wednesday as on Tuesday, i.e., that the origin of time is immaterial, the function f must have the form $f\,(E_j(t_1), t-t_1)$ and depend only on the difference between t and t_1. In the special case when the relation is linear and isotropic

$$P = \varepsilon_0 \int_{-\infty}^{\infty} f(\tau)E(t-\tau)\, d\tau, \tag{12.2}$$

and

$$f(\tau) = 0, \qquad \tau < 0. \tag{12.3}$$

If now we express $P(t)$ and $E(t)$ as Fourier integrals so that

$$P(t) = \int_{-\infty}^{\infty} P(\omega) \exp(j\omega t)\, d\omega, \qquad E(t) = \int_{-\infty}^{\infty} E(\omega) \exp(j\omega t)\, d\omega, \tag{12.4}$$

we must have, since $P(t)$ and $E(t)$ are real variables,

$$P(-\omega) = P^*(\omega), \qquad E(-\omega) = E^*(\omega). \tag{12.5}$$

If then we express the relation between $P(\omega)$ and $E(\omega)$ using a frequency-dependent complex susceptibility as

$$P(\omega) = \varepsilon_0\chi(\omega)E(\omega) \tag{12.6}$$

we have

$$\chi(\omega) = \chi^*(-\omega) = \int_{-\infty}^{\infty} f(\tau)\exp(-j\omega\tau)\,\mathrm{d}\tau. \tag{12.7}$$

For proof of this and the following results the reader should consult either Landau and Lifshitz (1960) or Faulkner (1969).

If we let

$$\chi(\omega) = \chi'(\omega) - j\chi''(\omega) \tag{12.8}$$

we see from (12.7) that χ' is an even and χ'' an odd function of ω, but in addition, because $f(\tau) = 0$ for $\tau < 0$ there are further restrictions, known as the Kramers–Kronig relations,

$$\chi''(\omega) = -\frac{1}{\pi}\int_{-\infty}^{\infty}\frac{\chi'(\omega_1)\,\mathrm{d}\omega_1}{\omega_1-\omega} \tag{12.9a}$$

and

$$\chi'(\omega) = \frac{1}{\pi}\int_{-\infty}^{\infty}\frac{\chi''(\omega_1)\,\mathrm{d}\omega_1}{\omega_1-\omega}, \tag{12.9b}$$

where the principal value integral is defined by

$$\int_{-\infty}^{\infty}\ldots\,\mathrm{d}\omega_1 = \lim_{\varepsilon\to 0}\left\{\int_{-\infty}^{\omega-\varepsilon}\ldots\,\mathrm{d}\omega_1 + \int_{\omega+\varepsilon}^{\infty}\ldots\,\mathrm{d}\omega_1\right\}. \tag{12.10}$$

These relations are a direct result of causality and have important consequences.

If χ is real at all frequencies so that $\chi'' \equiv 0$ we see that, from (12.9b), χ' is also zero; equally if χ' is independent of ω we find from (12.9a) that $\chi'' = 0$. Thus, all real media must have a frequency-dependent susceptibility which, in at least some frequency range, is complex. The average rate at which energy is dissipated in unit volume is the average of $E\dot{D}$ which is

$$W = \tfrac{1}{2}\mathscr{R}E^*j\omega D = \tfrac{1}{2}\omega\varepsilon_0\chi''EE^*, \tag{12.11}$$

thus all real media are dissipative at some frequency. These remarks apply with equal force to magnetic media.

In an anisotropic, but still linear, medium

$$D_i(\omega) = \varepsilon_0\varepsilon_{ij}(\omega)E_j(\omega) \tag{12.12}$$

or

$$P_i(\omega) = \varepsilon_0\chi_{ij}(\omega)E_j(\omega), \tag{12.13}$$

and the average dissipation is

$$W = \tfrac{1}{2}\mathscr{R}j\omega E_i^* D_i,$$

which gives

$$W = \tfrac{1}{2}\mathscr{R}j\omega\varepsilon_0 E_i^*\varepsilon_{ij}E_j. \tag{12.14}$$

When the implied sum over i and j has been performed we see that this will be zero only if

$$\varepsilon_{ij}^* = \varepsilon_{ji}, \tag{12.15}$$

and so this relation characterizes a lossless medium. It cannot, because of the Kramers–Kronig relations, hold at all frequencies but there are usually extended ranges of frequency in which it is almost exact.

Optical Activity

We have already seen, in the last chapter, that in isotropic media a complex value of ε leads to attenuation of waves, but in anisotropic media this is not necessarily so if (12.15) is obeyed. Consider a medium in which ε_{ij} has the form

$$\varepsilon_{ij} = \begin{pmatrix} 1 & j\alpha & 0 \\ -j\alpha & 1 & 0 \\ 0 & 0 & 1 \end{pmatrix},$$

and a field with components

$$E_x = \mathscr{R}(E_0 \exp j\omega t) = E_0 \cos \omega t$$

$$E_y = \mathscr{R}(jE_0 \exp j\omega t) = -E_0 \sin \omega t$$

is applied. This is a field rotating counter-clockwise about the z axis and the resulting displacements are easily seen to be

$$D_x = \varepsilon_0(1-\alpha)E_0 \cos \omega t = \varepsilon_0(1-\alpha)E_x, \qquad D_y = \varepsilon_0(1-\alpha)E_y$$

whereas for a clockwise rotating field we obtain $D_x = \varepsilon_0(1+\alpha)E_x$, $D_y = \varepsilon_0(1+\alpha)E_y$. The medium is therefore characterized by a definite sense of rotation. If the medium as a whole is stationary this is only possible if its orientation can be specified by a vector with the transformation properties of angular momentum or magnetic moment. Thus this constitutive relation is only possible for a magnetized medium. It in fact describes the Faraday effect in which α is proportional to both ω and an applied field B_z. A completely magnetic equivalent to this relation occurs in discussing magnetic resonance phenomena. For a full discussion see the early chapters of *Principles of Nuclear Magnetism* by Abragam (1961).

If we are dealing with waves it is possible for the parameter to depend on the wave vector $\boldsymbol{\beta}$. Thus, if the waves propagate as $\exp j(\omega t - \beta z)$ we might have a term in the constitutive relation such as

$$P_x = -\varepsilon_0 \gamma \frac{\partial E_y}{\partial z}$$

giving $\alpha = \beta\gamma$. In this case the fields rotating clockwise about the direction of propagation of the waves see a dielectric constant $(1+\alpha)$, and those rotating counter-clockwise see $(1-\alpha)$. They propagate with different velocities and this describes an optically active medium. We note that in the Faraday effect the rotation is defined with respect to an applied field **B**, but in optical activity it is with respect to the direction of propagation. A wave traversing an optically active medium once in each direction finally emerges un-rotated. If the rotation is due to the Faraday effect the double-path results in a double rotation.

For a medium to display optical activity it must possess a screw-sense or hand. This is a property which cannot easily be discussed in terms of point symmetry groups and is not connected with crystalline symmetry in the usual sense. It will occur whenever the molecular constituents define either three polar vectors **A**, **B**, **C** for which $\mathbf{A} \cdot (\mathbf{B} \wedge \mathbf{C})$ is non-zero and has a definite sign, or one polar and one axial vector **F** and **G** such that **F . G** is non-zero with a definite sign. These vectors may arise from the arrangement of the units in a crystal lattice as in $NaClO_3$ or from the molecular units themselves as in *l*-tartaric acid. In the latter case liquids and solutions may exhibit optical activity. For a very complete discussion see Agranovich and Ginzburg (1966); briefer treatments can be found in Nye (1957) or Born and Wolf (1959).

Birefringence

If the medium is effectively lossless and ε_{ij} or χ_{ij} are real, they must be symmetric tensors. It is then always possible to find a principal axis system in which they are diagonal, i.e.,

$$\varepsilon_{ij} \rightarrow \varepsilon_i \delta_{ij} = (1+\chi_i)\delta_{ij}. \tag{12.16}$$

The number of independent diagonal elements also depends on the point-group symmetry of the medium. In liquids and cubic crystals $\varepsilon_1 = \varepsilon_2 = \varepsilon_3$. In crystals with a three-fold or higher axis of symmetry, which we take as the 3 axis, $\varepsilon_1 = \varepsilon_2$ but ε_3 may be different. In optics such a medium is described as optically uni-axial. Crystals of lower symmetry with three different diagonal values are known as bi-axial.

The point-group symmetry of a crystal determines whether anisotropy may occur but does not guarantee that it will occur. Any arrangement, however anisotropic, of isotropic atoms, ions, or molecules will not lead to anisotropy unless the arrangement itself destroys the isotropy of the individual units. Since, on the whole, most atoms, ions, and molecular units have a high degree of symmetry, even in solids, large anisotropies in the electronic, high frequency, polarizability are rare. On the other hand, the low frequency ionic polarizability, which is more closely related to the atomic arrangement in a crystal, can be much larger. For example in KH_2PO_4, the optical (electronic) anisotropy is roughly 1·1/1 while the low frequency (ionic) anisotropy is 2/1.

In magnetic media the individual moments of the paramagnetic ions or atoms are often strongly influenced by the indirect effects of the inter-atomic fields and this can lead to very marked anisotropy. Although in paramagnetism the overall magnitude of χ_m is usually small, anisotropies of over $10^3/1$ in χ_m are not unusual. The effects are of most interest in low temperature physics and can be used, for example, to construct a spin refrigerator.

Wave propagation in anisotropic media is only simple if the electric field coincides in direction with a principal axis of the susceptibility or dielectric constant tensor. To indicate the complexity of the general problem we consider some of its main features. If we neglect magnetization, the field equations for a single frequency component are

$$\nabla . \mathbf{H} = \nabla . \mathbf{D} = 0, \qquad \nabla \wedge \mathbf{H} = j\omega\mathbf{D}, \qquad \nabla \wedge \mathbf{E} = -j\omega\mu_0\mathbf{H}. \tag{12.17}$$

It follows that **H** and **D** are perpendicular to the direction of propagation and each other, but that **E**, though perpendicular to **H**, is not necessarily perpendicular to the direction of propagation. Thus Poynting's vector **E** ∧ **H** need not coincide with the normals to the wave fronts. If we take the direction of propagation as the z axis, which is not necessarily a principal axis, so that the fields vary as $\exp j(\omega t-\beta z)$ we obtain the relations

$$\nabla . \mathbf{D} = 0 \rightarrow \varepsilon_{zx}E_x+\varepsilon_{zy}E_y+\varepsilon_{zz}E_z = 0, \tag{12.18a}$$

$$\nabla \wedge \mathbf{E} = -j\omega\mu_0\mathbf{H} \rightarrow \omega\mu_0 H_x = -\beta E_y, \qquad \omega\mu_0 H_y = \beta E_x, \qquad H_z = 0, \tag{12.18b}$$

$$\nabla \wedge \mathbf{H} = j\omega\mathbf{D} \rightarrow \varepsilon_{xx}E_x+\varepsilon_{xy}E_y+\varepsilon_{xz}E_z = \frac{\beta}{\omega\varepsilon_0} H_y = \frac{\beta^2}{k^2} E_x, \tag{12.18c}$$

$$\varepsilon_{yx}E_x+\varepsilon_{yy}E_y+\varepsilon_{yz}E_z = \frac{\beta^2}{k^2} E_y, \tag{12.18d}$$

where $k^2 = \omega^2\mu_0\varepsilon_0$. These equations lead to a secular equation which is quadratic in β^2. There are, therefore, two forward and two reverse waves. A prescribed field at a boundary of the medium in general excites both types of forward wave.

If z happens to be a principal axis we have $\varepsilon_{xz} = \varepsilon_{zx} = \varepsilon_{yz} = \varepsilon_{zy} = 0$, and from (12.18a) $E_z = 0$. The waves are now transverse and $\mathbf{E} \wedge \mathbf{H}$ is in the direction of propagation. These waves are "ordinary waves". If z is not a principal axis we have a wave with $E_z \neq 0$ which is known as an "extraordinary" wave. In a uni-axial crystal with $\varepsilon_1 = \varepsilon_2$ any direction in the 1 2 plane is a principal axis and the situation is somewhat simpler.

Wave propagation in anisotropic media has a well-deserved reputation for difficulty and most texts on crystal optics devote considerable attention to developing graphical or geometric constructions to avoid solving the wave equations. It is, unfortunately, a topic of considerable technical importance.

Piezo-electricity

The constitutive relation of almost any medium is changed by strain but in some media strain, by itself, leads to a dielectric polarization and a uniform field leads to strain. Media with this property are piezo-electric. We shall discuss some aspects of piezo-electricity in a later chapter and here we consider only the polarization due to strain. The reader unfamiliar with elasticity may either omit this section or refer to Landau and Lifshitz (1970). The strain in a medium, which we denote by S_{jk}, is a second rank symmetric polar tensor, i.e., it has the transformation properties of an outer product of displacement vectors. If we neglect the distinction between covariant and contravariant tensors the relation between S_{jk} and the polarization must be of the form

$$P_i = d_{ijk}S_{jk}, \tag{12.19}$$

and involve a third rank tensor coefficient d_{ijk}. Under the inversion operation P_i changes sign but S_{jk} does not and so d_{ijk} must change sign. In a medium with a centre of symmetry d_{ijk} must be the same after inversion and so can only be zero. Thus, piezo-electricity only occurs in acentric (non-centrosymmetric) crystals or media whose preparation has produced an artificial acentricity (some ceramics). In crystallography the presence of piezo-electricity is used as a test for acentricity. The non-zero coefficients in the tensor d_{ijk} are determined by the point-group symmetry of the crystal. For example, in cubic crystals such as GaAs, of symmetry $\bar{4}$ 3m, the only non-zero coefficients have $i \neq j \neq k$ and are all equal to d_{123}. The application of an oscillatory field to a piezo-electric crystal induces an oscillatory strain which in turn leads to a polarization. If the frequency of the field coincides with a mechanical resonance at ω_0 the relation between P and E will show a resonance peak. Thus we find an effective susceptibility of the general form

$$\chi(\omega) = \frac{A}{j+Q(\omega_0-\omega)}$$

where ω_0/Q is the width of the resonance. There is an absorption peak at $\omega = \omega_0$. In crystals such as quartz with a very high Q, several resonances near ω_0 may be observed corresponding to complex electro-mechanical interactions. A typical 30 MHz quartz crystal for radio use will often show two strong, and two or more weak, resonances near 30 MHz. Apart from their use as resonant elements, piezo-electric media have a considerable technical utility as electro-mechanical transducers. We may also note that piezo-electric resonances are a potential source of spurious lines in magnetic resonance experiments. (See Cady (1962) for a general reference.)

Pyro-electricity

We shall be discussing the temperature dependence of the constitutive relations in several of the later chapters; there is, however, one aspect, pyro-electricity, which follows appropriately on piezo-electricity. In some crystals a change in the temperature produces a spontaneous polarization **P** along a definite axis of the crystal. The relation is of the form

$$\mathbf{P} = \pi T$$

and can obviously only occur in crystals whose symmetry allows us to associate a polar vector π with the structure. A typical example is ZnO of point-group symmetry 6 mm.

Non-linear Effects

All the effects that we have so far discussed are linear and we now turn to non-linear phenomena. Here another distinction between electricity and magnetism becomes apparent. With the exception of ferroelectric media most dielectrics are sensibly linear. A field strong enough to produce appreciably non-linear phenomena is also strong enough to disrupt the medium. Thus, in dealing with dielectrics, it is usually sufficient to expand the polarization **P** as a power series and discard all except the first few terms. By contrast in magnetism, even the rather modest fields available in the laboratory are, especially at low temperatures, or in resonance phenomena, sufficient to produce non-linear effects of appreciable magnitude. Furthermore, the only magnetic media with appreciable values of χ_m are either ferro-, anti-ferro- or ferrimagnetic and very definitely non-linear. We shall discuss these effects in connection with hysteresis.

The most interesting consequences of dielectric non-linearity are observed in optics. A typical power series expansion of the polarization is of the form

$$P_i = \varepsilon_0\{\chi_{ij}E_j + d_{ijk}E_jE_k + g_{ijkl}E_jE_kE_l + \text{etc.}\}. \tag{12.20}$$

The non-linear coefficients d and g will usually be both frequency dependent and complex, but we neglect this complication. The reader seeking further complexity will be rewarded by a study of the text by Bloembergen (1965).

The lowest order non-linearity is associated with the third rank tensor d_{ijk} and must therefore vanish in centro-symmetric media. Since in addition, when we can neglect dispersion, the polarization P_i is connected with an energy density U by

$$P_i = -\frac{\partial U}{\partial E_i}$$

this leads to further relations amongst the components of the tensor d_{ijk} such as $d_{ijk} = d_{ikj} = d_{kji}$, etc. These generally reduce the number of independent non-zero components of d_{ijk} quite considerably. (Kleinman, 1962.)

If the field $\mathbf{E}(t)$ contains components such as $\mathbf{E}(1)\cos\omega_1 t$ and $\mathbf{E}(2)\cos\omega_2 t$ there will be terms in $\mathbf{P}(t)$ at the frequencies $\omega_1 \pm \omega_2$, $2\omega_1$, $2\omega_2$ and d.c. These correspond to the phenomena of frequency mixing, second harmonic generation and rectification. When these effects occur at optical frequencies the experimental consequences can be quite striking. If on the other hand one of the frequencies, say ω_1, is an optical frequency and the other ω_2 is zero, there is an additional term in $\mathbf{P}(t)$ at ω_1 proportional to $\mathbf{E}(1)$ and $\mathbf{E}(2)$; thus, omitting tensor indices,

$$P(t) = \varepsilon_0\{\chi E_1 \cos\omega_1 t + 2dE_2E_1\cos\omega_1 t\}.$$

This is equivalent to a change of $2dE_2$ in χ and leads to a refractive index linearly dependent on the field E_2. This linear electro-optic, or Pockels, effect can be used to modulate or deflect beams of light.

If we have a single crystal in which d is non-zero an intense beam of light of frequency ω will generate the second harmonic at 2ω. If the fundamental wave propagates as $\exp j(\omega t - \beta z)$ the non-linear polarization propagates as $\exp 2j(\omega t - \beta z)$ and acts as a source term for fields at 2ω. These fields, however, propagate freely as $\exp j(2\omega t - \gamma z)$ where, due to dispersion, γ is not exactly 2β. If $\varepsilon(\omega)$ is the dielectric constant at ω and $\varepsilon(2\omega)$ at 2ω we have

$$(2\beta)^2 = (2\omega)^2\mu_0\varepsilon_0\varepsilon(\omega),$$

$$\gamma^2 = (2\omega)^2\mu_0\varepsilon_0\varepsilon(2\omega).$$

The field equations are

$$\nabla\cdot\mathbf{H} = 0, \qquad \nabla\wedge\mathbf{E} = -\mu_0\dot{\mathbf{H}}$$

$$\varepsilon(2\omega)\varepsilon_0\nabla\cdot\mathbf{E} = -\nabla\cdot\mathbf{P}\exp 2j(\omega t - \beta z),$$

$$\nabla\wedge\mathbf{H} = \varepsilon(2\omega)\varepsilon_0\dot{\mathbf{E}} + \frac{\partial}{\partial t}\mathbf{P}\exp 2j(\omega t - \beta z),$$

and yield the wave equation

$$\nabla^2\mathbf{E} - \mu_0\varepsilon_0\varepsilon(2\omega)\mathbf{E} = \left(\mu_0\frac{\partial^2}{\partial t^2} - \frac{1}{\varepsilon_0\varepsilon(2\omega)}\nabla\nabla\cdot\right)\mathbf{P}\exp 2j(\omega t - \beta z).$$

If we suppose that $\mathbf{P}$ has a single transverse component, say $P_x = P$, then, with $E_x = E$, we find that the solution, appropriate to the initial condition $E_x = 0$ at $z = 0$, is

$$E(t, z) = \mathscr{R}\left\{\frac{4\omega^2\mu_0 P}{4\beta^2 - \gamma^2}(\exp(-2j\beta z) - \exp(-j\gamma z))\exp(2j\omega t)\right\}.$$

The intensity of the wave is $S = \frac{1}{2}(\varepsilon\varepsilon_0/\mu_0)^{\frac{1}{2}}E_x^2$ and this yields

$$S = 2\left(\frac{\varepsilon\varepsilon_0}{\mu_0}\right)^{\frac{1}{2}}\left(\frac{P}{\varepsilon\varepsilon_0}\right)^2\frac{\beta^2}{\beta^2 - (\frac{1}{2}\gamma)^2}\sin^2\left(\beta - \frac{\gamma}{2}\right)z.$$

Thus, in general S oscillates periodically along the path of the wave. If, however, $\beta = \gamma/2$ we obtain

$$S = \tfrac{1}{2}\left(\frac{\varepsilon\varepsilon_0}{\mu_0}\right)^{\frac{1}{2}}\left(\frac{P}{\varepsilon\varepsilon_0}\right)^2\beta^2 z^2 \qquad (12.21)$$

and the intensity grows quadratically with distance, and can attain a considerable magnitude.

In some media it is possible to choose the directions of polarizations of the primary wave and the secondary wave so that birefringence exactly cancels the dispersion. This is known as phase-matching, and is essential if any considerable fraction of the fundamental power is to be converted to the harmonic. In the phase-matched case we have

$$\frac{S(2\omega)}{S(\omega)} = \left(\frac{E(\omega)d}{\varepsilon}\right)^2 \beta^2 z^2.$$

Typical values of d/ε are about 10^{-11} mV^{-1} and if, in a crystal 1 cm long at $\lambda = 6000$ Å, we expect to obtain significant power conversion, we require $E(\omega) \approx 10^5$ V m^{-1}. This corresponds to $S(\omega) \approx 25$ watts mm^{-2} which is quite feasible with a laser source.

In media with a centre of symmetry $d_{ijk} = 0$ and the first non-linear term is $\varepsilon_0 g_{ijkl} E_j E_k E_l$. This leads to a quadratic electro-optic effect. In some media this is referred to as the Kerr effect (see below). If $g_{iiii} > 0$ as is usually the case, a field $E \cos \omega t$ leads to a term $\frac{3}{4}\varepsilon_0 g E^2 \cdot E \cos \omega t$ which appears as an increase in the linear susceptibility, or refractive index, proportional to the intensity. It is responsible for self-focusing in intense beams of light.

In general we expect the terms in the expansion of P in powers of E to be in the ratio

$$g:d:\chi :: \frac{1}{E_a^2} : \frac{1}{E_a} : 1$$

where E_a is an atomic or molecular field of the order of 1 volt per Å or 10^{10} V m^{-1}. This expectation is more or less realized in practice. In some media, notably polar liquids such as nitro-benzene, g is somewhat larger than this estimate predicts. In these media the molecules are acentric and individually exhibit a linear electro-optic effect, but in the absence of a field are oriented at random. The field, therefore, not only produces the electro-optic effect but also has to align the molecules. We now expect

$$g : \chi :: \frac{1}{E_a}\frac{p}{kT} : 1$$

where p is the molecular dipole moment. Since p/kT is about 10^{-8} mV^{-1} this effect, the Kerr effect, is some hundred times larger than the simple quadratic effect.

Hysteresis

Finally, we consider media which exhibit hysteresis so that, even for static fields, the polarization **P** or magnetization **M** is not a single valued function of the field. In a medium displaying dispersion and loss the magnetization **M** is also not a single valued function of **H** and lags in phase behind **H**. If we plot **M** against **H** at a finite frequency we obtain the curve shown in Fig. 12.1. However, as $\omega \to 0$ the ellipse degenerates to a straight line. In a material displaying hysteresis as distinct from dispersion the M, H curve encloses a finite area even for static fields.

In general there is no unique hysteresis loop for a ferromagnetic material. Only if the applied field H takes the material well into positive and negative saturation will we be able to restrict our attention to a single curve such as that shown in Fig. 12.2.

In all other cases we shall have to consider subsidiary hysteresis loops lying within the main loop. This situation is familiar in technical magnetism and the reader will find a useful account given by Hoselitz (1952). We may at this point perhaps remark that Fig. 12.2 illustrates behaviour very far removed from the simple relation $M = \chi H$.

Ferroelectric or ferromagnetic media can be used as memory elements in computers. It

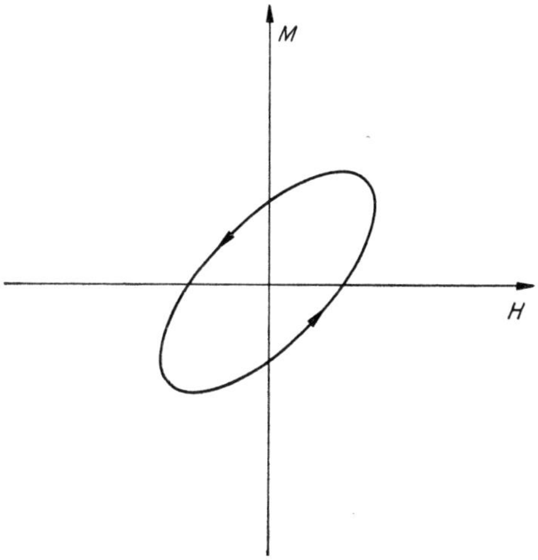

FIG. 12.1. *M*, *H* curve for a dispersive medium.

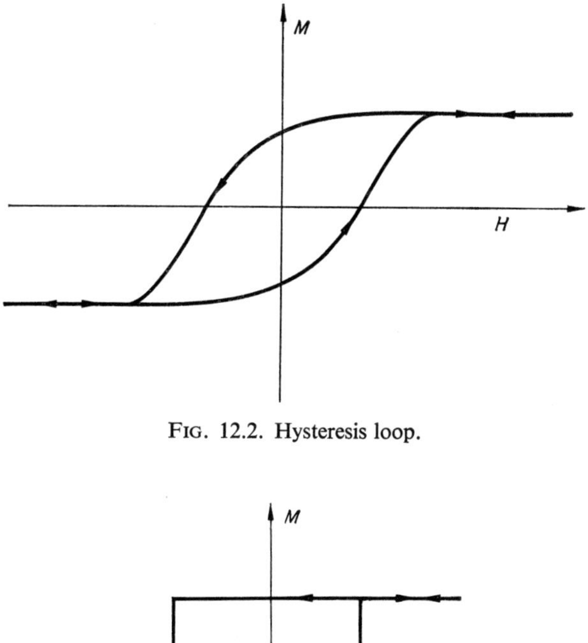

FIG. 12.2. Hysteresis loop.

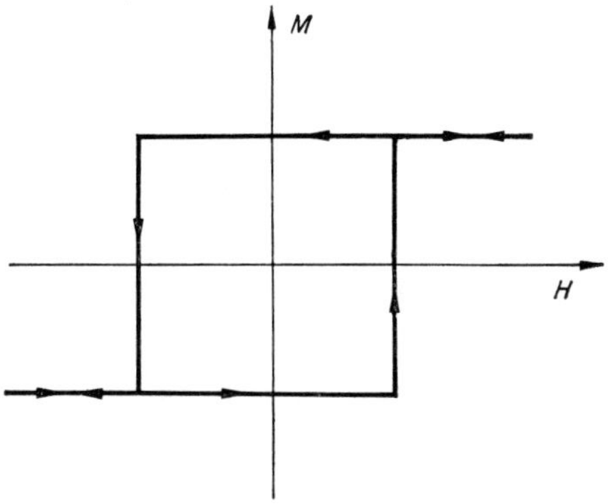

FIG. 12.3. Square hysteresis loop.

is then desirable to have a "square" hysteresis loop of the form shown in Fig. 12.3. We may note that, whatever the form of the intrinsic hysteresis loop of the material, the measured loop will only look like Fig. 12.3 if the experimental sample is either a closed core or an ellipsoid, so that M and H can simultaneously be uniform.

The saturation magnetization of technical materials is usually less than

$$\frac{2}{4\pi \times 10^{-7}}\ \mathrm{A\ m^{-1}}\ \text{(equivalent to 2 tesla or 20 kilogauss)}$$

and the coercive force less than

$$\frac{0{\cdot}2}{4\pi \times 10^{-7}}\ \mathrm{A\,.\,m^{-1}}\ \text{(2000 gauss)}$$

so that fields available in the laboratory are adequate to take a material round its complete hysteresis loop. This is not always the case in ferroelectricity. We must in any case distinguish carefully between ferroelectric and polar media. In a crystal of a polar symmetry class, there may be a permanent polarization built into the structure. It can only be reversed by a field strong enough to disrupt the crystal structure. In a ferroelectric the polarization is, at least in principle, reversible even though the fields required may be inconveniently high. It is worth noting that the built-in moment of a polar crystal is exceedingly high. The typical permanent molecular dipole moment corresponds to one electronic charge displaced by 1 Å. The usual unit, the debye, is 4·8 times less and is 10^{-18} esu or $3{\cdot}33 \times 10^{-30}$ coulomb m. Suppose we have a medium with units of moment 1 debye arranged on a cubic lattice of size 10 Å, then $P \approx 3 \times 10^{-3}$ coulomb m^{-2}. For a sphere of this material of radius 1 m the dipole moment is roughly $p = 10^{-2}$ coulomb m. The field at a distance r is

$$E \approx \frac{p}{4\pi\varepsilon_0 r^3} \approx \frac{10^8}{r^3}\ \mathrm{V\ m^{-1}}.$$

At a distance of 5 metres this would be enough to ionize air! Even if we use unrationalized, but more rational, units the results are quite impressive. The field at the surface of the body is still $10^6\ \mathrm{V\ cm^{-1}}$. Of course in actual crystals these effects are neutralized by extraneous surface charge and usually the only outwardly detectable effect of a permanent polarization is a transient change in **P** when the temperature is changed (pyro-electricity).

Problems

1. Verify that if χ' is independent of ω, $\chi'' = 0$.

2. In a crystal in which the 3-axis is an axis of six-fold symmetry the principal values of the dielectric constant tensor are $\varepsilon_1 = \varepsilon_2$, ε_3. A plane wave propagates in a direction making an angle θ with the 3 axis and lying in the 23 plane. Show that the propagation constants of the ordinary and extraordinary waves are given by

$$\beta_0^2 = \varepsilon_1 k^2, \qquad \beta_e^2 = k^2 \frac{\varepsilon_1\varepsilon_3}{\varepsilon_1 + (\varepsilon_3 - \varepsilon_1)\cos^2\theta}.$$

3. A certain crystal is known to belong to either the point group 3 or the group $\bar{3}$. Devise either an optical or an electric test which would distinguish between these groups.

4. ZnS exists in two crystal forms of symmetry $\bar{4}\,3\,m$ and $6\,mm$. How may the forms be distinguished by an electrical test?

5. A ferroelectric medium has a remanent polarization whose external effects are completely neutralized by extraneous surface charge. How may the polarization be detected by a purely electric process?

6. A crystal whose optical absorption depends on the direction of linear polarization of the light is said to be dichroic. Why is it possible for thick crystals to show strong dichroism without being strongly birefringent?

CHAPTER 13

Microscopic Aspects of the Dielectric Constitutive Relation

Introduction

The dielectric constitutive relation is primarily concerned with the relation between the polarization **P** and the macroscopic electric field **E** in a material medium; the effects of quadrupolarization **Q** are usually insignificant. The core of any microscopic interpretation of dielectric behaviour is a discussion of the response of the atomic or molecular units of a medium to an electric field and the nature of this response determines the general features of the dielectric behaviour. This is, unfortunately, not the only problem connected with a microscopic theory, for the field which acts on an atomic unit is a field at a point specified with microscopic precision and is not, therefore, necessarily the same as the macroscopic field **E** in the medium. This field, as we have seen, is specified by a process explicitly designed to eliminate all reference to the coordinates of the microscopic particles of the medium. The relation between this effective local field $\mathbf{E}^{\text{loc}}$ and the macroscopic field E is, as we shall see, rather complex, and no generally satisfactory method of dealing with the problem is known. The details of the relation in addition have a rather considerable effect on the calculated values of the dielectric properties and, as a result, attempts to give a precise, quantitative interpretation of experimental results in microscopic terms are rather unrewarding. There is little point in relating an easily measurable experimental quantity to a complex array of atomic parameters few of which can be calculated accurately or are accessible to direct measurement, especially when the final stages of the calculation contain a significant local field correction factor of doubtful accuracy. The function of a microscopic theory of dielectric behaviour is, therefore, less to provide us with an accurate predictive theory than to supply us with a means of organizing experimental data, a method of making rough qualitative predictions and an indication of possible semi-empirical rules and regularities in dielectric phenomena. It is in this spirit that we intend to approach a microscopic interpretation of the various varieties of dielectric behaviour.

In any microscopic theory of dielectric phenomena we begin by identifying within the medium a set or sets of atomic or molecular sub-units in terms of whose response we propose to interpret the overall dielectric behaviour. Thus, in discussing, for example, crystalline NaCl we identify three sets of units, first of all the ionic charges in a unit cell, secondly the individual Na^+ ions and thirdly the individual Cl^- ions. The overall dielectric response will be the sum of the induced ionic polarization and the electronic polarizations of the two types of ions. As we shall see these three types of response, although they contribute additively to **P**, cannot be treated separately. The induced ionic polarization is to some extent determined by the electronic polarization and vice versa. Nevertheless, at some stage in the calculation we reach a point at which we have N_s sub-units of type s in unit volume of moment $\mathbf{p}(s)$ and then proceed to express **P** as

$$\mathbf{P} = \sum N_s \mathbf{p}(s). \tag{13.1}$$

The next step in the calculation is to relate the local field $\mathbf{E}^{\text{loc}}$ acting on the units to the macroscopic field $\mathbf{E}$ in the medium. In some especially simple cases, as we shall see, we can write $\mathbf{E}^{\text{loc}} = \mathbf{E}+\gamma/\varepsilon_0\mathbf{P}$ where γ is a dimensionless constant, but in general the relation will be more complex and, in some cases, it may turn out to be hardly worth while attempting to reduce the response of many interacting sub-units to a sum of the responses of a set of equivalent non-interacting units to an effective local field. The final stage in the calculation is, in most cases, the most interesting and significant one of obtaining, from a consideration of their internal structure, the response of the individual units to the local field which acts upon them. By and large it is at this stage that the different types of dielectric phenomena peculiar to different types of media become evident and can be related to the detailed microscopic structure of the medium.

In elementary atomic physics and solid state physics a great deal of attention is paid to systems which are either especially simple such as the hydrogen atom, or highly idealized, and this is liable to convey the impression that the dielectric response of atomic or molecular units should be calculable, with some precision from their structure. This is, unfortunately, not the case, since the polarizability of an atomic or molecular structure often depends critically and in an unpredictable way on rather fine details of the electronic configuration. A nice example is provided by the two simple oxides quartz, SiO_2, and rutile, TiO_2, whose dielectric behaviours are grossly dissimilar. Quartz is an *almost* isotropic crystal of electronic susceptibility near unity and low frequency susceptibility near 3. Rutile has a markedly anisotropic susceptibility. The electronic susceptibility is near 7 and the total low frequency susceptibility near 100. There are indeed few cases other than perhaps the lighter noble gases (but not their condensed phases) where even the most elaborate calculations yield results comparable in accuracy with those obtained by the simplest experimental methods.

In most of this chapter we shall be concerned with the response of particular types of microscopic sub-units to the local electric field but we begin by considering the relation of the local electric field to the macroscopic field. As we have remarked, this problem arises because the field acting on a microscopic unit, and effective in polarizing it, is a field at a point specified with microscopic precision, whereas the macroscopic electric field is intentionally defined in such a way that it makes no explicit reference to microscopic coordinates, and is indeed an average field over a loosely defined region including not only many atomic sites, but also the regions of space between these sites.

The Local Field

To start with we shall consider a finite material body in which the atoms have a specified impressed polarization so that, at the site $\mathbf{r}(n)$ there is a unit with a dipole moment $\mathbf{p}(n)$ and no other electric moment. This is a highly artificial situation and we shall, at a later point, have to consider how far it is relevant to real dielectric media. The polarization $\mathbf{P}(\mathbf{R})$ is obtained by forming the expression $\sum_n \mathbf{p}(n)\delta(\mathbf{r}(n)-\mathbf{R})$ and then truncating its spatial Fourier spectrum in such a way that all explicit reference to the atomic coordinates $\mathbf{r}(n)$ is removed from $\mathbf{P}(\mathbf{R})$. Once this has been done, or assumed to have been done, the resulting macroscopic electric field $\mathbf{E}(\mathbf{R})$ can be obtained from

$$\mathbf{E}(\mathbf{R}) = \nabla_1 \int_{\text{body}} \frac{\nabla_2 \cdot \mathbf{P}(\mathbf{R}_2)\,d^3\mathbf{R}_2}{4\pi\varepsilon_0 R_{12}}. \tag{13.2}$$

Microscopic Aspects of the Dielectric Constitutive Relation

As long as the electric effects due to each of the atomic sub-units of the medium can be regarded as equivalent to those of point dipoles at the atomic sites, the exact field at the atomic site $\mathbf{r}(n)$ due to all dipoles except that at $\mathbf{r}(n)$ is

$$E_i^{\text{loc}}(\mathbf{r}(n)) = \sum_{n'\neq n} \Gamma_{ij}(n, n')p_j(n') \tag{13.3}$$

where

$$\Gamma_{ij}(n, n') = \frac{3r_i(n, n')r_j(n, n') - r^2(n, n')\delta_{ij}}{4\pi\varepsilon_0 r^5(n, n')} \tag{13.4}$$

with

$$\mathbf{r}(n, n') = \mathbf{r}(n) - \mathbf{r}(n'). \tag{13.5}$$

We now assume, for simplicity, that the polarization within the body is uniform and that the dipole moments at each atomic site are identical. This first of all allows us to express (13.2) in terms of an integral over the surface of the body as

$$\mathbf{E}(\mathbf{R}_1) = -\nabla_1 \int_{\text{surface}} \frac{P_n(\mathbf{R}_2)\,\mathrm{d}S_2}{4\pi\varepsilon_0 R_{12}} \tag{13.6}$$

and secondly allows us to simplify (13.3). We have, if p_j is the jth component of the uniform dipole polarization,

$$E_i^{\text{loc}}(\mathbf{r}(n)) = p_j \sum_{n'\neq n} \Gamma_{ij}(n, n'). \tag{13.7}$$

If we now take a surface of macroscopic dimensions surrounding the point $\mathbf{r}(n)$ but within the body (thus dividing the body into a near region B and a distant region C) we have

$$E_i^{\text{loc}}(\mathbf{r}(n)) = p_j \sum_{n'\neq n}^{B} \Gamma_{ij}(n, n') + p_j \sum_{n'}^{C} \Gamma_{ij}(n, n'). \tag{13.8}$$

The expression $\sum_{n'\neq n}^{B}\Gamma_{ij}(n, n')$ is a symmetric tensor and, for any given distribution of atomic sites $\mathbf{r}(n')$, we can choose the surface of B as an ellipsoid in such a way that the sum is identically zero. The shape of the ellipsoidal surface, i.e., the coefficients a_{ij} in its equation $a_{ij}r_i r_j = 1$, is determined by the lattice of the atomic sites. The remaining sum over the region C in (13.8) is over a region at a macroscopic distance from $\mathbf{r}(n)$ and can be evaluated macroscopically from

$$\mathbf{E}^{\text{loc}}(\mathbf{r}(n)) = \nabla_n \int_{\text{region } C} \frac{\nabla_{n'} \cdot \mathbf{P}(\mathbf{r}(n'))\,\mathrm{d}^3\mathbf{r}(n')}{4\pi\varepsilon_0 r(n, n')},$$

where $\mathbf{r}(n')$ and $\mathbf{r}(n, n')$ are regarded as continuous variables. This is, however, just $\mathbf{E}(\mathbf{R}_1 = \mathbf{r}(n))$ given by (13.2), but less the contribution, calculated macroscopically, which arises from the near region B. We therefore have

$$\mathbf{E}^{\text{loc}} = \mathbf{E} - \nabla \int_{\text{region } B} \frac{\nabla \cdot \mathbf{P}\,\mathrm{d}^3\mathbf{R}}{4\pi\varepsilon_0 R}. \tag{13.9}$$

This can be evaluated by standard methods for the ellipsoidal region B and we obtain

$$E_i^{\text{loc}} = E_i + \frac{1}{\varepsilon_0}\gamma_{ij}P_j, \tag{13.10}$$

where γ_{ij} is the shape factor for the ellipsoidal boundary of B, and this, we recall, is determined by the lattice on which the dipoles are located. If the atoms are distributed either on a cubic lattice or isotropically and randomly in space the ellipsoid is a sphere, and γ_{ij} becomes a scalar, $\gamma = \frac{1}{3}$. We then obtain the Lorentz formula

$$\mathbf{E}^{\text{loc}} = \mathbf{E} + \frac{1}{3\varepsilon_0}\mathbf{P}. \tag{13.11}$$

For a more general crystal lattice γ_{ij} remains a symmetric tensor whose diagonal elements satisfy $\gamma_{11}+\gamma_{22}+\gamma_{33} = 1$. For lattices which occur in practice it is likely that the diagonal elements are not far from $\frac{1}{3}$ and the off diagonal elements small, but the widespread assumption that (13.11) rather than (13.10) is generally valid is probably unfounded.

The assumptions made explicitly in the derivation of (13.10) are (a) that the effects of the atomic sub-units are completely described by their dipole moments and (b) that the moments of each unit are identical in magnitude and direction. To this we must add a further implicit assumption which is that, if the dipoles are randomly distributed in space, we need only consider the expectation value of $\mathbf{E}^{\text{loc}}$ for this random distribution.

If we have two atomic units of linear dimensions a_0 a distance d apart, the dipole electric field at one unit due to the other is of order a_0/d^3 while the 2^l pole field is of order a_0^l/d^{l+2}. Thus the dipole field will only be the dominant term if a_0/d is small. This is certainly the case in gases where $a_0/d \approx 1/30$ at s.t.p., but in solids and liquids where $a_0/d \sim 1$ it is a very dubious approximation. The usefulness of the dipole approximation depends mainly on the fact that we are only interested in the dipole polarization induced by the local field and that, due to symmetry, the higher multipole fields do not induce an appreciable dipole moment. Little attention appears to have been devoted to higher order multipole effects largely because the dipole term is already sufficiently complicated and uncertain to make the inclusion of higher order terms a pointless exercise.

The second assumption, that the moments at each lattice site are identical, is less critical. In the chapter on statistical mechanics we consider the consequences of relaxing this assumption. We shall see that they are not important. We may, however, remark that the usual statement, in elementary texts, that we can consider all the moments on each lattice site to be identical when we are dealing with induced polarization, but not when we are dealing with the polarization of molecules with permanent moments, is false. The expectation value of the moments at each site may well be identical but in both cases individual moments have a considerable statistical dispersion about this mean. In permanent dipolar polarization this dispersion is of thermal origin. In induced electronic polarization it is a quantum mechanical effect and, far from being small, is the origin of the van der Waals forces. In some crystals these forces are the dominant term in the cohesive energy of the crystal.

The final, implicit assumption that, if the atomic units are randomly distributed in space we need only consider a mean isotropic distribution, is not satisfied in liquids and gases at high density. Fluctuations in the distance between the units then have small, but detectable, effects on the dielectric properties. We shall not, however, consider this further here.

Let us now look at the consequences of the relation (13.11) for dielectric theory. For brevity and simplicity we shall confine our attention to cubic or isotropic media where the simpler form

$$\mathbf{E}^{\text{loc}} = \mathbf{E} + \frac{\gamma}{\varepsilon_0}\mathbf{P} \tag{13.12}$$

is valid. Suppose that, as a result of an atomic calculation, we have found that the dipole moment of an atomic unit in a field $\mathbf{E}^\circ$ is $= \varepsilon_0 \alpha \mathbf{E}^\circ$ and that there are N such units in unit volume, we then have

$$\mathbf{P} = \varepsilon_0 N\alpha \mathbf{E}^\circ. \tag{13.13}$$

If now we make the further assumption that the effective field which polarizes the atoms is $\mathbf{E}^{loc}$ we have

$$\mathbf{P} = \varepsilon_0 N\alpha \mathbf{E} + N\alpha\gamma \mathbf{P}, \tag{13.14}$$

so that

$$\mathbf{P} = \frac{\varepsilon_0 N\alpha \mathbf{E}}{1-\gamma N\alpha}. \tag{13.15}$$

Thus the dielectric susceptibility is

$$\chi = \frac{N\alpha}{1-\gamma N\alpha}. \tag{13.16}$$

Suppose that, as in a non-polar gas such as Xe at s.t.p., $N\alpha \approx 10^{-3}$, then the value of χ is not very sensitive to the value of γ. On the other hand, in solids and liquids when $N\alpha$ is of the order of unity the effect of the value of γ is appreciable. Thus if $N\alpha = 2$ and $\gamma = \frac{1}{3} = 0{\cdot}33$ we have $\chi = 6$ whereas if $\gamma = 0{\cdot}30$ we have $\chi = 5$. Now, even in a cubic crystal, γ will only be exactly $\frac{1}{3}$ if the atoms are well-separated entities. This may be approximately the case in NaCl but is far from the case in AgCl where there is appreciable overlap of the electronic wave-functions. As the electronic distribution associated with each atom or ion becomes more extended in space we expect the effective local field to approach more and more closely to the macroscopic field, and γ to decrease. We would certainly be unable to decide on an exact value for γ to within 10% in a crystal such as AgCl. Indeed, in a crystal such as diamond where there is appreciable electronic overlap over several atoms, we might reasonably expect the electrons to respond to the average macroscopic field and the effective value of γ to be near zero. Further, given a complicated enough overlapping electronic structure, there is no reason to suppose that polarization of the electrons of one atom could not have an inhibiting effect on its neighbours and lead to negative effective values of γ. Thus, if experimentally we know that $\chi = 6$, we cannot be sure whether this implies $\gamma = \frac{1}{3}$ and $N\alpha = 2$ or perhaps $\gamma = \frac{1}{4}$ and $N\alpha = 2{\cdot}4$. For this reason comparison between atomic theory and experiment is rather unrewarding.

Clearly if $N\alpha > 3$ and $\gamma = \frac{1}{3}$ the dielectric properties of the medium will be anomalous. We expect a spontaneous polarization in zero field. This is, indeed, a part of the mechanism which leads to ferro-electricity. It is, however, not quite so simple as this indicates. Water contains molecules with a permanent dipole moment and ought to have $\gamma \approx \frac{1}{3}$. On the other hand, measurements of the moment and the application of elementary statistical mechanics lead to $N\alpha \sim 9$ at 300 K. Despite this, water is neither ferroelectric nor spontaneously polarized. The resolution of this paradox is due to Onsager (1936) and the paradox arises because we have assumed that the field effective in orienting the dipoles is $\mathbf{E}^{loc}$ given by (13.11). Now $\mathbf{E}^{loc}$ given by (13.11) is the field at the site of a molecule when all the molecules including the one in question have the same specified moment and direction. The direction of the molecules, other than the central one, is however in part determined by the orientation of the central one. Thus the field these molecules exert on the central one contains a component $\mathbf{E}^{reaction}$ which depends on the orientation of the central dipole and rotates with it. This field can have no effect on the orientation of the central dipole and should be subtracted from $\mathbf{E}^{\text{loc}}$. Thus we have

$$\mathbf{P} = N\alpha\varepsilon_0\mathbf{E}^\circ,$$

with

$$\mathbf{E}^\circ = \mathbf{E}^{\text{loc}} - \mathbf{E}^{\text{reaction}}.$$

If we have a single dipole $\mathbf{p}$ at the centre of a macroscopic spherical cavity of radius a in a medium of susceptibility χ a straightforward macroscopic calculation yields for the reaction field

$$\mathbf{E}^{\text{reaction}} = \frac{2\chi}{3+2\chi}\frac{\mathbf{p}}{4\pi\varepsilon_0 a^3}.$$

Onsager assumed that this would also hold for a microscopic dipole in a microscopic cavity, and that he could determine the radius of the cavity a from the density of dipoles N by

$$N \cdot \frac{4\pi}{3} a^3 = 1.$$

Since $N\mathbf{p} = \mathbf{P}$ we have $\mathbf{P} = 3\mathbf{p}/4\pi a^3$ and so

$$\mathbf{E}^{\text{reaction}} = \frac{1}{3\varepsilon_0}\mathbf{P} \cdot \frac{2\chi}{3+2\chi}$$

and

$$\mathbf{E}^\circ = \frac{1}{3\varepsilon_0}\mathbf{P}\left(1 - \frac{2\chi}{3+2\chi}\right) = \frac{1}{\varepsilon_0}\frac{\mathbf{P}}{3+2\chi}.$$

Thus we obtain

$$\mathbf{P} = \varepsilon_0 N\alpha\mathbf{E} + \frac{N\alpha}{3+2\chi}\mathbf{P},$$

which gives the relations

$$N\alpha = \frac{\chi(1+\frac{2}{3}\chi)}{1+\chi} \tag{13.17}$$

and

$$\chi = \tfrac{3}{4}\{N\alpha - 1 + (1 + \tfrac{2}{3}N + N^2\alpha^2)^{\frac{1}{2}}\}. \tag{13.18}$$

If we compare these with the Lorentz relations

$$N\alpha = \frac{3\chi}{3+\chi}$$

$$\chi = \frac{N\alpha}{1-\frac{1}{3}N\alpha}$$

we see that, in the Onsager theory, χ remains finite for all values of $N\alpha$ and that as $N\alpha \to \infty$, $\chi \to \frac{3}{2}N\alpha$.

Although the Onsager theory does not give particularly close quantitative agreement with experiment it is clearly of importance in explaining why polar liquids do not exhibit spontaneous polarization. For a more detailed account the reader should refer to either Smythe (1955) or Frohlich (1958).

For small values of $N\alpha$ equation (13.18) yields

$$\chi \sim N\alpha(1 + \tfrac{1}{3}N\alpha)$$

which, to this order, coincides with the Lorentz expression, thus the Onsager and Lorentz treatments only lead to different results when $N\alpha$ is not too small.

When we are dealing with induced polarization, rather than polarization due to the orientation of molecules with permanent dipole moments, the reaction field is effective in producing polarization and Onsager's refinement of the Lorentz approach is unnecessary. Thus, in general, in media displaying induced electronic or ionic polarization, but not orientation polarization, the first approximation to the local field correction will always be of the form

$$E_i^{\text{loc}} = E_i + \frac{1}{\varepsilon_0}\gamma_{ij}P_j,$$

even if the tensor factor is not quantitatively given by a shape factor associated with the lattice.

In some media, e.g. piezo-electrics or non-linear media, the polarization of the individual atoms is essentially an impressed polarization due to strain or a non-linear interaction. This polarization then reacts with the linear polarizability of the medium to produce an additional effect. Suppose that the directly impressed polarization density is π and the resulting polarization $\mathbf{P}$ we then have

$$\mathbf{P} = \pi + \varepsilon_0 N\alpha \mathbf{E}^{\text{loc}}$$

or

$$\mathbf{P} = \pi + N\alpha\gamma\mathbf{P}$$

so that the net polarization is

$$\mathbf{P} = \frac{\pi}{1 - N\alpha\gamma}. \qquad (13.19)$$

If $N\alpha\gamma$ approaches unity this can be very substantially larger than the direct effect. We shall consider an example of this type of effect later, in connection with non-linear phenomena.

The Basic Polarization Mechanisms

When atoms or molecules are subjected to an electric field the forces acting on the negative electrons and the positive nuclei are in opposite directions, and, as a result, there is a net relative displacement of the centres of gravity of the positive nuclear charge and the negative electronic charge. The true mass centre of gravity of the system is however unchanged and so, since nuclei are much heavier than electrons, the predominant displacement is that of the electronic charge. The induced dipolar polarization due to this process is therefore known as electronic polarization. Because it involves only the motion of light electrons, whereas all other processes involve complete atoms or ions, it is the only process effective at the highest, i.e. optical, frequencies. Optical phenomena are almost exclusively due to electronic polarization.

An electric field applied to an ionic crystal such as NaCl not only displaces electrons and nuclei relative to each other in the individual ions but also displaces the Na^+ ions relative to the Cl^- ions. The resulting polarization is known as ionic polarization. Because it involves ions, generally with masses more than 10^4 greater than the mass of the electron, it is not effective at optical frequencies. Ionic polarization is essentially a low frequency phenomenon and in few media is it significant at frequencies much above 10^{14} Hz or wavelengths shorter than 3 microns.

Many molecules, e.g. HCl, possess permanent electric dipole moments of the order of 1 debye ($3{\cdot}3 \times 10^{-30}$ coulomb m) (or one electron at 1/4·8 Å). In a gas or a liquid these molecules are free to rotate and an applied field, competing with the forces of thermal disorder, tends to line the moments parallel to the field and produce a net polarization. This dipolar polarization is temperature dependent and, like ionic polarization, essentially a low frequency effect. As we shall see later it is, in any case, closely related to ionic polarization, for obviously a crystal of HCl formed by condensation of the gas will exhibit ionic polarization. In subsequent sections of this chapter we treat each of these basic mechanisms in more detail. In later sections we deal with the combined effects of electronic and ionic polarization (since all ionic media also exhibit electronic polarization), the microscopic interpretation of non-linear effects, the additivity of contributions from individual structural units to the overall behaviour and the effects of a small concentration of mobile electrons.

Electronic Polarization

Before we look at the quantum mechanical aspects of these phenomena it is useful to consider several simple classical models.

An atom consists of a massive positive nucleus surrounded by a cloud of negative electrons. It is not unreasonable to suppose that in an applied field $\mathbf{E}$ the electrons redistribute themselves in such a way as to cancel the effects of $\mathbf{E}$ within the atom. In other words, an atom of radius a behaves like a conducting sphere of radius a and in a field $\mathbf{E}$ acquires an induced dipole moment equal to $4\pi a^3\varepsilon_0\mathbf{E}$. Thus, if we write $\mathbf{p} = \alpha\varepsilon_0\mathbf{E}$ the polarizability is $\alpha = 4\pi a^3$.

The electrons are bound by the coulomb field of the nucleus and the attractive electric force acting on the electrons at a radius r is $Ze^2/4\pi\varepsilon_0 r^2$ where Z is the effective nuclear charge. This is balanced by centrifugal forces but again it is not unreasonable to suppose that, since the actual orbit is one of stable equilibrium, the additional force for a displacement δr is of the order $(Ze^2/4\pi\varepsilon_0 r^2)/(\delta r/r)$. Thus, in a field E the displacement is of the order of $4\pi\varepsilon_0 r^3/Ze$, and so with Z electrons of charge e the induced dipole moment is again $4\pi\varepsilon_0 r^3\mathbf{E}$.

If there are N atoms in unit volume, $\mathbf{P} = N\alpha\varepsilon_0 E$ and the susceptibility is $\chi = N\alpha = 4\pi Na^3$. The volume occupied by each atom, i.e. $1/N$, is $(4\pi/3)a^3$ and so we obtain $\chi \approx 3$ or, if the filling factor is somewhat lower, i.e., there are empty spaces or small voids between the atoms, $\chi \approx 1$ to 3. In fact the vast majority of transparent solids have a value of χ due to electronic polarization which lies between these limits, i.e., corresponds to a refractive index between 1·4 and 2.

If the electrons in an atom are in equilibrium in zero field the energy of the atom must be a minimum with respect to any displacement. Thus, the energy must increase quadratically for small displacements. In other words, the electrons are bound in an approximately harmonic potential well, with a restoring force proportional to displacement. If we write this force as $-kx$ it would appear from our earlier argument that $k \approx Ze^2/4\pi\varepsilon_0 r^3$ so that the natural resonance frequency ω_e of the electrons about equilibrium is given by

$$\omega_e^2 \approx \frac{Ze^2}{4\pi\varepsilon_0 r^3 m}. \tag{13.20}$$

Planck's constant $\hbar$ is in fact hidden in this result for r is of the order of the Bohr radius

$a_0 = 4\pi\varepsilon_0(\hbar^2/me^2)$, i.e. 0·53 Å. A classical equation of motion for an electron in a field $E(t) = E\exp(j\omega t)$ would be

$$\ddot{x}+\omega_e^2 x = \frac{e}{m}E(t).$$

The effects of radiation, and the residual effects of the actual anharmonicity of the potential which couple the electronic motion to other forms of motion in a classical model, would add a term to this equation which, to a first approximation, looks like a frictional term. Thus we have

$$\ddot{x}+\omega_e\Delta\dot{x}+\omega_e^2 x = \frac{e}{m}E\exp(j\omega t).$$

The response at ω is

$$x = \frac{\frac{e}{m}E\exp(j\omega t)}{\omega_e^2-\omega^2+j\omega\omega_e\Delta}.$$

With N atoms in unit volume the polarization is Nex and

$$\chi = \frac{\omega_p^2}{\omega_e^2-\omega^2+j\omega\omega_e\Delta} \tag{13.21}$$

where the plasma frequency ω_p is defined by

$$\omega_p^2 = \frac{Ne^2}{\varepsilon_0 m}. \tag{13.22}$$

The static susceptibility is

$$\chi_0 = \frac{\omega_p^2}{\omega_e^2} \tag{13.23}$$

and we can express the high frequency susceptibility as

$$\chi(\omega) = \frac{\chi_0}{1-\frac{\omega^2}{\omega_e^2}+j\frac{\omega}{\omega_e}\Delta}. \tag{13.24}$$

At frequencies well above the resonance frequency ω_e the susceptibility becomes small. In most cases ω_e lies in the near ultra-violet region of the spectrum and the electronic susceptibility in the visible is not far from the static value. It falls rapidly to near zero in the far ultra-violet and soft X-ray regions.

Clearly $\chi(\omega)$ is complex and we write it as $\chi = \chi'-j\chi''$; we have

$$\chi' = \frac{(1-\omega^2/\omega_e^2)\chi_0}{(1-\omega^2/\omega_e^2)^2+(\omega/\omega_e\Delta)^2} \tag{13.25a}$$

and

$$\chi'' = \frac{\omega/\omega_e\Delta\chi_0}{(1-\omega^2/\omega_e^2)^2+(\omega/\omega_e\Delta)^2}. \tag{13.25b}$$

The imaginary part χ'', and therefore the dissipation or attenuation, rises to a peak at the resonance frequency ω_e where it is

$$\chi'' \approx \frac{\chi_0}{\Delta}. \tag{13.26}$$

The fractional width of the absorption region is Δ and the width in frequency $\omega_e/2\pi\Delta$ Hz. Because χ_0 is expected to be near unity and Δ is at least likely to be less than unity we expect χ'' to be large and therefore $\tan\delta = \chi''/1+\chi'$ to be large. The absorption at resonance (in a solid) is therefore exceedingly intense and the solid is not only opaque but, if the absorption occurs in the visible, has a metallic appearance because the large value of χ'' makes a large change in the wave impedance and leads to specular reflection. This behaviour is observed in solid dyes. The weak colours characteristic of inorganic materials, such as copper sulphate, have a different origin. They are due to weak, almost forbidden, absorption lines which correspond to resonances making almost no contribution to the dielectric properties.

The quantum mechanical expression for the electronic polarizability of an atom is (neglecting terms due to collisions which correspond to the frictional terms in the classical equation)

$$\alpha = \frac{2e^2}{\hbar\varepsilon_0}\sum_n \frac{\omega_n x_{0n} x_{n0}}{\omega_n^2-\omega^2}, \tag{13.27}$$

where ex_{0n} is the matrix element of the electric dipole moment operator between the ground state 0 and the state n whose energy is $\hbar\omega_n$. This expression is of little practical use since, except for the very simplest atoms, we have no way of calculating the matrix elements with any precision nor are they easy to measure. There are, however, two approximate methods of estimating α. The first is useful in estimating the static susceptibility as $\omega \to 0$ and the second, based on a sum rule, is useful in discussing the frequency response or dispersion.

If the atom has no permanent dipole moment $x_{00} = 0$ and the static polarizability is

$$\alpha = \frac{2e^2}{\hbar\varepsilon_0}\sum_{n\neq 0}\frac{x_{0n}x_{n0}}{\omega_n} = \frac{2e^2}{\hbar\varepsilon_0}\sum_n \frac{x_{0n}x_{n0}}{\omega_n}. \tag{13.28}$$

In this sum the frequency ω_n has a rather limited range of variation but the matrix elements fluctuate wildly with the summation index n. Thus we can replace ω_n by an average value ω_e and

$$\alpha = \frac{2e^2}{\varepsilon_0\hbar\omega_e}\sum_n x_{0n}x_{n0}. \tag{13.29}$$

The sum is now just $(x^2)_{00}$ the expectation value of x^2 in the ground state which we write as $\langle x^2\rangle$. Thus, to this approximation

$$\alpha = \frac{2e^2}{\varepsilon_0\hbar\omega_e}\langle x^2\rangle. \tag{13.30}$$

If the atom is approximately spherical $\hbar\omega_e \approx e^2/8\pi\varepsilon_0 r$ and $\langle x^2\rangle \sim \frac{1}{3}r^2$ so that $\alpha \sim (16\pi/3)r^3$. This is, to within a factor $\frac{3}{4}$, the same as the expression $4\pi r^3$ for the classical models. Of somewhat more interest is the light equation (13.30) sheds on the origin of anisotropy in the electronic polarizability, and thus on the dielectric constitutive relation at high frequencies, and on birefringence. We see that if the atom is not spherical, α will be greatest when **E** lies in the direction in which the atom has its greatest spatial extent. This also applies to molecules and, for example, explains why in iodoform CHI_3, in which the almost planar molecules are arranged normal to the c axis of the crystal, the refractive index is larger when **E** is polarized normal to the axis than when it is parallel to the axis. In a molecule such as $CH_3(CH_2)_nCOOH$, there is little electronic communication between groups more than one bond apart and each group responds independently and almost isotropically to a field. By contrast in conjugated molecules, e.g., $CH_3—CH═CH—CH═CH_2$, containing alternating single and double bonds, the π electrons move freely through the structure and not only is the anisotropy large but also the resonance frequency is low. Many dyes contain this type of structure. For a discussion of this type of behaviour the reader should consult Platt (1964). The qualitative notion that the polarizability is related to the spatial extent of the electronic distribution in the ground state is of some importance since it relates optical properties to molecular structure as it is envisaged in chemistry.

Inorganic molecular groupings rarely lead to large anisotropies; an exception is tellurium (see below) but some conjugated organic molecules, notably carotenoids, have anisotropies exceeding 5/1. We would expect many organic media to be exceedingly anisotropic were it not for the tendency of highly anisotropic molecules to form crystals with several cancelling molecules in each unit cell.

Gyrotropy, or optical activity, is a rather more subtle form of anisotropy which can be explained in these terms. Consider a molecule whose electronic distribution has the shape shown in Fig. 13.1.

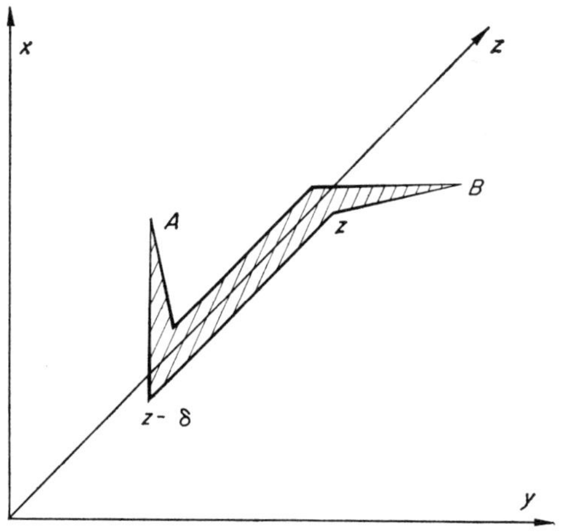

FIG. 13.1. An optically active molecule.

In a field $E_x(z)$ the limb A acquires a polarization proportional to $E_x(z-\delta)$ and, as a result, electrons are expelled from A towards B, where they produce a negative component p_y. Thus the constitutive relation will contain terms of the form

$$P_y(z) = -N\beta\varepsilon_0 E(z-\delta),$$

and

$$P_x(z) = -N\beta\varepsilon_0 E_y(z+\delta),$$

or

$$P_y(z) = -N\beta\varepsilon_0 E_x(z) + N\beta\varepsilon_0 \delta \frac{\partial E_x}{\partial z},$$

$$P_x(z) = -N\beta\varepsilon_0 E_y(z) - N\beta\varepsilon_0 \delta \frac{\partial E_y}{\partial z}.$$

This is just what we require for optical activity and the rotation of the plane of polarization of light of vacuum wavelength λ advancing along the z axis is

$$\frac{\mathrm{d}\theta}{\mathrm{d}z} = \tfrac{1}{2}N\beta\delta\left(\frac{2\pi}{\lambda}\right)^2. \tag{13.31}$$

The term $N\beta$ corresponds to an off-diagonal term in the ordinary anisotropic susceptibility and might be expected to be of the order of 10^{-2} to 10^{-1}. At $\lambda = 6000$ Å we have $\mathrm{d}\theta/\mathrm{d}z \approx 10^{12}\delta$ radians m^{-1}. If $\delta \approx 1$ Å or 10^{-10} m we get $\mathrm{d}\theta/\mathrm{d}z \approx 100$ rad m^{-1} which is of the general order of rotations observed in simple solids. The rotation in liquids is generally less because of the random orientation of the molecules.

In order to exhibit a large optical activity a medium must contain molecules with a pronounced screw sense in which electronic communication between extremes of the molecule is good. This favours conjugated organic molecules. A notable example is the benzene derivative hexa-helicene whose six rings have the configuration of a lock-washer.

Our general arguments suggest that the electronic susceptibility χ will be of the order of 3η where η is the fraction of the volume of the medium actually occupied by atoms. This is slightly misleading, for in many media the electronic wave functions overlap several atoms. In this case the values of r which determine α and those which determine the packing may be quite different. This is the case in media such as ZnS, GaP, or Ge, and for example in tellurium the electronic susceptibility is not only large but markedly anisotropic with principal values of 15 and 35.

An alternative approximation to the atomic polarizability is based on the sum rule for the dipole matrix elements. Because electrons are fermions the electronic properties of atoms on molecules are dominated by single electron excitations and, in this case, the fundamental commutation rule for coordinates and momenta leads directly to a sum rule

$$\sum_n \omega_n x_{0n} x_{n0} = \frac{\hbar}{2m}. \tag{13.32}$$

If we define the dimensionless oscillator strength f_n associated with the transition x_{0n} by

$$f_n = \frac{2m\omega_n}{\hbar} x_{0n} x_{n0}, \tag{13.33}$$

the sum rule becomes

$$\sum_n f_n = 1, \tag{13.34}$$

and we have

$$\alpha = \frac{e^2}{\varepsilon_0 m}\sum_n \frac{f_n}{\omega_n^2 - \omega^2}, \tag{13.35}$$

so that

$$\chi = \frac{Ne^2}{\varepsilon_0 m}\sum_n \frac{f_n}{\omega_n^2 - \omega^2}. \tag{13.36}$$

This is like the sum of a set of classical oscillators with resonance frequencies ω_n and "strengths" f_n less than unity.

Dissipation plays no part in quantum mechanics but when we consider a large assembly of atomic systems, anharmonic interactions described in a higher order of perturbation theory lead to an equivalent behaviour and so we can replace (13.36) by

$$\chi = \frac{Ne^2}{\varepsilon_0 m} \sum_n \frac{f_n}{\omega_n^2 - \omega^2 + j\omega\omega_n \Delta_n}. \tag{13.37}$$

Each of the terms in (13.37) is known as a Sellmeier term and in practice, except near a resonance, the dielectric properties of even quite complex media can be accurately represented by a very small number of Sellmeier terms, usually only one for each identifiable atomic species. Thus, in KCl two terms, one for the K^+ and one for the Cl^- ions, are adequate.

Near a resonance the absorption is intense, usually of the order of several powers of e^{-1} per wavelength if f_n is at all large. Thus even if $f \approx 10^{-4}$ the absorption length for an optical resonance is less than 1 mm. The pale colours of transition group compounds are, as we have already remarked, due to highly forbidden resonances with very low oscillator strengths. Only in the ultraviolet, or with dye crystals in the visible, do we observe the gross effects of an allowed resonance.

The polarizability of a single electron in an atom of effective nuclear charge Z is

$$\alpha \approx \frac{2e^2}{3\hbar\omega_e \varepsilon_0} r^2 \approx kZ^{-4}, \tag{13.38}$$

thus for the isoelectronic ions K^+ and Cl^- with $Z \approx 19-10 = 9$ and $Z \approx 17-10 = 7$ we expect the ratio of the polarizabilities to be $(7/9)^4 \approx 0{\cdot}36$. The experimental ratio is 0·33.

In general, in ionic media the main contribution to the electronic susceptibility comes from the negative ions. In organic media the most effective components are regions of the molecule which display alternating single and double bonds.

We have seen that, although the gross dielectric susceptibility of a medium comes only from a few resonances, with high oscillator strengths near unity, even a weak resonance of low strength can cause appreciable attentuation. It follows that very small concentrations of impurities with appreciable oscillator strengths can make an otherwise transparent medium opaque, lossy, or coloured. Further, if the impurities are exceedingly anisotropic we can have a highly dichroic medium (anisotropic attentuation). This is used to good effect in commercial polarizers such as Polaroid. Since the best optical glass is claimed to have an optical attenuation of less than 0·2 db per metre it follows that tiny traces of impurity can have a catastrophic effect on its properties.

Ionic Polarization

The 111 planes of KCl contain only one type of ion K^+ or Cl^- and, if crystals of KCl grew with 111 faces, the crystal would consist of layers of Cl^- and K^+ ions. One face would contain only negative ions and the opposite face only positive ions. This would give the crystal an enormous built-in permanent ionic polarization with **P** of the order of 1 coulomb m^{-2}. The field at the surface of the crystal would be of the order of 10^{10} V m^{-1} and there would be a very large electrostatic energy associated with this field. Since KCl is held together mainly by the electrostatic forces between ions this configuration would be unstable. For this reason KCl crystals do not grow with 111 faces but only with faces such as 100 which contain equal numbers of positive and negative ions. In some crystals, however, where the binding is predominantly covalent, and electrostatic forces play only a subsidiary role, the situation is different. For example, hexagonal BeO forms crystals with faces consisting only of O or Be atoms. The residual ionic charge, though small, leads to large values of **P**. In practice the effects cannot be observed directly because the migration of even a small number of mobile carriers through the crystal or the accumulation of even a monatomic layer of ions from the atmosphere is sufficient to neutralize the effect completely. On the other hand, if the crystal is heated and the lattice expands, or there is a slight change in the charge distribution, a transient unbalanced polarization is observed. This pyroelectric effect is confined to crystals of the polar symmetry classes and in some cases, e.g., tri-glycine sulphate, it is large enough to be used to detect thermal radiation.

In other crystals which, although they have no polar axis, lack a centre of inversion symmetry, mechanical stress or strain can produce a distortion of the ionic charge distribution which leads to a net polarization. If the applied stress is S_{jk} the polarization P_i is given by $P_i = d_{ijk}S_{jk}$. In the converse effect a field E_i produces a strain σ_{jk} given by $\sigma_{jk} = d_{ijk}E_i$ and in chapter 15 we show that the two coefficients are the same. We might expect a field of the order of the inter-ionic field, i.e., about 1 volt per angstrom or 10^{10} V m^{-1} to produce unity strain, thus the general order of magnitude of those components of d_{ijk} allowed by crystal symmetry is 10^{-10} m V^{-1}. Actual values are usually, but not invariably, somewhat less due to geometric factors, but some media, e.g. NH_4 H_2PO_4 (A.D.P.), have coefficients of this magnitude. The polarization for a given strain is of the order of $Bd\sigma$ where B is the bulk modulus, usually 10^{10} newtons m^{-2} or more. Thus $P \approx \sigma$ and the associated electric field produced by strain is $E \sim P/\varepsilon\varepsilon_0 \approx 10^{10}\sigma$. A displacement x therefore generates a voltage of the order of $10^{10}x$. In other words, 1 Å displacement leads to about a volt. Thus piezo-electric effects are not small. The fact that the crystal in a gramophone pick-up generates almost a volt although it is exceedingly loosely coupled to the motion of the stylus is an obvious example.

We have introduced this section by discussing two somewhat esoteric effects to emphasize that ionic polarization is a rather general and complex effect and that its possibilities have been by no means exhausted when we have discussed the normal linear response to which we now turn our attention.

When we discuss the electronic polarizability of atoms we are to some extent misled by the notion that we might be able to perform an exact, or at least reasonably precise, quantum mechanical calculation of α. We are therefore inclined to regard classical or semi-classical models as rather poor substitutes for a proper calculation. This is quite illogical for, as long as we insist on deriving a coherent classical response (polarization) to a coherent classical field, quantum mechanics cannot lead to qualitatively different results from classical mechanics. In the case of ionic polarization we are subject to no such temptation. We know *a priori* that the forces which bind ions in a solid lattice are altogether too complicated for any exact treatment to be possible, and we therefore are prepared to use classical language from the start.

We know that the ions have a stable configuration in the lattice. Thus each ion sits in a potential well and its equilibrium is a position of minimum potential. For small displacements relative to its neighbours its potential energy will be a power series in the components of the displacement starting with a quadratic term. For small displacements the ion will obey Hooke's law. Of course all the ions are coupled together and for some purposes it will be useful, if not necessary, to consider the normal modes of vibration of the crystal as a whole, i.e., lattice waves, rather than vibrations of individual ions, but even then the problem is a classical one. The interaction potential between neighbouring ions is certainly not exactly harmonic or quadratic and, if we excite vibrations with a field of frequency ω, these anharmonic terms will generate new frequencies. The anharmonic coupling of the coherent driven vibrations with thermally excited vibrations will also generate new frequencies. In the theory of thermal conductivity we describe these processes in terms of collisions between phonons but for our purposes their effect is different. They represent a mechanism which transfers energy from a coherently excited mode of vibration to a vast number of incoherent modes. In other words, they represent a dissipation process.

In discussing ionic polarization we are then confronted with the classical problem of discussing the response of charged ions in a lattice interacting with their neighbours through predominantly harmonic forces and subject to essentially viscous dissipation. To fix our ideas we shall discuss ionic polarization in NaCl since this illustrates most of the principles involved.

The lattice modes in a crystal such as NaCl, with two types of atom, can be classified as transverse or longitudinal modes on the one hand, and as acoustic or optical modes on the other. In an acoustic mode the ions of both types oscillate more or less in phase, exactly in phase in the long wavelength limit. In an optical mode their oscillations are more or less out of phase, exactly so in the long wavelength limit. At a later stage we shall have to consider the effects of a finite wavelength but for the moment it is unnecessary. The reason for this is simply that the velocity of light, even in a dielectric medium, is much greater than the velocity of sound or lattice waves. Thus even at the highest frequencies the wavelength of an electromagnetic disturbance in a crystal will be several thousand angstroms while the lattice spacing, which sets the wavelength scale for lattice modes, is only a few angstroms. Thus we are only concerned with the long wave limit of the lattice modes. In this case it is sufficient to consider the response of a single pair of ions.

We begin by making a rough estimate of the force constant restoring ions to their equilibrium positions. Because Laplace's equation $\nabla^2\phi = 0$ does not allow the electrostatic potential to take absolute maximum or minimum values a lattice of structurelesss point charges cannot be stable. The stability of the NaCl lattice must be due to the internal electrostatic structure of the ions themselves, and this of course is itself only stable for quantum mechanical reasons. Nevertheless, the ordinary coulomb force $F = q^2/4\pi\varepsilon_0 r^2$ between two ions of charge q a distance r apart must be a dominant term in the force and we might expect the rate of change of the force with displacement to be of the order of $\partial F/\partial r$, i.e. about $q^2/2\pi\varepsilon_0 r^3$. With ions of mass $m \sim 5\times 10^{29}$ kg this gives a resonance frequency of $\nu = 1/2\pi(q^2/2\pi\varepsilon_0 mr^3)^{\frac{1}{2}} \approx 10^{13}$ Hz if we take r to be 5 Å. This corresponds to a wavelength of 30 microns, i.e., in the infra-red.

If the ionic charges are q and $-q$, the ionic masses M^+ and M^- and the restoring force for a small relative displacement $x^+ - x^-$ of the two ionic sublattices is $K(x^+ - x^-)$, the equations of motion for the ions are

$$M^+\ddot{x}^+ + K(x^+ - x^-) = qE(t) = q\mathscr{R}(E\exp(j\omega t))$$
$$M\ddot{x}^- + K(x^- - x^+) = -qE(t).$$

These yield the equation

$$\frac{\partial^2}{\partial t^2}(x^+ - x^-) + \frac{K}{M}(x^+ - x^-) = \frac{q}{M}E(t)$$

where M is the reduced mass defined by

$$\frac{1}{M} = \frac{1}{M^+} + \frac{1}{M^-}. \qquad (13.39)$$

If there are N ions of each type in unit volume

$$P = Nq(x^+ - x^-), \tag{13.40}$$

and the polarization satisfies the equation

$$\ddot{P} + \omega_i^2 P = \frac{Ne^2}{M} E(t) \tag{13.41}$$

where the ionic resonance frequency is

$$\omega_i = \left(\frac{K}{M}\right)^{\frac{1}{2}}. \tag{13.42}$$

If the effects of anharmonic terms in dissipating energy amongst other lattice modes are included as a damping term we get

$$\ddot{P} + \omega_i \Delta_i P + \omega_i^2 P = \frac{Ne^2}{M} E(t), \tag{13.43}$$

and this yields

$$\chi(\omega) = \frac{Ne^2}{\varepsilon_0 M} \frac{1}{\omega_i^2 - \omega^2 + j\omega\omega_i\Delta_i}. \tag{13.44}$$

We can express this in terms of the static susceptibility

$$\chi_0 = \frac{Ne^2}{\varepsilon_0 M \omega_i^2} = \frac{Ne^2}{\varepsilon_0 K} \tag{13.45}$$

as

$$\chi(\omega) = \chi_0 \frac{\omega^2}{\omega_i^2 - \omega^2 + j\omega\omega_i\Delta_i}. \tag{13.46}$$

If $K \approx e^2/\pi\varepsilon_0 r^3$ then $\chi_0 \sim \pi N r^3$ and, with $2N$ ions altogether in unit volume, $Nr^3 \approx \frac{1}{2}$ and $\chi_0 \approx 1{\cdot}5$. A local field correction with $\gamma = \frac{1}{3}$ would increase χ_0 to 3. In fact, the observed ionic susceptibilities of alkali halide crystals range from 2·2 in KI to 5·3 in LiF. The much larger values of χ_0 which occur in complex ionic crystals such as KH_2PO_4 are, as we shall see later (pp. 138 ff.), due to interaction between electronic and ionic polarization.

As ω increases, χ increases to a maximum as ω approaches ω_i and in this region is complex and leads to attenuation. For $\omega > \omega_i$ the ionic susceptibility is negative and decreases in magnitude as ω becomes large. By the time optical frequencies are reached χ_{ionic} is small.

The general behaviour of the ionic polarization is similar to that of the electronic polarization but with a resonance frequency some 100 times lower. In the region where χ_{ionic} is appreciable, $\chi_{\text{electronic}}$ is almost constant; this is usually expressed by writing $\chi_{\text{electronic}}$ as χ_∞, the implication being that χ_∞ is the susceptibility at a frequency well above the ionic resonance. The total dielectric constant in this range is

$$\varepsilon(\omega) = 1 + \chi_\infty + \frac{(\chi_{dc} - \chi_\infty)\omega_i^2}{\omega_i^2 - \omega^2}$$

or

$$\varepsilon(\omega) = \varepsilon_\infty + \frac{(\varepsilon_{dc} - \varepsilon_\infty)\omega_i^2}{\omega_i^2 - \omega^2}. \tag{13.47}$$

This is negative over the frequency range from $\omega = \omega_i$ to $\omega = \omega_i\,(\varepsilon_{dc}/\varepsilon_\infty)^{\frac{1}{2}}$. In this frequency interval the crystal cannot propagate electromagnetic waves and in addition the wave impedance $Z = (\mu_0/\varepsilon\varepsilon_0)^{\frac{1}{2}}$ is pure imaginary. If $Z = jx$ the fraction of the incident radiation from vacuum reflected at the surface is

$$\left|\frac{(\mu_0/\varepsilon_0)^{\frac{1}{2}} - jx}{(\mu_0/\varepsilon_0)^{\frac{1}{2}} + jx}\right|^2 = 1.$$

Thus in this range an ionic crystal reflects strongly. This is, of course, the typical behaviour observed in the infra-red, there is a range of frequencies over which ionic crystals are opaque.

Dipolar Polarization

If we have a gas, such as HCl or water vapour, of molecules each of which has a permanent electric dipole moment p the energy of the dipoles in a field $\mathbf{E}$ will be $-\mathbf{p}.\mathbf{E} = -pE \cos\theta$ where θ is the angle between $\mathbf{p}$ and $\mathbf{E}$. At a temperature T the molecules are distributed in energy according to the Boltzmann factor and so the expectation value of $\cos\theta$ is

$$\langle\cos\theta\rangle = \frac{\int_0^\pi \cos\theta \exp(pE\cos\theta/kT)\sin\theta\, d\theta}{\int_0^\pi \exp(pE\cos\theta/kT)\sin\theta\, d\theta}.$$

The magnitude of molecular dipole moments is about 1 debye or $3{\cdot}33\times10^{-30}$ coulomb m and at room temperature $kT \approx 4\times10^{-21}$ joules so that, in fields appreciably less than the breakdown strength of the gas, $pE \ll kT$. The integrals then yield

$$\langle\cos\theta\rangle \sim \frac{1}{3}\frac{pE}{kT}$$

and so, with N molecules in unit volume, the polarization parallel to E is

$$P = Np\,\langle\cos\theta\rangle = \frac{Np^2E}{3kT}$$

and the susceptibility is

$$\chi = \frac{Np^2}{3\varepsilon_0 kT}. \tag{13.48}$$

This result is known as the Langevin–Debye law and, although the derivation given is to be found in most texts without comment (a notable exception being van Vleck, 1932), it is, as we shall see, rather misleading, either from a classical or a quantum mechanical point of view. The result is, however, correct.

Classically, if we have an assembly of freely rotating dipoles, their response to an applied field E is quite different depending on whether their rotational energy is greater or less than pE, and also on whether their axis of rotation is parallel to or normal to $\mathbf{E}$. If the axis is parallel to $\mathbf{E}$ then, even if the energy is greater than pE, there is a net polarization parallel to $\mathbf{E}$. We have only to think of a ball being swung round on a string in the earth's gravitational field to see why this is so. The string describes the surface of a cone with the ball below the centre of rotation rather than in a horizontal plane. If, however, the axis is normal to $\mathbf{E}$ then, when pE is more than the energy of rotation, the dipoles oscillate about the direction of $\mathbf{E}$ while if the energy of rotation is more than pE they continue to rotate and spend more time with $\mathbf{p}$ opposed to $\mathbf{E}$ than with $\mathbf{p}$ parallel to $\mathbf{E}$. The net effect of $\mathbf{E}$ in this case is a negative polarization. When an average is taken over all possible orientations of the axis of rotation relative to **E** the average polarization due to molecules with rotational energy greater than pE is zero. The polarization is solely due to molecules with less rotational energy.

In quantum mechanics the corresponding result is that in weak fields, i.e., those for which pE is less than the energy of the first excited rotational level, only those molecules in the ground state, with zero rotational quantum number, contribute to **P**. The polarization of the ground state molecules is essentially an induced effect and can be obtained from the standard quantum mechanical formula for the polarizability,

$$\alpha(\omega) = \frac{2}{\varepsilon_0 \hbar} \sum_n \frac{\omega_n (p_x)_{0n} (p_x)_{n0}}{\omega_n^2 - \omega^2} \tag{13.49}$$

where $(p_x)_{0n}$ is the dipole moment matrix element between the ground state and the nth rotational level. For a rigid rotating dumbbell molecule $(p_x)_{0n}$ is zero except when n corresponds to the first excited state. For this system, since $\sum_n (p_x)_{0n}(p_x)_{n0} = \langle p_x^2 \rangle$ is the expectation value of p_x^2 in the ground state which is $\frac{1}{3}p^2$, we know the value of the matrix element and so

$$\alpha(\omega) = \frac{2}{3\varepsilon_0 \hbar} \frac{\omega_1 p^2}{\omega_1^2 - \omega^2}. \tag{13.50}$$

The static value is

$$\alpha_0 = \frac{2}{3\varepsilon_0} \frac{p^2}{\hbar \omega_1},$$

and, since $\hbar\omega_1 = \hbar^2/I$ where I is the moment of inertia of the molecule,

$$\alpha_0 = \frac{2}{3} \frac{p^2 I}{\varepsilon_0 \hbar^2}.$$

The number of molecules in the ground state is found (see chapter 16) from statistical mechanics to be $N' = N(\hbar^2/2IkT)$ so that

$$\chi_0 = N'\alpha_0 = \frac{Np^2}{3\varepsilon_0 kT}$$

which coincides with our earlier result.

The frequency dependence is given by (13.50), although collisions occurring at a rate $\omega_1 \Delta$ where $\Delta \ll 1$ produce damping and

$$\chi(\omega) = \frac{\chi_0 \omega_1^2}{\omega_1^2 - \omega^2 + j\omega\omega_1\Delta}.$$

This more or less describes the situation in a gas. There is intense absorption as ω approaches ω_1. At higher frequencies absorption also occurs due to transitions between higher rotational levels, e.g., $n = 2$ to $n = 3$, for, at room temperature at least, many of the lower rotational levels are appreciably populated.

In a liquid, collisions are so frequent that any resonant response is smeared out, indeed the collision frequency is such that a classification in terms of rotational levels is almost meaningless. The polarization essentially obeys a relaxation equation

$$\tau\dot{P} + P = \bar{P} = \frac{Np^2}{3\varepsilon_0 kT} E = \varepsilon_0 \chi_0 E$$

so that

$$\chi(\omega) = \frac{\chi_0}{1+j\omega\tau}. \tag{13.51}$$

The loss angle is $\delta \approx \omega\tau$ and so, even if $\tau \approx 10^{-12}$ s, the loss is appreciable at radio frequencies. Most dipolar media are indeed rather lossy; this is also true of plastics. For example, acrylic plastics with dipolar groupings are very much more lossy than hydrocarbon plastics such as polystyrene or polyethylene.

It is, perhaps, worth pointing out that, since the dipolar polarization in a gas such as HCl is due to induced polarization of the ground state, there is a smooth transition from this type of polarization to ionic polarization, as the gas is condensed to form an ionic or partially ionic crystal.

Electronic and Ionic Polarization

In a crystal such as NaCl the response to an electric field contains contributions from the ionic polarization of the lattice and also from the electronic polarization of the ions. Furthermore, a distortion of the ionic lattice will itself also distort the electron distributions of the atoms and influence the electronic polarization. In addition, the electronic polarizability of the ions has a direct effect on the forces resisting distortion of the lattice. As a result the actual rcsponse to a field will be exceedingly complicated. We can, however, obtain considerable insight into the qualitative nature of the dielectric response if we make the sweeping assumption that all these effects are described by the local field correction, and that the same local field acts on both the ions and the electrons in these ions. Thus basically we assume that if α_i is the ionic polarizability and α_e the electronic polarizability and there are N_i ionic units and N_e electronic units in unit volume,

$$P = (N_i\alpha_i + N_e\alpha_e)\varepsilon_0 E^{\text{loc}} \tag{13.52}$$

and

$$E^{\text{loc}} = E + \frac{\gamma}{\varepsilon_0}P. \tag{13.53}$$

Before we explore the consequences of this assumption, however, it will be useful to discuss the description of ionic displacements in terms of lattice waves or phonons.

We begin by considering a one-dimensional chain of atoms, in which two types of atom with masses m and μ alternate in the chain and a harmonic restoring force acts to restore the separation of the atoms to its equilibrium value l, which is half the period of the structure. If a_n is the displacement of the nth mass m and α_n of the nth mass μ, which follows the nth mass m in the chain, the equations of free motion are

$$m\ddot{a}_n = k(\alpha_n + \alpha_{n-1} - 2a_n)$$

$$\mu\ddot{\alpha}_n = k(a_{n+1} + a_1 + - 2\alpha_n)$$

where k is the force constant. If a wave propagates in the chain so that $a_n = a \exp j(\omega t - 2\beta nl)$ and $\alpha_n = \alpha \exp j(\omega t - 2\beta nl)$ we obtain the equations

$$(2k - m\omega^2)a = k\alpha(1 + \exp(2j\beta l),$$

$$(2k - \mu\omega^2)\alpha = ka(1 + \exp(-2j\beta l),$$

and the secular equation

$$\omega^2 = \frac{k}{M}\left\{1 \pm \left(1 - 4\frac{m\mu}{(m+\mu)^2}\sin^2 \beta l\right)^{\frac{1}{2}}\right\} \tag{13.54}$$

where $1/M = 1/m + 1/\mu$ defines the effective mass.

In solids the typical lattice spacing is a few angstroms whereas the wavelengths of the electromagnetic waves to be considered are several thousand angstroms. Thus we are only concerned with the limit $\beta l \ll 1$. The secular equation then has roots $\omega \to 0$ and $\omega \to (2k/M)^{\frac{1}{2}}$ as $\beta l \to 0$. The low frequency, or acoustic, branch has $a = \alpha$ and the atoms of mass m and μ on neighbouring lattice sites oscillate in phase. If the atoms have opposite charges $\pm q$ this leads to no net polarization. In the high frequency, or optical, branch $ma = -\mu\alpha$ and a net polarization $q(a-\alpha)$ is associated with the wave. Only the optical branch interacts with an electric field of long wavelength. These results are valid whether the wave represents longitudinal motion of the atoms or motion transverse to the length of the array. The force constants k may be different in the two cases and so the longitudinal wave cut-off frequency ω_l may differ from the transverse cut-off ω_τ. However, in a three-dimensional lattice a transverse displacement relative to one line of atoms is a longitudinal displacement relative to another line and it will certainly simplify the results if we assume that $\omega_l = \omega_\tau = \omega_i$.

If an electric field E^{loc}, which varies as $\exp j(\omega t - \beta z)$ with $\beta l \ll 1$, interacts with the atoms the equations of motion are

$$\ddot{a}_n = \frac{2k}{m}(\alpha_n - a_n) + \frac{q}{m}E^{\text{loc}},$$

$$\ddot{\alpha}_n = \frac{2k}{\mu}(a_n - \alpha_n) - \frac{q}{\mu}E^{\text{loc}},$$

and so the polarization $P_i = Nq(a_n - \alpha_n)$ satisfies

$$\ddot{P}_i + \omega_i^2 P_i = \frac{Nq^2}{M}E^{\text{loc}}. \tag{13.55}$$

If we let $\chi_i^* = Nq^2/\varepsilon_0 M\omega_i^2$ be the ionic susceptibility, uncorrected for local fields and at zero frequency, and χ_e^* be the similar, but frequency-independent, electronic susceptibility, then at zero frequency

$$P = P_e + P_i = \varepsilon_0\chi_e^* E^{\text{loc}} + P_i = \varepsilon_0(\chi_e^* + \chi_i^*)E^{\text{loc}},$$

or

$$P = \varepsilon_0(\chi_e^* + \chi_i^*)\left(E + \frac{\gamma}{\varepsilon_0}P\right).$$

Thus

$$\chi_0 = \varepsilon(0) - 1 = \frac{P}{\varepsilon_0 E} = \frac{\chi_e^* + \chi_i^*}{1 - \gamma(\chi_e^* + \chi_i^*)}. \tag{13.56}$$

We see that χ_0 is (if γ is positive) larger than the sum of

$$\chi_e = \frac{\chi_e^*}{1 - \gamma\chi_e^*} \quad \text{and} \quad \chi_i = \frac{\chi_i^*}{1 - \gamma\chi_i^*}.$$

We now consider the difference in the effects of the longitudinal and transverse optical modes and, to start with, we omit the local field correction. For either mode

$$\ddot{P}_i + \omega_i^2 P_i = \varepsilon_0\omega_i^2\chi_i^* E, \tag{13.57}$$

and

$$P = \varepsilon_0 \chi_e^* E + P_i, \tag{13.58}$$

and for either wave we must have $\nabla . \mathbf{D} = \nabla . (\varepsilon_0 \mathbf{E} + \mathbf{P}) = 0$. For the transverse waves this is an identity and so, for these waves,

$$P = \varepsilon_0 \chi_e^* E + \frac{\varepsilon_0 \chi_i^* \omega_i^2 E}{\omega_i^2 - \omega^2}, \tag{13.59}$$

or

$$\varepsilon(\omega) = 1 + \chi_e^* + \chi_i^* \frac{\omega_i^2}{\omega_i^2 - \omega^2}. \tag{13.60}$$

For longitudinal waves propagating as $\exp j(\omega t - \beta z)$, however, $\nabla . \mathbf{D} = (\partial D_z / \partial z) = -j\beta (\varepsilon_0 E + P)$ and because β, though small on a lattice scale, is finite we must have $\varepsilon_0 E = -P$. Equation (13.58) therefore yields $(1 + \chi_e^*) P = P_i$ and so

$$\varepsilon_0 E = -\frac{P_i}{1 + \chi_e^*}. \tag{13.61}$$

When this is inserted in equation (13.57) we obtain the homogeneous equation

$$\ddot{P}_i + \omega_i^2 \left(1 + \frac{\chi_i^*}{1 + \chi_e^*} \right) P_i = 0. \tag{13.62}$$

Thus the longitudinal phonons are not coupled to the field and also the longitudinal phonon cut-off frequency is raised from ω_i to ω_l where

$$\omega_l^2 = \omega_i^2 \left(\frac{1 + \chi_e^* + \chi_i^*}{1 + \chi_e^*} \right) = \omega_i^2 \frac{\varepsilon(0)}{\varepsilon(\infty)}. \tag{13.63}$$

This is known as the Lyddane–Sachs–Teller relation. Notice that $\varepsilon(\omega) < 0$ and the material is opaque for $\omega_i < \omega < \omega_l$.

If the local field correction is included we have

$$\ddot{P}_i + \omega_i^2 P_i = \varepsilon_0 \omega_i^2 \chi_i^* (E + \frac{\gamma}{\varepsilon_0} P), \tag{13.64}$$

$$P = \varepsilon_0 \chi_e^* (E + \frac{\gamma}{\varepsilon_0} P) + P_i, \tag{13.65}$$

and for longitudinal waves alone $\varepsilon_0 E = -P$. For transverse waves

$$\ddot{P}_i + \omega_\tau^2 P_i = \varepsilon_0 \omega_i^2 \frac{\chi_i^*}{1 - \gamma \chi_e^*} E, \tag{13.66}$$

where the new transverse optical mode limit ω_τ is given by

$$\omega_\tau^2 = \omega_i^2 \frac{1 - \gamma(\chi_e^* + \chi_i^*)}{1 - \gamma \chi_e^*}. \tag{13.67}$$

The optical dielectric constant is

$$\varepsilon(\infty) = 1 + \frac{\chi_e^*}{1 - \gamma \chi_e^*} \tag{13.68}$$

and the static dielectric constant is still given by (13.56), i.e.,

$$\varepsilon(0) = 1+\frac{\chi_e^*+\chi_i^*}{1-\gamma(\chi_e^*+\chi_i^*)}. \tag{13.69}$$

For the longitudinal waves with $\varepsilon_0 E = -P$ we obtain

$$\ddot{P}_i+\omega_l^2 P_i = 0$$

where now, with the local field correction included, we have

$$\omega_l^2 = \omega_i^2 \frac{1+(1-\gamma)(\chi_i^*+\chi_e^*)}{1+(1-\gamma)\chi_e^*}. \tag{13.70}$$

The ratio

$$\frac{\omega_l^2}{\omega_\tau^2} = \frac{[1+(1-\gamma)(\chi_i^*+\chi_e^*)][1-\gamma\chi_e^*]}{[1+(1-\gamma)\chi_e^*][1-\gamma(\chi_i^*+\chi_e^*)]} = \frac{\varepsilon(0)}{\varepsilon(\infty)} \tag{13.71}$$

is, however, still given by the Lyddane–Sachs–Teller relation. This relation is therefore unaffected by the local field correction and, again, the medium is opaque when $\omega_\tau < \omega < \omega_l$. We leave it to the reader to verify that equations (13.67), (13.68), and (13.69) together with $\chi_i^* = q^2N/\varepsilon_0 M\omega^2$ yield

$$\varepsilon(\omega) = \varepsilon(\infty)+(\varepsilon(0)-\varepsilon(\infty))\frac{\omega_\tau^2}{\omega_\tau^2-\omega^2}. \tag{13.72}$$

The local field correction decreases both ω_l and ω_τ but the effect is most marked on ω_τ. Indeed if $\gamma(\chi_e^*+\chi_i^*) \to 1$ we see from (13.66) that $\omega_\tau^2 \to 0$ while ω_l^2 remains finite and of the same order as ω_i^2. In this case not only does $\omega_\tau^2 \to 0$ but also $\varepsilon(0) \to \infty$. Thus the interaction between electronic and ionic polarizability may make the unpolarized lattice with $P = 0$ unstable. The medium will then distort to form a ferroelectric structure. If for some hypothetical ionic lattice $\omega_\tau \to 0$ and $\varepsilon(0) \to \infty$ then this lattice is unstable and the real crystal lattice is distorted, and quite probably corresponds to a non-zero value of P. In this distorted lattice $\omega_\tau \neq 0$, and the low frequency susceptibility and dielectric constant, though large, are not infinite. In addition, though the medium may display rather high dielectric loss at radio frequencies, the real part of ε is not negative and the reflection at the surface is not total.

A change in temperature not only changes the lattice spacing but as a consequence may also change both γ and the intrinsic polarizabilities, thus it is quite possible for a particular medium to exist in a spontaneously polarized form at one temperature and a normal form at another temperature. Also, if the lattice distortion in the polarized form is small, a strong enough applied field may be able to switch the lattice to a new configuration and the relation between P and E will display hysteresis. We have therefore a qualitative interpretation of the behaviour of ferroelectric media. It should not be taken too seriously for the assumption that the interaction between ionic and electronic polarization is described by the local field implies that only dipolar interactions between neighbouring ions are significant. In solids, with ions in relatively close proximity to each other, this is unlikely to be a very good approximation.

In an ionic crystal both the electronic and ionic polarizations are produced by the same, or approximately the same, local field. It follows that the ratio of the electronic to ionic contributions to P is

$$\frac{P_e}{P_i} = \frac{\chi_e^*}{\chi_i^*} \tag{13.73}$$

and is independent of the local field correction. Thus if in some medium χ_e obtained from the optical refractive index is say 2 and the d.c. susceptibility is say 100 it does not follow that 98% of the d.c. polarization is ionic. It is much more likely that χ_e^* and χ_i^* are comparable and the two contributions of much the same magnitude. The d.c. value χ_0 is large because $\gamma(\chi_e^* + \chi_i^*)$ is near unity.

Non-linear Effects

We have so far only considered anharmonicity as a potential dissipative mechanism but clearly, if we drive a medium with a large enough electric field, the anharmonic terms in both the electronic and ionic restoring forces will lead to a non-linear response. If the applied field is at a high frequency this leads to effects such as second harmonic generation. Provided that we are dealing only with optical frequency fields the non-linear response is associated solely with the electronic polarizability and we begin by considering this case. It is possible, and quite illuminating, to treat the response of an atomic system to a strong electric field in terms of a classical anharmonic oscillator model, and before we look at the quantum mechanical theory, we do this. For simplicity we consider only a one-dimensional analysis in which an atom with a natural resonance frequency ω_0 is subjected to a field $E(t)$ containing two frequencies. We include the anharmonic term in the equation of motion of the electron which is, therefore,

$$\ddot{x} + \omega_0^2 x + g x^2 = \frac{q}{m} E(t), \tag{13.74}$$

and we express $E(t)$ as $E_\beta \exp(j\beta t) + E_\beta^* \exp(-j\beta t) + E \exp(j\gamma t) + E_\gamma^* \exp(-j\gamma t)$ which is real. For small amplitudes the response is almost linear and x contains terms of the form

$$x_\beta = \frac{q/m \,.\, E_\beta \exp(j\beta t)}{\omega_0^2 - \beta^2},$$

and

$$x_\gamma = \frac{q/m \,.\, E_\gamma \exp(j\gamma t)}{\omega_0^2 - \gamma^2}.$$

In the next approximation, in which we insert these linear terms into the anharmonic term in (13.74) we obtain, for example, a term at the sum frequency $\alpha = \beta + \gamma$ varying as $\exp(j\alpha t)$ and of course other terms involving $E_\gamma^* E_\beta^*$ varying as $\exp(-j\alpha t)$. The term in question is clearly

$$x_\alpha = -\frac{2g(q/m)^2 E_\beta E_\gamma \exp(j\alpha t)}{(\omega_0^2 - \alpha^2)(\omega_0^2 - \beta^2)(\omega_0^2 - \gamma^2)}, \tag{13.75}$$

and the corresponding polarization is $P_\alpha = Nqx_\alpha$. Thus, if we were to write $P_\alpha = \varepsilon_0 d_{\alpha\beta\gamma} E_\beta E_\gamma$ with an implied sum over all values of β and γ for which $\beta + \gamma = \alpha$, and to allow β and γ to take negative values with the convention that $E_{-\beta} = E_\beta^*$, we could express the results as

$$d_{\alpha\beta\gamma} = -Ng\frac{q^3}{m^2} \cdot \frac{1}{(\omega_0^2 - \alpha^2)(\omega_0^2 - \beta^2)(\omega_0^2 - \gamma^2)}. \tag{13.76}$$

The linear susceptibility χ_β is given by

$$\chi_\beta = \frac{Nq^2}{\varepsilon_0 m} \cdot \frac{1}{\omega_0^2 - \beta^2}$$

and so we can express (13.75) as

$$d_{\alpha\beta\gamma} = \chi_\alpha \chi_\beta \chi_\gamma \left(\frac{-gm\varepsilon_0^2}{N^2 q^3} \right).$$

The anharmonic term in the potential energy is $\frac{1}{3}gmx^3$ and it is reasonable to assume that, when $x \approx r \approx (1/N)^{\frac{1}{3}}$, the mean atomic separation, this term is comparable with the electrostatic potential energy $q^2/(4\pi\varepsilon_0 r)$, thus $gm \approx \frac{1}{4}(q^2/\varepsilon_0 r^4)$. With this assumption

$$d_{\alpha\beta\gamma} = \chi_\alpha \chi_\beta \chi_\gamma \Delta, \tag{13.77}$$

where

$$\Delta = -\frac{\varepsilon_0 r^2}{4q} \approx E_A^{-1}, \tag{13.78}$$

where E_A is a field of the order of 1 V per Å or 10^{10} V m^{-1}. If we include a local field correction in equation (13.74) this will appear twice in the expression for x_α and there will be a further local field correction in the relation between P_α and Nqx_α (see equation (13.19)). Since each of these corrections appears once in χ, χ_β or χ_γ, equation (13.77) is not altered. Although χ varies appreciably from material to material and χ^3 even more so, e.g., $\chi^3 \approx$ 4000 for tellurium and $\chi^3 \approx 4$ for KH_2PO_4, the term Δ is unlikely to vary by more than a factor 10 and so, apart from symmetry restrictions or geometric effects, the non-linear susceptibility d is mainly determined by χ. Equation (13.77) expresses an empirical rule first discovered by Miller (1964).

A quantum mechanical treatment of the non-linear polarizability, keeping in the frequency dependence, can be given (see, e.g., Bloembergen, 1965) but the results are unwieldy and therefore we consider only the low frequency limit, where the frequencies involved are reasonably well below any electronic resonances or absorption lines. In this case we can derive the polarizability of an atom from the expression for the energy of the ground state in an electric field. The quadratic term in this expression is, to a first approximation,

$$W_2 = -\frac{e^2}{\hbar\omega_0} X_{ij} E_i E_j$$

where the tensor X_{ij} is given in terms of the expectation values of the displacement vector components x_i in the ground state as

$$X_{ij} = \langle (x_i - \langle x_i \rangle)(x_j - \langle x_j \rangle) \rangle.$$

It is the second order semi-invariant moment of the charge distribution. The energy $\hbar\omega_0$ is the average energy difference between the excited states and the ground state. The induced atomic moment is

$$p_i = -\frac{\partial W}{\partial E_i} = \frac{e^2}{\hbar\omega_0}(X_{ij} + X_{ji})E_j$$

and since $X_{ji} = X_{ij}$, the polarizability is

$$\alpha_{ij} = \frac{2e^2}{\varepsilon_0 \hbar\omega_0} X_{ij}.$$

The third order term in W is approximately

$$W_3 = -\frac{e^3}{(\hbar\omega_0)^2} X_{ijk} E_i E_j E_k \tag{13.79}$$

and so

$$p_i = -\frac{\partial W_3}{\partial E_i} = \frac{e^3}{(\hbar\omega_0)^2}(X_{ijk} + X_{jki} + \text{etc.}) E_j E_k,$$

where the third order semi-invariant is

$$X_{ijk} = \langle (x_i - \langle x_i \rangle)(x_i - \langle x_j \rangle)(x_k - \langle x_k \rangle) \rangle. \tag{13.80}$$

Since $X_{ijk} = X_{jki}$, etc., we have for the non-linear polarizability

$$\alpha_{ijk} = \frac{3e^3}{\varepsilon_0 (\hbar\omega_0)^2} X_{ijk}. \tag{13.81}$$

The non-linear polarizability is essentially proportional to the third order distortion of the shape of the electronic ground state.

In a crystal, if there are N^s atoms, ions or groups of non-linear polarizability α^s_{ijk}, the non-linear susceptibility is

$$d_{ijk} = \sum_s N^s \alpha^s_{ijk} \tag{13.82}$$

but, in forming this sum, we have to remember that the axes x_i, x_j and x_k are axes in space, not principal axes for each group. Thus the evaluation of (13.82) is usually a lengthy business. In crystals with a centre of symmetry or a nearly centro-symmetric structure d_{ijk} will be zero or small even if individual groups have large values of α^s_{ijk}. This is, for example, the case in the langbeinite double sulphates such as $K_2Mg_2(SO_4)_3$. The sum will also be small, even if the crystal is markedly acentric, unless at least some of the structural units have large values of α_{ijk}. Only highly acentric lattices containing acentric units yield large values of d_{ijk} and even then, as equation (13.77) indicates, the linear susceptibility χ must also be large. In recent years considerable progress has been made in relating the non-linear coefficient d_{ijk} to molecular structure (see Jeggo and Boyd, 1970), but we shall not pursue this further here.

If the applied field contains, in addition to a field $E^\beta \exp(j\beta t)$, a d.c. or low frequency field E° the response of the medium will involve the ionic polarizability. Thus, in the Pockel's or linear electro-optic effect, where the term in the polarization at the frequency β proportional to $E^\beta E^\circ$ appears as a field dependent term in the refractive index, we cannot ignore ionic effects. The main influence of ionic polarization can, however, be rather simply deduced if we write equation (13.76) in terms of the electronic polarizations P^β_e and P°_e as

$$x_\alpha = -\frac{2g}{N^2 q^2} \frac{P^\beta_e P^\circ_e}{\omega_0^2 - \alpha^2}.$$

The optical term P^β_e is unaltered by ionic effects but P°_e, the d.c. term, can be obtained from (13.73) as $P^\circ_e = P^\circ \chi^*_e / (\chi^*_e + \chi^*_i)$. Thus the non-linear response is proportional to the total static polarization P° and therefore to the total static dielectric susceptibility χ_0. Miller's rule, equation (13.76), is therefore valid in this case if the appropriate total susceptibility χ_0 is used for χ_γ as $\gamma \to 0$, and this is in agreement with experimental results. There is a substantial literature on the subject of non-linear dielectric effects in optics and the book by

Bloembergen (1965) is a useful introduction. The paper by Robinson (1967) deals in a less formal way with their relation to molecular structure and also contains a guide to the notation.

Additivity Rules

Although *ab initio* calculations of the electronic susceptibility of even the simplest media are not very successful, there is at least one semi-empirical rule which is sometimes helpful in estimating this quantity or the refractive index $n = \varepsilon^{\frac{1}{2}}$.

If we have a medium containing N^s units of polarizability α^s the susceptibility corrected for local field effects using the Lorentz formula $E^{\text{loc}} = E + 1/3\varepsilon_0 P$ is

$$n^2 - 1 = \chi = \frac{\sum_s N^s \alpha^s}{1 - \frac{1}{3}\sum\limits_s N_s \alpha_s} \tag{13.83}$$

so that

$$\tfrac{1}{3}\sum_s N^s \alpha^s = \frac{n^2-1}{n^2+2} \tag{13.84}$$

or, in cgs units,

$$\frac{4\pi}{3}\sum_s N^s \alpha^s = \frac{n^2-1}{n^2+2}. \tag{13.85}$$

If we let L be Avogadro's number the quantity $r_s = \frac{1}{3}L\alpha^s$ (or $r_s = (4\pi/3)L\alpha^s$) is known as the molar refractivity of the species s. Thus if the effects are additive as implied by (13.83) and the molar volume occupied by the species s is V_s we have

$$\frac{n^2-1}{n^2+2} = \sum_s \frac{r_s}{V_s}. \tag{13.86}$$

Values of r_s can be deduced for different species, e.g., the alkali and halide ions in alkali–halide crystals and the refractive index of say KCl related to those of NaCl, NaF and KF. This is not very successful with ionic crystals but in organic media if the basic structural units are taken as bonds, e.g., C—H, C—C, C—N, N—H, C=C, etc., a rather small number of basic values of r_s can be used to make surprisingly accurate predictions for complex compounds. For example, with bond refractivities obtained from simpler compounds, e.g. CH_4, the predicted value for CH_3POCl_2 is $r = 24{\cdot}93$ while the value obtained from the measured refractive index is 24·96. There is a very full discussion of the basis and use of this rule by le Fevre (1965). The fact that the rule works so well is probably the strongest experimental evidence for the validity of the Lorentz local field formula on which it is based.

Dielectric Behaviour and Conduction Electrons

If a medium contains N mobile electrons in unit volume the average equation of motion of an electron can be written as

$$\ddot{x} + \frac{\dot{x}}{\tau} = \frac{q}{m}E \tag{13.87}$$

where τ is an effective collision time. At low frequencies when $\omega\tau \ll 1$ the effect of the electrons is primarily described by the conduction current density

$$J = Ne\dot{x} = \frac{Nq^2\tau}{m} E.$$

At higher frequencies we have

$$J = \frac{Nq^2\tau}{m(1+j\omega\tau)} E,$$

and the electrical conductivity $J/E = \sigma(\omega)$ can be expressed as

$$\sigma(\omega) = \frac{\sigma_0}{1+j\omega\tau}. \tag{13.88}$$

It is, however, a matter of choice in macroscopic problems whether we regard the medium as possessing a complex conductivity or whether we integrate (13.87) once again and obtain

$$x = \frac{q\tau/m}{j\omega - \omega^2\tau},$$

and

$$P = -\frac{(Nq^2/m)E}{\omega^2 - j(\omega/\tau)},$$

and regard the medium as having a complex susceptibility

$$\chi = -\frac{Nq^2}{\varepsilon_0 m} \cdot \frac{1}{\omega^2 - (j\omega/\tau)} = -\frac{\omega_p^2}{(\omega^2 - j\omega/\tau)}$$

or

$$\chi = -\frac{\omega_p^2 \omega^2}{\omega^4 + (\omega^2/\tau^2)} - j\frac{\omega\omega_p^2/\tau}{\omega^4 + (\omega^2/\tau^2)}.$$

At very high frequencies, e.g. at infra-red frequencies, in a semiconductor where ω is greater than both ω_p and $1/\tau$ the most significant term is the imaginary part of χ which leads to attenuation.

This is

$$-j\chi'' = -j\frac{\omega_p^2}{\omega^3\tau} = -j\frac{\sigma_0}{\omega\varepsilon_0\omega^2\tau^2}. \tag{13.89}$$

If the normal dielectric constant is ε the propagation constant of a wave is

$$\beta = \omega(\mu_0\varepsilon_0)^{\frac{1}{2}}\left(\varepsilon - j\frac{\sigma_0}{\omega\varepsilon_0\omega^2\tau^2}\right)^{\frac{1}{2}}$$

$$\approx \omega(\mu_0\varepsilon\varepsilon_0)^{\frac{1}{2}} - \tfrac{1}{2}j\frac{\sigma_0(\mu_0/\varepsilon\varepsilon_0)^{\frac{1}{2}}}{\omega^2\tau^2},$$

and the intensity of a wave decays as $\exp(-\alpha z)$ where

$$\alpha = \frac{\sigma_0(\mu_0/\varepsilon\varepsilon_0)^{\frac{1}{2}}}{\omega^2\tau^2}. \tag{13.90}$$

This attenuation, inversely proportional to ω^2, due to mobile carriers exerts a dominant influence on the properties of, for example, silicon in the far infra-red. Similar effects may also be observed in plasmas.

Conclusion

We have now surveyed a selection of the features of dielectric behaviour for which a qualitative interpretation can be given in terms of the microscopic atomic structure of a medium. We have seen that, in the absence of a local field correction or interactions, both electronic and ionic polarizability would contribute terms of the order of unity to χ. The fact that interactions can lead to values of χ as high as 10^4, or even to ferroelectricity, serves to emphasize the complexity of dielectric phenomena and the importance of the local field correction. It also serves to discourage us from placing too great reliance on predictions based on the properties of non-interacting atoms, even if these have been obtained as the result of a flawless and virtuoso theoretical calculation. It is with some relief that in the next chapter we turn to the relative simplicity of magnetic phenomena. Except at very low temperatures magnetic interactions are trivial and utterly negligible in comparison with lattice energies. It is only when electrostatic forces and the Pauli principle conspire to produce exchange forces that we observe phenomena comparable in complexity with those in the simplest dielectrics.

Problems

1. Atomic magnetic moments are of the order of 1 Bohr magneton, 10^{-23} J $\times$ T^{-1} whereas molecular electric moments are of the order of 1 debye, $\frac{1}{3}\times 10^{-29}$ m coulomb. Compare the interaction energy of two magnetic atoms with that of two dipolar molecules when, in both cases, the separation is 3 Å. Express the two energies in the form $k\theta$ where $k = 1{\cdot}38\times 10^{-23}$ J deg^{-1} is Boltzmann's constant.

2. Derive equation (13.4).

3. Show that the field in a spherical cavity in a medium of polarization **P**, where the field **E** is parallel to **P**, is $\mathbf{E}+(1/3\varepsilon_0)\mathbf{P}$.

4. A dipole of moment **p** is placed at the centre of a spherical cavity in a medium of linear susceptibility χ. Show that in addition to the direct dipole field there is a uniform field in the cavity given by $2\chi\mathbf{p}/4\pi\varepsilon_0 a^3(3+2\chi)$where a is the radius of the cavity.

5. Derive equation (13.48), the Langevin–Debye law.

6. The cubic crystal $NaClO_3$ rotates the plane of polarized light equally for any direction of propagation. Obtain the form of its dielectric constitutive relation. What is the point group symmetry of the crystal?

7. The optical attenuation of the best quality glass is reputed to be 20 db km^{-1}. If this is due to an impurity with a broad absorption band in the visible with an oscillator strength near unity, estimate the concentration of the impurity.

8. In NaCl why would you expect the dispersion of the dielectric constant in the visible to be predominantly associated with the Cl^- ions?

9. Show that if a medium has $\varepsilon(\omega) < 0$ it will reflect incident e.m. radiation of frequency $\omega/2\pi$.

Microscopic Aspects of the Dielectric Constitutive Relation

10. The refractive index of NaCl is 1·55 and its static dielectric constant 6. Assuming that the local field correction factor $\gamma = \frac{1}{3}$, find χ_e^* and χ_i^*. How much larger would χ_e^* have to be to lead to ferroelectric instability?

11. Verify equation (13.72).

12. A molecule in the form of a rigid linear dumbbell with a dipole moment parallel to the axis of the dumbbell rotates with kinetic energy W about an axis normal to the dumbbell. An electric field E such that $pE < W$ is applied in a direction normal to the axis of rotation. Describe in classical terms the subsequent motion and show that the time average of the component of the dipole moment parallel to E is negative.

13. A crystal exhibits a strong linear electro-optic effect. Why is it likely to be piezo-electric?

14. A crystal exhibits a linear electro-optic effect and the non-linear susceptibility $\chi^{(2)}$ has a component $\chi_{333}^{(2)}$. What property must the point group of the crystal possess?

CHAPTER 14

Microscopic Aspects of the Magnetic Constitutive Relation

Diamagnetism and Paramagnetism

Although all solids and liquids exhibit appreciable dielectric effects at all frequencies from the ultra-violet down to d.c., few media have correspondingly large magnetic properties. Indeed, with the exception of three phenomena, ferromagnetism, magnetic resonance, and magneto-caloric effects at low temperatures, magnetism is primarily of interest as an experimental diagnostic tool in atomic and, especially, solid state physics. Furthermore, since magnetic interactions play little or no part in determining the structure of matter, and magnetic phenomena can be regarded as an isolated property superimposed on a given structural background, it is possible to give a briefer and somewhat less discursive treatment of magnetism than is the case with dielectrics. We may also add that, because of its importance to the solid-state physicist, magnetism is given a much more thorough coverage in the general literature devoted to electromagnetic properties.

We begin by considering an electron of charge e and mass m bound in a potential V, e.g., the coulomb potential of a nucleus, and acted on by fields $\mathbf{H} = 1/\mu_0 \, \nabla \wedge \mathbf{A}$ and $E = -\dot{\mathbf{A}} - \nabla\phi$. The Hamiltonian energy function (see, e.g., Goldstein, 1959) is

$$\mathscr{H} = \frac{1}{2m}(\mathbf{p}-e\mathbf{A})^2 + e\phi + V, \tag{14.1}$$

where $\mathbf{p}$ is the canonical momentum. If $\mathbf{E}$ and $\mathbf{H}$ are static uniform fields parallel to the z axis and we use a gauge in which $\nabla \cdot \mathbf{A} = 0$ we have $\phi = -Ez$ and $A = \mu_0 H/2\,(-y, x, 0)$ and we can write (14.1) as

$$\mathscr{H} = \frac{p^2}{2m} + V - \frac{e\mu_0}{2m} H l_z - eE_z + \frac{e^2\mu_0^2}{8m} H^2(x^2+y^2), \tag{14.2}$$

where

$$l_z = xp_y - yp_x \tag{14.3}$$

is the z component of the angular momentum $\mathbf{l} = \mathbf{r} \wedge \mathbf{p}$. The quantum mechanical Hamiltonian is obtained by replacing the electronic variables by their equivalent operators, and the ground state energy to second order in E and H obtained from (14.2) is

$$W = W_0 - \frac{e\mu_0}{2m} H\langle l_z\rangle - eE\langle z\rangle - \frac{e^2\mu_0^2}{4\hbar\omega_0 m^2} H^2\langle l_z^2\rangle$$
$$-\frac{e^2}{\hbar\omega_0} E^2\langle z^2\rangle + \frac{e^2\mu_0^2}{8m} H^2\langle x^2+y^2\rangle \tag{14.4}$$

where $\langle l_z\rangle$, $\langle z\rangle$, etc., are ground state expectation values and $\hbar\omega_0$ is an average excited state energy.

Now, whereas it is impossible for $\langle x^2\rangle$, $\langle y^2\rangle$ or $\langle z^2\rangle$ to vanish, atomic ground states frequently have zero angular momentum and $\langle l_z\rangle = \langle l_z^2\rangle = 0$. Further, if the system has a centre of symmetry $\langle z\rangle = 0$ and we are left with

$$W-W_0 = -\frac{e^2}{\hbar\omega_0}E^2\langle z^2\rangle+\frac{e^2\mu_0^2}{8m}H^2\langle x^2+y^2\rangle.$$

For near spherical system, $\langle x^2\rangle = \langle y^2\rangle = \langle z^2\rangle = \frac{1}{3}\langle r^2\rangle$ and so,

$$W-W_0 = -\frac{e^2}{3\hbar\omega_0}E^2\langle r^2\rangle+\frac{e^2\mu_0^2}{12m}H^2\langle r^2\rangle. \tag{14.5}$$

The induced electric moment π and magnetic moment μ are given by

$$\pi = -\frac{\partial W}{\partial E}, \qquad \mu = -\frac{1}{\mu_0}\frac{\partial W}{\partial H}, \tag{14.6}$$

and so the electric and magnetic polarizabilities are

$$\alpha_e = \frac{2}{3}\frac{e^2}{\varepsilon_0\hbar\omega_0}\langle r^2\rangle \tag{14.7a}$$

$$\alpha_m = -\frac{e^2\mu_0}{6m}\langle r^2\rangle. \tag{14.7b}$$

With N atoms in unit volume we have

$$\chi_e = N\alpha_e, \qquad \chi_m = N\alpha_m$$

and the ratio is

$$\frac{\chi_m}{\chi_e} = -\frac{\hbar\omega_0\mu_0\varepsilon_0}{4m} = -\frac{\hbar\omega_0}{4mc^2}. \tag{14.8}$$

Since $\chi_e \approx 1$ in most solids and $\hbar\omega_0 \approx 10$ eV while $4mc^2 \sim 2\times10^6$ eV, χ_m is of the order of -5×10^{-6}, which is negligible in its macroscopic effects however exciting it may be to the solid-state physicist. Substantial magnetic effects can, therefore, only arise in systems of atoms, ions or molecules which have ground states of non-zero angular momentum. In this case the dominant magnetic term in $W-W_0$ is $-e\mu_0/2m\, H\langle l_z\rangle$. The angular momentum is quantized in units of $\hbar$ and so a natural unit for atomic moments is the Bohr magneton

$$\beta = \frac{e\hbar}{2m} = 9{\cdot}27\times10^{-24} \approx 10^{-23}\text{ joule tesla}^{-1}. \tag{14.9}$$

In many cases an atomic system has a ground state of definite angular momentum $J\hbar$ and degeneracy $(2J+1)$ and no further excited states within an energy comparable with either the thermal energy kT or any practicable magnetic energy $\beta\mu_0H$. It is then permissible, for magnetic discussions, to neglect all the terms in the Hamiltonian $\mathscr{H}$ except

$$\mathscr{H} = -\frac{e\hbar\mu_0}{2m}\mathbf{H}\,.\,\mathbf{J} = -\mu_0\beta\mathbf{H}\,.\,\mathbf{J}. \tag{14.10}$$

We can include the effects of spin magnetism by allowing J to take half integral values and introducing a dimensionless g factor of the order of unity so that

$$\mathscr{H} = -g\beta\mu_0\mathbf{H}\,.\,\mathbf{J}. \tag{14.11}$$

We will not repeat the familiar argument that now leads to Curie's law

$$\chi = \frac{C}{T} = \frac{\mu_0 g^2 J(J+1)\beta^2 N}{3kT}. \tag{14.12}$$

If we take $N \approx 10^{29}\ \mathrm{m}^{-3}$, the typical density of atoms in solids, we obtain a Curie constant C near unity so that $\chi \approx 1/T$. However, in most paramagnetic media only a few atoms in the formula unit, e.g., the Ce^{+++} ions in $Ce_2Mg_3(NO_3)_{12}{\cdot}24H_2O$, are magnetic and N is more often of the order of $10^{27}\ \mathrm{m}^{-3}$, giving $\chi \approx 10^{-2}/T$. Thus, except at low temperatures, the positive paramagnetic susceptibility is, like the negative temperature-independent diamagnetic susceptibility, too small to qualify as a gross macroscopic property. Except for resonance phenomena, which we shall discuss shortly, the only significant macroscopic effects associated with paramagnetism are those in which it is used in a thermodynamic cycle to produce low temperatures.

In a solid, torques and couples act on the atoms as the result of electrostatic forces. The magnetic moment and angular momentum are now no longer free, and new terms appear in the Hamiltonian to describe these torques. The form of these terms is restricted by the symmetry of the atomic site and the requirement that $\mathscr{H}$ be time-reversal invariant. Since the couples are already, because of their electrostatic origin, time-reversal invariant, whereas $\mathbf{J}$ being an angular momentum variable is not, the new terms must contain only even powers of $\mathbf{J}$. Further, since for a given $\mathbf{J}$ there are only $(2J+1)^2-1$ independent combinations of the components of $\mathbf{J}$, the number of new terms is often quite small. If, for example, $J = \frac{1}{2}$ there are no new terms. A typical example of an effective Hamiltonian is

$$\mathscr{H} = g\beta \mathbf{H} \,.\, \mathbf{J} + J_i D_{ij} J_j$$

where the principal axes of D_{ij} are related to the crystal axes.

In many cases these additional terms split the $2J+1$ states into well-separated groups of levels and very often, in the important case when J is half-integral, there is an isolated doublet ground state with all other levels at much higher energies. In this case it is often advantageous to consider this doublet as arising from a system with a fictitious effective angular moment $\hbar S = \frac{1}{2}\hbar$. If we neglect hyperfine structure the most general possible effective spin Hamiltonian is

$$\mathscr{H} = \mu_0 \beta H_i g_{ij} S_j \tag{14.13}$$

where the principal axes of the g tensor are related to crystal axes. (This is, of course, only an adequate Hamiltonian if $\mu_0 g\beta H$ is small compared with the energy of the next group of levels.) Especially in the rare-earth salts the g tensor can be extremely anisotropic; for example, it is possible to have a set of principal (diagonal) values $g_{11} = g_{22} = 0$, $g_{33} = 15$. Since

$$\chi_{ii} = \mu_0 g_{ii}^2 s(s+1) N \frac{\beta^2}{3kT} \tag{14.14}$$

the susceptibility, though not large, can be exceedingly anisotropic. This property is made use of in a number of cryogenic applications.

A local field correction with

$$H^{\mathrm{loc}} = H + \gamma M$$

leads to a corrected susceptibility of the form

$$\chi = \frac{C/T}{1-\gamma C/T},$$

which we can express in the form of the Curie–Weiss law

$$\chi = \frac{C}{T-\theta} \tag{14.15}$$

with a Curie–Weiss temperature

$$\theta = \gamma C. \tag{14.16}$$

Clearly the behaviour of χ is singular when $T \to \theta$ but this is not the explanation of ferromagnetism since, with $\gamma \approx \frac{1}{3}$ and $C \sim 1°$, the critical temperature is below 1 K whereas many ferromagnetic media have Curie points above 300 K. In any case the Lorentz local field treatment is inappropriate in this case and the Onsager, or a more sophisticated approach, does not lead to a singularity in χ.

The Frequency Dependence of Paramagnetism

We now turn to the dynamic behaviour of paramagnetism and its frequency dependence. The equation of motion of a single isolated paramagnetic atom of angular momentum or spin $\mathbf{J}\hbar$ and magnetic moment $\boldsymbol{\mu} = g\beta\mathbf{J}$ which must (because of the Wigner–Eckart theorem) be parallel to $\mathbf{J}$ in a magnetic field $\mathbf{H}$ is

$$\dot{\mathbf{J}} = \mu_0 g\beta \mathbf{J} \wedge \mathbf{H}, \tag{14.17}$$

or

$$\dot{\boldsymbol{\mu}} = \frac{\mu_0 g\beta}{\hbar} \boldsymbol{\mu} \wedge \mathbf{H}. \tag{14.18}$$

Thus if we have N non-interacting spins in unit volume so that $\mathbf{M} = N\boldsymbol{\mu}$ we have

$$\dot{\mathbf{M}} = \frac{\mu_0 g\beta}{\hbar} \mathbf{M} \wedge \mathbf{H}. \tag{14.19}$$

Suppose that the field $\mathbf{H}$ is parallel to the z axis and is raised from zero to its final value. According to (14.19) $\dot{M}_z = 0$ and so no net magnetization results. This is quite different from any of the cases of dielectric polarization. In paramagnetism, a net magnetization can only result if either the spins interact one with another or with an external lattice. Interactions thus play a crucial role in the approach of $\mathbf{M}$ to its equilibrium value $\chi\mathbf{H}$.

We consider first an isolated set of interacting spins. In zero field the interactions determine a set of energy levels for the system as a whole and the occupation of these levels is determined by statistical thermodynamics. If we apply a field slowly the levels change in energy and their occupation numbers also change. The magnetization of the system then follows the slow increase of the field. In the language of thermodynamics this is a reversible isentropic change and the energy and temperature of the system change. If, however, we change the field so rapidly that the interactions are not strong enough, either to maintain the coupling of the spin system as a whole, or to ensure that the occupation of the levels follows the changing field and level structure, each spin will eventually, at increasing rates of change of the field, behave independently and no net magnetization will occur. The details of the response at intermediate rates will depend on the nature of the interactions and may be either a damped resonant response or a relaxation type of response. Thus, in a field $H(t) = \mathscr{R}[H \exp(j\omega t)]$ we expect either

$$\chi(\omega) = \frac{\chi(0)}{1 - \frac{\omega^2}{\omega_0^2} + j\omega\tau_0}, \tag{14.20}$$

or

$$\chi(\omega) = \frac{\chi(0)}{1 + j\omega\tau}. \tag{14.21}$$

The values of the characteristic frequencies ω_0 or $1/\tau$ will depend on the strength of the interactions. In simple paramagnetic media they lie in the radio or low microwave frequency part of the spectrum, i.e., below about 10^{10} Hz, and paramagnetism is an ineffective mechanism in the infra-red and visible regions of the spectrum. In ferromagnetic and antiferromagnetic media with Curie or Neel temperatures θ, we expect $\hbar\omega_0$ to be of the order of $k\theta$ and thus ω_0 should lie in the infra-red. This expectation is only partly fulfilled as, in these media, more complex phenomena associated with long-range magnetic order are liable to intervene at lower frequencies.

If the spin system is coupled to a lattice, or non-magnetic heat sink, there will be energy interchange between the two systems. In paramagnetic media (though not necessarily in exchange coupled media) this interchange will be characterized by a second relaxation time, usually referred to as the spin–lattice relaxation time τ_1 to distinguish it from our earlier spin–spin relaxation time, which we now denote by τ_2. If χ_T is the isothermal static susceptibility and χ_S is the adiabatic or isentropic susceptibility we have

$$\chi(\omega) = \frac{\chi_S}{1 + j\omega\tau_2} + \frac{\chi_T - \chi_S}{1 + j\omega\tau_1}. \tag{14.22}$$

Since in many paramagnetic media $\tau_2 < 10^{-8}$ s, while $\tau_1 > 10^{-3}$ s, we can obtain τ_T from low frequency or static measurements and τ_S from measurements at, say, 10^6 Hz. We shall consider this topic in more detail in chapter 15.

Whatever the mechanism involved in the approach of the magnetization to its equilibrium value the shortest characteristic time is certainly longer than 10^{-13} s, and so, at optical frequencies, the only remaining magnetic effect is a weak diamagnetism, so weak indeed that it is not only swamped by dielectric effects but cannot readily be separated from electric quadrupole effects.

Magnetic Resonance

In a static magnetic field H_0 parallel to the z axis the equations of motion of the magnetization can be written as

$$\dot{M}_x + \frac{M_x}{\tau_2} = \frac{\mu_0 g\beta}{\hbar} H_0 M_y \tag{14.23a}$$

$$\dot{M}_y + \frac{M_y}{\tau_2} = -\frac{\mu_0 g\beta}{\hbar} H_0 M_x \tag{14.23b}$$

$$\dot{M}_z + \frac{M_z}{\tau_1} = \frac{\bar{M}_z}{\tau_1}. \tag{14.23c}$$

In these equations, due to Bloch, τ_2 is the transverse or spin–spin relaxation time and τ_1

the longitudinal or spin–lattice relaxation time. The transverse time τ_2 enters into the equations for $\dot{M}_x$ and $\dot{M}_y$ because changes in M_x and M_y do not involve energy leaving or entering the spin system. *Per contra* changes in M_z are accompanied by energy exchange with the lattice and so involve τ_1. The equilibrium solution of equations (14.22), neglecting relaxation, corresponds to a precession of **M** about the direction of **H** at an angular rate $\omega_0 = \mu_0 g\beta H_0/\hbar$. If $\mu_0 H_0 = 1$ tesla this rate corresponds to a frequency of about 3×10^{10} Hz, that is, in the microwave region.

If in addition to H_0 we apply a small perpendicular oscillatory field $H_1(t) = \mathscr{R}[H_1 \exp(j\omega t)]$ along the x axis, the approximate solution of the equations of motion (see Abragam, 1961) is

$$M_x = \mathscr{R}\frac{\omega_0\gamma H_1 \exp(j\omega t)}{\omega_0^2-\omega^2+j\dfrac{\omega}{\tau_2}}M_z, \tag{14.24a}$$

$$M_y = \mathscr{R}\frac{j\omega\gamma H_1 \exp(j\omega t)}{\omega_0^2-\omega^2+j\dfrac{\omega}{\tau_2}}M_z, \tag{14.24b}$$

and

$$M_z \approx \bar{M}_z = \chi(0)H_0, \tag{14.24c}$$

where $\gamma = g\beta\mu_0/\hbar$.

At resonance when $\omega = \omega_0$, we obtain

$$M_x = \mathscr{R}\{-j\gamma H_1\tau_2 M_z \exp(j\omega t)\},$$

$$M_y = \mathscr{R}\{\gamma H_1\tau_2 M_z \exp(j\omega t)\},$$

and so $\chi_{xx} = -j\gamma M_z\tau_2$ is pure imaginary, corresponding to resonant absorption, and χ_{xy} leads to a transverse magnetization.

The relaxation time τ_2 in the presence of a strong field H_0 is often quite long and, if we write the resonant susceptibilities as

$$-j\chi''_{xx} = -j\chi(0)\,\omega_0\tau_2,$$

$$\chi_{xy} = \chi(0)\,\omega_0\tau_2,$$

we see that, even if $\chi(0)$ is small, χ''_{xx} and χ_{xy} can be large. Thus if $\chi(0) \approx 10^{-4}$, $\omega_0 \approx 2\times10^{11}$ sec^{-1} and $\tau_2 \approx 10^{-7}$ s we have $\chi''_{xx} = \chi_{xy} \sim 2$. Near a resonance paramagnetic effects are large enough to be of macroscopic significance. A number of microwave devices, e.g., masers, isolators and gyrators, utilize magnetic resonance phenomena in paramagnetic (or ferrimagnetic) media. The value of χ_{xy}, even well away from resonance, can be quite appreciable, though imaginary, and this leads to strong gyratory effects. In some media the rotation of the plane of polarization can exceed 1 radian per wavelength. This may be compared with the much smaller effects associated with optical activity and even the Faraday effect.

Ferromagnetism

We now turn to the magnetic phenomena of most significance in macroscopic electromagnetism, that is ferromagnetism and ferrimagnetism. In media where there is appreciable overlap of the electronic wave functions associated with neighbouring atoms, electrostatic forces may favour either anti-symmetric or symmetric combinations of the individual wave functions. Since the overall state function, including spin, must be anti-symmetric this influences the relative spin orientations of neighbouring atoms. Electrostatic interaction energies range from perhaps 0·1 to 10 eV per atom and so it is not surprising that this exchange interaction can compete with thermal disorder at temperatures up to 1000 K where $kT \approx 0{\cdot}1$ eV. In most cases the favoured arrangement is anti-ferromagnetic with alternate spins anti-parallel but in a few cases, especially in metals, the lowest energy is associated with a ferromagnetic array of parallel spins and moments. If there are N spins per unit volume of moment β this leads to a saturation magnetization $M = N\beta$. In iron $N \approx 10^{29}\ \text{m}^{-3}$, $N\beta \approx 10^{6}$ joule tesla^{-1} m^{-3} and $\mu_0 N\beta \sim 1$ tesla. The experimental value is about 2 tesla. In some media, notably the ferrites such as $Fe_3O_4 = FeO{\cdot}Fe_2O_3$ the interaction between one group of ions Fe^{2+} and the other group Fe^{3+} is strongly antiferromagnetic. Because there are twice as many ions of the second type there is a net magnetization whose saturation value is about $\frac{1}{3}N\beta$. This type of medium is called a ferrimagnet.

In ferromagnets and ferrimagnets the strong exchange interaction tends to produce an ordered state with a large net magnetization. The direction of the magnetization is fixed in the crystal lattice by the anisotropy of the lattice. As a result there is a large stored energy of the order of $\mu_0 M^2$, per unit volume of material, associated with the field. The system can therefore reduce its total energy by breaking up into a number of oppositely magnetized domains. Roughly speaking if a body of volume V breaks up into n domains the field energy is reduced by a factor $1/n$, but of course at the same time there is an increase in energy due to magnetic disorder at the domain boundaries. The balance between these two effects determines the stable domain size. The relation is not simple and depends on both the shape of the body and the crystalline anisotropy which favours directions certain of magnetization. Further if the body is an aggregate of microcrystals each microcrystal will often be a single domain.

In a field two processes contribute to the appearance of a net magnetization, either the orientation of single domains changes or favourably oriented domains grow by domain wall movement at the expense of unfavourably oriented domains. The first process is resisted by crystalline anisotropy and so only occurs in relatively high fields. In an unstrained single crystal of suitable geometry there is essentially no barrier to domain wall movement and this process is effective in low fields. In favourable cases initial susceptibilities as large as 10^6 may be observed. This is, however, exceptional and the presence of strain will result in much lower values. In aggregates, of course, the boundaries of the microcrystals hinder domain growth. If the grain of the structure is exceedingly small domain wall movement is almost entirely inhibited and only domain reversal is possible. In this case the medium displays a low initial susceptibility, considerable hysteresis and a high coercive force.

There are also resonance effects associated with ferromagnetic and ferrimagnetic media. In ferromagnetic metals these are of little technical interest because electromagnetic fields can barely penetrate the medium but in ferrimagnetic media, many of which are good insulators, the effects are large and can be used to construct a variety of important microwave devices. An insulating ferrimagnetic material is, of course, also useful as a core for high frequency transformers.

The Difference between Dielectric and Magnetic Phenomena

We conclude this chapter by considering the differences between dielectric and magnetic phenomena. First of all we have the obvious fact that, except in ferromagnetic and ferrimagnetic media and in resonance phenomena, magnetic effects are much smaller than dielectric effects, and secondly that magnetism is essentially a low frequency phenomenon. Further, the dynamical processes involved in magnetization are quite different from those involved in dielectric polarization and this, combined with the relative freedom of motion of atomic magnetic moments, leads to resonance phenomena with no dielectric counterpart. A difference of another kind arises from the fact that electrostatic interactions in a solid are comparable with the interactions which determine the structure whereas magnetic interactions are by comparison trivial. Thus, if we magnetize a medium to saturation, there is usually no visible or audible effect on the medium, whereas if we subject a medium to an electric field which is strong enough to produce an appreciable displacement of the microscopic charges, the result is usually catastrophic. As a result we are frequently concerned with magnetic media whose response is saturated and non-linear, whereas non-linearity in a dielectric medium is something of a curiosity. Finally, dielectric phenomena are easily obscured by the effects of mobile charge whereas magnetic effects are not similarly affected, thus whereas a polarized dielectric soon either loses its polarization or has it neutralized by extraneous surface charge, a permanent magnet can retain its magnetization indefinitely. We may also remark that magnetism is essentially a quantum mechanical effect. In any completely classical theory the overall magnetic behaviour of atoms gives vanishing effects. For a full discussion of this topic the reader is referred to van Vleck (1932) which, despite its age, is still the most generally useful text.

Problems

1. In equation (14.1) set $V = \phi = 0$ and $\nabla . \mathbf{A} = \partial A/\partial t = 0$. Verify that it leads to the Lorentz equation of motion.

2. Derive equation (14.2).

3. Use second order perturbation theory to obtain equation (14.4).

4. The ionization potential of neon is 21·5 eV and the refractive index of the gas at s.t.p. is 1·00007. What is the diamagnetic susceptibility of the gas?

5. Derive Curie's law, equation (14.12). Show that, if χ_T is defined as $(\partial M/\partial H)_T$, it tends to zero in large fields. How large must the field be in a typical paramagnetic medium at 1 K?

6. The paramagnetic susceptibility of an anisotropic crystal is C/T measured along the axis of symmetry and zero in a plane normal to this axis. What is the susceptibility of a powdered sample?

7. Chrome alum is a cubic crystal. Neglecting local field corrections the susceptibility is C/T. Show that, using the Lorentz local field correction, the magnetic moment of a spherical sample of volume V in a weak field H is CHV/T. Is this result modified if one uses the Onsager local field theory?

8. Estimate the spin–spin relaxation time for dipoles of moment 1 Bohr magneton arranged on a cubic lattice of spacing 10 Å.

9. A cavity resonator of volume V has a quality factor Q_0. A small crystal of paramagnetic material of volume v and imaginary susceptibility $-j\chi''$ is placed in the cavity. What is the approximate change in the quality factor? Take as an example $V = 10\ \text{cm}^3$, $v = 1\ \text{mm}^3$, $Q_0 = 10^4$, $\chi'' = 10^{-2}$.

10. A small sphere of iron consists of a single domain. Estimate the stored energy in the external magnetic field as a function of the radius of the sphere.

CHAPTER 15

Thermodynamics

Introduction

There are a number of reasons why the thermodynamic aspects of electromagnetic phenomena are important. First of all some important topics, the production of low temperatures by adiabatic demagnetization is an obvious example, clearly call for a thermodynamic treatment. Secondly, the generality of thermodynamic results, which are independent of any particular model or microscopic theory, means that they can be used as a check on the more fanciful speculations of theorists and the more hopeful assumptions of inventors. Thirdly, thermodynamics often yields results which are difficult to obtain from a microscopic theory or relates parameters of technical interest to others which are easier to measure. Fourthly, the machinery of statistical mechanics which connects microscopic theory with macroscopic phenomena is based on thermodynamic concepts. Finally, the whole concept of an equilibrium state of a macroscopic system is based on thermodynamic ideas.

We begin with a brief summary of the classical thermodynamics of macroscopic systems. Thermodynamics is based on the recognition that energy can exist in two forms: coherent energy freely convertible to mechanical work and incoherent energy or heat. Whereas work can always be completely converted to heat, heat cannot in general be completely converted to work. A key concept is the notion of an equilibrium state of a macroscopic system: this is a state which persists in macroscopically unchanging form when all the external influences are kept constant. Amongst these external influences we may distinguish parameters such as pressure, volume, stress, or electric field. In addition to these parameters we find experimentally that if the system can exchange heat with its surroundings there is an additional purely thermal parameter, the temperature T, which must also be held constant. The internal constitution of a macroscopic system determines an equation of state which relates the possible values of the external parameters which correspond to an equilibrium state of the system. For example, for a given number of molecules of helium there is a relation between the volume V and the pressure p and temperature T. Thus if, for given values of p and T, the volume does not have the value determined by the equation of state, the system is not in equilibrium and V will change spontaneously until eventually it reaches the appropriate value.

The first law of thermodynamics can be reduced to the statement that to every recognizable and repeatable macroscopic state of a system, whether it is an equilibrium state or not, there corresponds a total energy function U with a definite value. In any change between recognizable macroscopic states the change δU in U is related to the work δW done on the system and the heat δQ added to the system by

$$\delta U = \delta W + \delta Q. \tag{15.1}$$

The recond law is only concerned with equilibrium states. Every equilibrium state of a system not only has a definite value of the total energy U but also of a second function of state, the entropy S. In a change between two equilibrium states at the same temperature T the heat taken in δQ satisfies

$$\delta Q \leqq T\,\delta S. \tag{15.2}$$

If the change proceeds in the opposite direction the heat taken in is $\delta Q' \leqq -T\,\delta S$. If the change is reversible, with no permanent change in the surroundings, we must have $\delta Q + \delta Q' = 0$ and this is only possible if $\delta Q = T\,\delta S$. Thus the equality sign in (15.2) applies to reversible changes.

The functions U and S both have definite values in any equilibrium state and so does the temperature. The function

$$F = U - TS \tag{15.3}$$

is therefore also a function of state. In a change between two equilibrium states

$$\delta F = \delta U - T\,\delta S - S\,\delta T$$

and, if we use (15.1) and (15.2), we obtain

$$\delta F \leqq -S\,\delta T + \delta W. \tag{15.4}$$

In a change at constant temperature with $\delta T = 0$ we have $\delta F \leqq \delta W$ or $-\delta W \leqq -\delta F$. Thus the maximum work that can be extracted from the system in an isothermal change is $-\delta F$. For this reason F is known as the free energy, or sometimes as the Helmholtz free energy to distinguish it from the Gibbs free energy $G = U - TS + pV$.

If a system is maintained at constant temperature and constrained in such a way, e.g., by holding the volume constant, that it can do no external work then any possible change satisfies $\delta F \leqq 0$. If a system is constrained in this way and is in equilibrium, and there is another equilibrium state compatible with the same constraints, then the system can only change to this other state if it has an equal or lower free energy. If the second state has a lower free energy the change is irreversible and the original state was not an equilibrium state contrary to our assumption. We conclude that if a system is maintained at constant temperature and constrained so that it cannot exchange work with its surroundings then F is a minimum for the equilibrium state. If α is an internal macroscopic parameter which can change without external work being performed then, in the equilibrium state,

$$\frac{\partial F}{\partial \alpha} = 0. \tag{15.5}$$

If a system is taken by a succession of reversible changes from a state (1) to a state (2) we have

$$S(2) = S(1) + \int_1^2 \frac{dQ}{T} \tag{15.6}$$

and

$$F(2) = F(1) - \int_1^2 S\,dT + \int_1^2 dW, \tag{15.7}$$

and we can use these relations to calculate or measure changes in S and F, or to relate S and F in an arbitrary state to their values in some standard state.

Because F is a function of state the differentials, e.g., $\partial F/\partial T$, of F with respect to variables such as T which define the state have a unique and definite meaning. For example, if we are considering a fluid for which $\delta W = -p\delta V$ then, in a change between two states defined by the values of T and V in the two states, we have

$$\delta F = -S\,\delta T - p\,\delta V. \tag{15.8}$$

The equality sign holds for any infinitesimal change between equilibrium states. We therefore have

$$S = -\left(\frac{\partial F}{\partial T}\right)_V \tag{15.9a}$$

$$p = -\left(\frac{\partial F}{\partial V}\right)_T \tag{15.9b}$$

and also

$$\left(\frac{\partial S}{\partial V}\right)_T = -\frac{\partial^2 F}{\partial T\,\partial V} = \left(\frac{\partial p}{\partial T}\right)_V. \tag{15.10}$$

Equation (15.10) is known as a Maxwell relation. It yields an experimentally rather inaccessible quantity $(\partial S/\partial V)_T$, of considerable technical interest to refrigerator designers, in terms of an easily measurable quantity $(\partial p/\partial T)_V$. We also have

$$U = F + TS = F - T\left(\frac{\partial F}{\partial T}\right)_V, \tag{15.11}$$

thus if F is known as function of V and T we can obtain S, p and U directly from F and the whole of the thermodynamic properties of the fluid are known. This is essentially the basis of a statistical mechanical treatment of fluids. The ultimate aim of the treatment is to obtain $F(V, t)$ from the microscopic or atomic structure of the fluid.

We note in this connection the crucial role played by the expression $-p\,\delta V$ for the work δW. Before we can make any extended use of thermodynamics we have to be able to express δW in terms of variables which have definite values in each equilibrium state. It is exactly at this point that electromagnetic thermodynamics parts company with fluid thermodynamics. The functions U, S and F have the same significance in every branch of thermodynamics; the appropriate expression for δW, however, varies from topic to topic. The central problem in electromagnetism is to find convenient forms for δW.

Electromagnetic Work

External agencies can do work on a system contained within a closed surface S either by the action of fields across the surface or by the action of mechanical or other non-electromagnetic forces across the surface. We are only concerned with the electromagnetic term, and electromagnetic thermodynamics is based on the expression

$$\frac{\mathrm{d}W}{\mathrm{d}t} = -\oint(\mathbf{E}\wedge\mathbf{H})\,.\,\mathrm{d}\mathbf{S} \tag{15.12}$$

for the rate at which external agencies do work on the system. We can, however, also express this in terms of conditions within the volume V of the system as

$$\frac{\mathrm{d}W}{\mathrm{d}T} = \int(\mathbf{E}\,.\,\mathbf{J} + \mathbf{E}\,.\,\dot{\mathbf{D}} + \mathbf{H}\,.\,\dot{\mathbf{B}})\,\mathrm{d}V. \tag{15.13}$$

If it were not for the term $\mathbf{E}\,.\,\mathbf{J}$ we could integrate this over a finite interval of time δt and obtain δW immediately. Indeed for systems in which there are no mobile currents and $\mathbf{J} = 0$ we have

$$\delta W = \int(\mathbf{E} \,.\, \delta\mathbf{D}+\mathbf{H} \,.\, \delta\mathbf{B})\, \mathrm{d}V. \tag{15.14}$$

We now consider the term $\mathbf{E} \,.\, \mathbf{J}$ separately. We may distinguish two essentially different cases. The first case occurs when there are superconductors in the system so that an equilibrium state with a finite value of $\mathbf{J}$ can occur. In all other cases $\mathbf{J}$ must be zero in an equilibrium state. This is obvious if there are ohmic conductors present for a non-vanishing value of $\mathbf{J}$ leads to irreversible heat generation and cannot correspond to an equilibrium state. However, even if $\mathbf{J}$ is associated with the free motion of particles in vacuum or the motion of macroscopic charged bodies, $\mathbf{J}$ must also vanish, for no macroscopic motion of particles or bodies occurs in equilibrium states.

If there are superconductors in the system we can certainly assume that the charge density ρ is zero and in this case we can derive $\mathbf{B}$ from a vector potential $\mathbf{A}$ as $\mathbf{B} = \nabla \wedge \mathbf{A}$ and $\mathbf{E}$ as $-\dot{\mathbf{A}}$ and use the gauge in which $\nabla \,.\, \mathbf{A} = 0$. We then have

$$\int \mathbf{E} \,.\, \mathbf{J}\, \mathrm{d}V = -\int \mathbf{J} \,.\, \dot{\mathbf{A}}\, \mathrm{d}V \tag{15.15}$$

so that the work corresponding to this term in a change between equilibrium states is

$$\delta W_J = -\int \mathbf{J} \,.\, \delta\mathbf{A}\, \mathrm{d}V. \tag{15.16}$$

In this case, although we can omit the term $\mathbf{E} \,.\, \delta\mathbf{D}$, we must retain the magnetic term. However, we also have $\mathbf{H} = \mu_0\mathbf{B}$, so that the thermodynamics of superconductors can be based on the equation

$$\delta W = \int\left(\frac{1}{\mu_0}\mathbf{B} \,.\, \delta\mathbf{B}-\mathbf{J} \,.\, \delta\mathbf{A}\right)\mathrm{d}V. \tag{15.17}$$

We next consider the case when there are no ohmic currents present but charges (mobile charges or charged bodies) are displaced during the change between equilibrium states. If the change is to be reversible the forces on these charges must be in balance during the change. If the only forces are those due to $\mathbf{E}$, then $\mathbf{E}$ must vanish and the term $\mathbf{E} \,.\, \mathbf{J}$ is eliminated. If, however, there are other forces, e.g. internal forces due to elastic constraints, then the time integral of $\int \mathbf{E} \,.\, \mathbf{J}\, \mathrm{d}V$ gives the work done against these forces and the increase δU_M in the internal stored mechanical potential energy. Thus, in this case,

$$\delta W = \int(\mathbf{E} \,.\, \delta\mathbf{D}+\mathbf{H} \,.\, \delta\mathbf{B})\, \mathrm{d}V+\delta U_M. \tag{15.18}$$

We might use this, for example, to discuss a system of charged electrodes on yielding supports. If the forces which balance $\mathbf{E}$ during the change are external, then additional external work is done on the system and this must be added to δW explicitly.

Finally, if ohmic currents flow during the change of state the change will be irreversible. However, if we consider the system shown in Fig. 15.1, and the external condition being changed is the voltage applied at A, which is raised from ϕ_1 to ϕ_2 in a time t, the system certainly has well-defined initial and final equilibrium states and we can calculate their

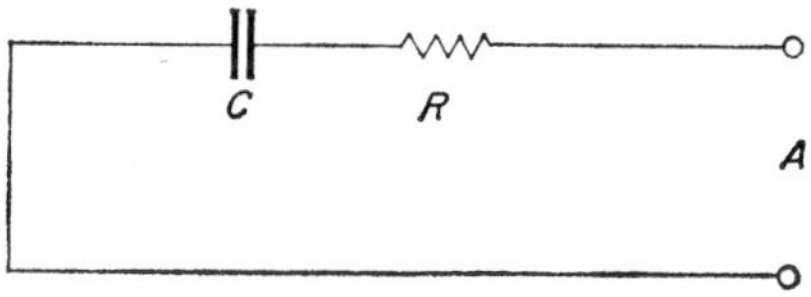

FIG. 15.1. A thermodynamic system.

difference in free energy, or the reversible work done by considering slow changes. Thus the energy dissipated as heat in R is $RC^2(\phi_2-\phi_1)^2t^{-1}$ and this can be as little as we please if t is long enough. This case is not essentially different from the previous case for, if ϕ changes slowly, the difference between ϕ and the potential q/c across the capacitance is negligible and the mobile charges in R are effectively acted on by a vanishingly small electric field.

In all the cases that we have to consider the stored mechanical energy term δU_M is either absent, or can be eliminated by considering first a rigid system and then finally relaxing the mechanical constraint and discussing the mechanical terms separately. For example, if a parallel-plate condenser has an elastic dielectric spacer, we can consider the electromagnetic properties of a system with a rigid dielectric and then separately discuss the mechanical work required to distort the dielectric. Henceforth, we shall omit all mechanical terms.

It is also usually possible to treat dielectric and magnetic phenomena in isolation. Thus, for superconductors we require equation (15.17), for dielectrics

$$\delta W = \int \mathbf{E} \, . \, \delta \mathbf{D} \, \mathrm{d}V, \tag{15.19}$$

and for magnetic media

$$\delta W = \int \mathbf{H} \, . \, \delta \mathbf{B} \, \mathrm{d}V. \tag{15.20}$$

Electric results can often be transposed to magnetic results by the substitutions $\mathbf{E} \to \mathbf{H}$, $\mathbf{D} \to \mathbf{B}$, $\mathbf{P} \to \mu_0\mathbf{M}$, but because of the different way electric and magnetic fields are related to their sources the equations may have a very different physical significance.

In the remaining sections of this chapter our main concern is to convert these three equations, either into useful forms for practical applications or to forms which make contact with the results obtained in statistical mechanics. In the course of this work we shall have frequent, if implicit, recourse to the notions of constant voltage and constant current generators. Because practical voltage and current generators are devices which are usually only in thermodynamic equilibrium when either the batteries are completely discharged, or there is a strike or fuel shortage at the power station, it is important to establish that idealized devices do not themselves violate thermodynamic laws. This will occupy us in the next section. The devices described there, though conceptually simple, are not, however, of much practical use. Our purpose in describing them is solely to show that the notion of a constant voltage or constant current generator in an equilibrium state is *not*, of itself, inconsistent with thermodynamics.

Voltage and Current Generators

Figure 15.2 shows a device in which a long vertical electrode, or plate, A supports a mass m and moves freely within a metal case. The charge on A is q and the capacitance between A and the case is $\gamma(x-x_0)$ where $x-x_0$ is the height of the mass m above some fixed reference point x_0. The reversible work required to raise the mass from x_0 to x with no charge on the plate is $mg(x-x_0)$ and the reversible work required to charge the plate at this position is $\frac{1}{2}(q^2/\gamma(x-x_0))$; thus, neglecting any constant thermal term, the free energy is

$$F(x, q, T) = +mg(x-x_0)+\tfrac{1}{2}\frac{q^2}{\gamma(x-x_0)}.$$

If the plate is freed the parameter x can change without external work being done and so, in the equilibrium state, $\partial F/\partial x = 0$ and this gives

$$x-x_0 = \frac{q}{(2mg\gamma)^{\frac{1}{2}}},$$

and

$$F(q, T) = +\left(\frac{2mg}{\gamma}\right)^{\frac{1}{2}} q.$$

For this system $dW = \phi\, dq$ and so

$$\phi = \frac{\partial F}{\partial q} = \left(\frac{2mg}{\gamma}\right)^{\frac{1}{2}} \tag{15.21}$$

which is constant, and in addition we now have

$$F = +\phi\, q. \tag{15.22}$$

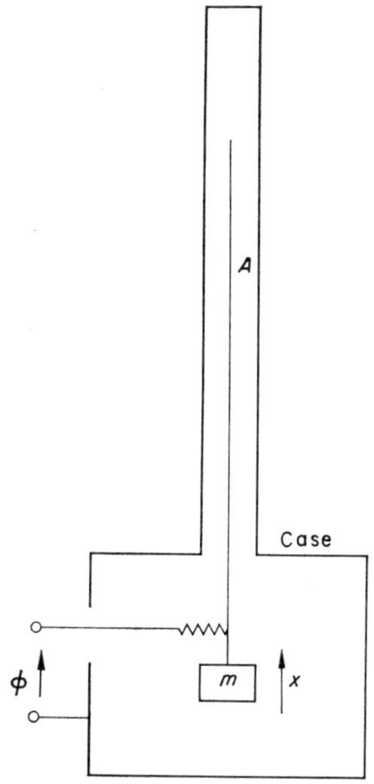

FIG. 15.2. A constant voltage generator.

By regarding this device as the prototype of all constant voltage sources we can avoid a discussion of, for example, chemical batteries. The system shown in Fig. 15.2 has definite equilibrium states and, if we introduce an infinitesimal viscous friction term, it will always come to equilibrium.

A constant current generator with a well-defined thermal equilibrium state can only be constructed from conductors of zero resistance. A superconductor is satisfactory but we note that we only use the property of zero resistance. The Meissner effect is not needed. A possible device connected to a load L is shown in Fig. 15.3. It consists of a vertical superconducting parallel wire line of inductance λ per unit length which supports a freely moving short circuit of mass m. If ψ is the flux linking the device, the work done to increase ψ by an external source connected across AB is $I\,\mathrm{d}\psi$ (see equation (15.16)) and we obtain for the free energy at fixed x

$$F = F_0+mg(x-x_0)+\int_0^{\psi} I\,\mathrm{d}\psi$$

or

$$F = F_0+mg(x-x_0) = \tfrac{1}{2}\frac{\psi^2}{\lambda(x-x_0)}.$$

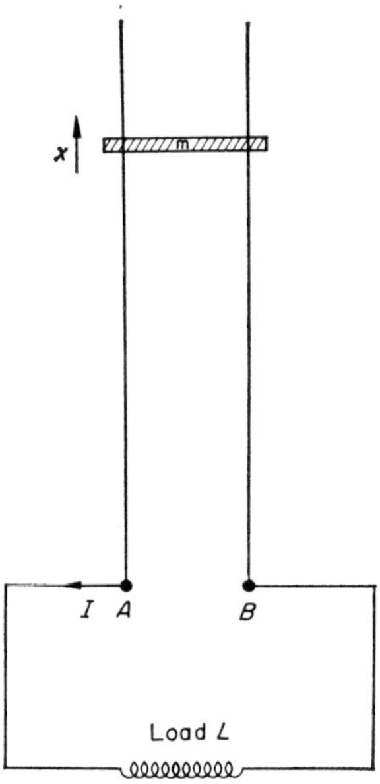

FIG. 15.3. A constant current generator.

If the constraint on x is relaxed $\partial F/\partial x = 0$ yields

$$x-x_0 = \frac{\psi}{(2mg\lambda)^{\frac{1}{2}}}$$

and

$$F = F_0+\left(\frac{2mg}{\lambda}\right)^{\frac{1}{2}}\psi.$$

Thus the current is

$$I = \frac{\partial F}{\partial \psi} = \left(\frac{2mg}{\lambda}\right)^{\frac{1}{2}} \qquad (15.23)$$

and

$$F = F_0+I\psi. \qquad (15.24)$$

The device therefore behaves as a constant current generator with a definite equilibrium state. We can ensure that it reaches equilibrium by including an infinitesimal viscous frictional term in the equation of motion of the sliding short.

Capacitors

Before we begin the formal development in terms of fields it is useful to look briefly at a rather elementary problem which clarifies some of our ideas about thermodynamic functions.

If we have a parallel-plate capacitor with a charge q and a potential difference ϕ, whose plates are held at a separation x by an externally applied force X, then, for a small change in the conditions,

$$dF = \phi\, dq + X\, dx. \qquad (15.25)$$

We can use this to calculate the free energy F in terms of F_0, the free energy at zero charge and separation x_0, as

$$F = F_0 + \tfrac{1}{2}\phi q, \qquad (15.26)$$

by first separating the plates to x which requires no work, since $X = 0$, and then charging the system. If the area of the plates is A we also have

$$q = \frac{\varepsilon_0 A}{x}\,\phi \qquad (15.27)$$

and so we can express F either in terms of q as

$$F = F_0 + \tfrac{1}{2}\frac{x}{\varepsilon_0 A}q^2, \qquad (15.28)$$

or in terms of ϕ as

$$F = F_0 + \tfrac{1}{2}\frac{\varepsilon_0 A}{x}\phi^2. \qquad (15.29)$$

If the plates are isolated so that q is a constant we can obtain the force X required to maintain the separation x from (15.25) as

$$X = \left(\frac{\partial F}{\partial x}\right)_q = \tfrac{1}{2}\frac{q^2}{\varepsilon_0 A}. \qquad (15.30)$$

If, however, the plates are maintained at constant potential by a battery we cannot obtain X from (15.25) for, as x changes, q changes and the free energy of the whole system changes, partly because the free energy of the capacitor changes, and partly because the free energy of the battery changes. We have in fact for the whole system

$$dF_{\text{total}} = \phi\, dq + X\, dx - d(\phi q) \qquad (15.31)$$

and this gives

$$dF_{\text{total}} = X\, dx - q\, d\phi. \qquad (15.32)$$

Now, if F_{ob} is the free energy of the undischarged battery,

$$F_{\text{total}} = F + F_{ob} - \phi q = F_o + F_{ob} - \tfrac{1}{2}\phi q$$

and we can express this in terms of ϕ using (15.29) as

$$F_{\text{total}} = F_o + F_{ob} - \tfrac{1}{2}\frac{\varepsilon_0 A}{x}\phi^2.$$

From (15.32) we have

$$X = \left(\frac{\partial F_{\text{total}}}{\partial x}\right)_\phi = \tfrac{1}{2}\frac{\varepsilon_0 A}{x^2}\phi^2 = \tfrac{1}{2}\frac{q^2}{\varepsilon_0 A}$$

which gives the same force as (15.30). We see that, apart from the constant term F_{ob} which is irrelevant, both F_{total} and dF_{total} can be expressed in terms of parameters X, x, ϕ and q which refer only to the state of the condenser. Thus the function

$$\tilde{F} = F - \phi q \qquad (15.33)$$

which satisfies

$$d\tilde{F} = X\,dx - q\,d\phi \tag{15.34}$$

and

$$\tilde{F} = F_0 - \tfrac{1}{2}q\phi \tag{15.35}$$

is, apart from a constant F_{ob}, the same as F_{total}. It is a perfectly good function of state for the capacitor alone and it is tempting to believe that it is the free energy function for a capacitor and a set of batteries. Thus if, as in Fig. 15.4, we have a capacitor which can be connected to either of two batteries A or B by a switch, we might imagine that the change in

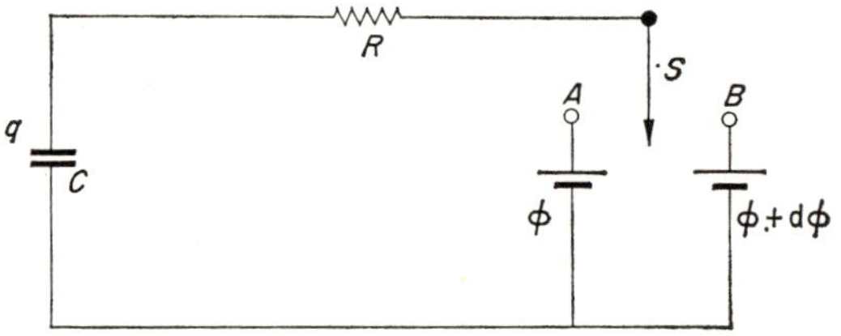

FIG. 15.4. A circuit with two batteries.

the total free energy of the system as S is changed from A to B could be expressed as $d\tilde{F} = -q\,d\phi$. This is not so. With the switch connected to A the total free energy is

$$F_{\text{total}} = F_o + F_{oA} + F_{oB} + \tfrac{1}{2}C\phi^2 - q\phi$$

or

$$F_{\text{total}} = F_o + F_{oA} + F_{oB} - \tfrac{1}{2}C\phi^2.$$

When the switch is changed to B the free energy of the battery A remains constant, the free energy of the condenser rises by $\tfrac{1}{2}C(\phi+d\phi)^2 - \tfrac{1}{2}C\phi^2 = C\phi\,d\phi + \tfrac{1}{2}C(d\phi)^2$ and the free energy of B falls by $-(\phi+d\phi)dq = -(\phi+d\phi)C\,d\phi = -C\phi\,d\phi - C(d\phi)^2$. Thus the total change in the total free energy is $-\tfrac{1}{2}C(d\phi)^2$. This is not equal to $dF = -q\,d\phi$. Indeed it is not even of first order in $d\phi$. For infinitesimal changes it is zero. Thus, although $\tilde{F} = F - q\phi$ is a well-defined function of state for the condenser, it is only equivalent to the total free energy of the condenser and its battery if the same battery is permanently connected to the plates. We can generalize this to a system of n electrodes. The function $\tilde{F} = F - \sum_n \phi_n q_n$, though a well-defined function of state, is only the total free energy of the system and its batteries if the same batteries remain permanently connected to the same electrodes. If we know the functional dependence of $\tilde{F}$ on the ϕ_n then the equation

$$d\tilde{F} = -\sum_n q_n\,d\phi_n + \text{mechanical and thermal terms} \tag{15.36}$$

allows us to calculate the charges as $q_n = -\partial\tilde{F}/\partial\phi_n$ but $d\tilde{F}$ given by (15.36) is not the change in the total free energy when the ϕ_n are changed by switching new batteries into the system.

Dielectric Systems

If we have a system of dielectric or polarized bodies, in a region where the field is $\mathbf{E}(\mathbf{r})$, the fundamental thermodynamic result for the difference between the free energies of two equilibrium states is

$$\delta F = -S\,\delta T + \int \mathbf{E}(\mathbf{r}) \,.\, \delta\mathbf{D}(\mathbf{r})\,dV. \tag{15.37}$$

In general this equation is thoroughly intractable and useless unless we know that **E** and **D** are uniform throughout the system. We shall therefore have to convert (15.37) to a more useful form. If, however, **E** and **D** are uniform and the system is homogeneous it is possible to define f as the free energy per unit volume, σ as the entropy per unit volume and write (15.37) as

$$\delta f = -\sigma\,\delta T + \mathbf{E}\,.\,\delta\mathbf{D}. \tag{15.38}$$

We then have

$$\sigma = -\left(\frac{\partial f}{\partial T}\right)_D$$

and

$$E_i = \left(\frac{\partial f}{\partial D_i}\right)_T$$

and these equations yield the rather pointless Maxwell relation

$$\left(\frac{\partial E_i}{\partial T}\right)_D = -\left(\frac{\partial \sigma}{\partial D_i}\right)_T.$$

The function

$$\tilde{f} = f - \mathbf{E}\,.\,\mathbf{D} \tag{15.39}$$

is a well-defined function of state, though not of course the free energy density, and it satisfies

$$\delta\tilde{f} = -\sigma\,\delta T - \mathbf{D}\,.\,\delta\mathbf{E}. \tag{15.40}$$

It yields the Maxwell relation

$$\left(\frac{\partial D_i}{\partial T}\right)_E = \left(\frac{\partial \sigma}{\partial E_i}\right)_T \tag{15.41}$$

which is, on the face of it, somewhat more useful for, if $D = \varepsilon(T)\varepsilon_0 E$, we have

$$\left(\frac{\partial \sigma}{\partial E}\right)_T = \varepsilon_0 E\left(\frac{\partial \varepsilon(T)}{\partial T}\right)_E \tag{15.42}$$

and, if ε decreases with increasing T, heat is evolved as E is increased.

The function

$$\tilde{f}^{\dagger} = \tilde{f} + \tfrac{1}{2}\varepsilon_0 E^2 \tag{15.43}$$

has the thermodynamic identity

$$\delta\tilde{f}^{\dagger} = -\sigma\,\delta T - \mathbf{P}\,.\,\delta\mathbf{E} \tag{15.44}$$

and yields the Maxwell relation

$$\left(\frac{\partial \sigma}{\partial E_i}\right)_T = \left(\frac{\partial P_i}{\partial T}\right)_E \tag{15.45}$$

which is identical with (15.41) since $\mathbf{D} = \varepsilon_0\mathbf{E} + \mathbf{P}$. For reasons which can only be regarded as metaphysical $\tilde{f}^{\dagger}$ is sometimes referred to as the free energy density of the dielectric medium. Great care has to be exercised in using equations (15.38)–(15.45) not so much because they refer only to homogeneous media in uniform fields as because σ is defined as the entropy per unit *volume*. In the thermodynamics of fluids and elastic media the entropy density nearly always refers to entropy per unit *mass* since the volume is a thermodynamic variable. An excellent and profound treatment of the use of these relations to discuss the thermodynamics of deformable media has been given by Landau and Lifshitz (1960).

Thermodynamics

In practical applications, and in theoretical physics, we are primarily interested in the thermodynamic behaviour of dielectric bodies in an impressed field due, either to a given set of fixed charges, or to a given set of electrodes with fixed charges, or to a given set of electrodes maintained at fixed potentials. It is therefore highly desirable to manipulate equation (15.37) into a form where the electric variable can be expressed in terms of $\delta\mathbf{E}^*$, the change in the applied field. This is not difficult but, with the exception of Landau and Lifshitz (1960), few books present the complete calculation. The treatment we shall give differs in some respects from that given by Landau and Lifshitz and also includes some aspects which they ignore.

First of all, we prove a general result. If $\mathbf{F}$ is a vector which can be expressed as $-\nabla\psi$ and $\mathbf{G}$ is a vector which satisfies $\nabla \cdot \mathbf{G} = 0$ in the interior of a volume, we have

$$\int \mathbf{F} \cdot \mathbf{G} \, dV = -\int (\mathbf{G} \cdot \nabla\psi) \, dV = -\int \nabla \cdot (\mathbf{G}\psi) \, dV = -\int \psi G_n \, dS.$$

If in addition ψ is specified on surfaces S_k, and G_n on these surfaces is equal to $-\omega$, a surface "charge", we have

$$-\int \psi G_n \, dS = \sum_k \psi_k \int_{S_k} \omega \, dS_k = \sum_k \psi_k \tau_k \tag{15.46}$$

where τ_k is the "charge" on the kth surface. If we apply this to the electric vectors $\mathbf{E}$ and $\mathbf{D}$ due to charges q_k on surfaces at potentials ϕ_k we obtain the relations

$$\int \mathbf{E} \cdot \mathbf{D} \, dV = \sum_k \phi_k q_k, \tag{15.47a}$$

$$\int \mathbf{E} \cdot \delta\mathbf{D} \, dV = \sum_k \phi_k \, \delta q_k, \tag{15.47b}$$

$$\int \mathbf{D} \cdot \delta\mathbf{E} \, dV = \sum_k q_k \, \delta\phi_k. \tag{15.47c}$$

We have derived (15.46) in an abstract form to emphasize that the set of equations (15.47) do *not* depend on $\mathbf{E}$ and $\mathbf{D}$ being aspects of the same field.

The thermodynamic identity for F can now be written as

$$\delta F = -S \, \delta T + \sum_k \phi_k \, \delta q_k \tag{15.48}$$

and, if we define $\tilde{F}$ as

$$\tilde{F} = F - \int \mathbf{E} \cdot \mathbf{D} \, dV, \tag{15.49}$$

we have

$$\delta \tilde{F} = -S \, \delta T - \sum_k q_k \, \delta\phi_k. \tag{15.50}$$

If a system of electrodes produces a field $\mathbf{E}^*$ in empty space and the total free energy of the system is F_{mt}^{total}, then when we introduce dielectric bodies into the system the new total free energy (provided the charge is isothermal and reversible) is

$$F^{\text{total}} = F_{mt}^{\text{total}} + F_0 + W, \tag{15.51}$$

where W is the work done to move the bodies and F_0 is the free energy of the bodies in zero field. It is not unreasonable to refer to the additional term $F_0 + W$ as the free energy of the bodies in the impressed field $\mathbf{E}^*$. We shall consider later what, if anything, this means.

We now define two new functions F^* and $\tilde{F}^*$. If F_{mt} is the free energy of a system of isolated electrodes at potentials ϕ_k^* and with charges q_k^* when no dielectric bodies are present, and F is the free energy of the same system of electrodes with the same charges but, of course, different potentials with a dielectric body or bodies present, then

$$F^* = F - F_{mt}. \tag{15.52}$$

Notice that F^* is specifically defined with reference to fields produced by fixed charges. It is equal to the sum of the free energies of the bodies outside the field and the work done to introduce the bodies isothermally and reversibly into the field keeping the charges constant.

If $\tilde{F}_{mt}$ refers to the empty system and $\tilde{F}$ to the same system with the same electrode potentials (but not necessarily the same charges), then

$$\tilde{F}^* = \tilde{F} - \tilde{F}_{mt}. \tag{15.53}$$

It is equal to the sum of the free energies of the bodies outside the field and the work required to introduce them into the field when the electrodes are held at constant potentials by batteries.

The function F^* depends on the constitution of the dielectric body and the charges q_k^* on the electrodes while $\tilde{F}^*$ depends on the bodies and the potentials ϕ_k^* of the electrodes. The difference δF^* between the values of F^* for different charges $q_k^* + \delta q_k^*$ is the additional work required to introduce the body into the new field. The difference $\delta \tilde{F}^*$ for different potentials has the same meaning. We have, however, the thermodynamic identities

$$\delta F = -S\,\delta T + \int \mathbf{E} \cdot \delta \mathbf{D}\, \mathrm{d}V,$$

$$\delta F_{mt} = \int \varepsilon_0 \mathbf{E}^* \cdot \delta \mathbf{E}^*\, \mathrm{d}V,$$

$$\delta \tilde{F} = -S\,\delta T - \int \mathbf{D} \cdot \delta \mathbf{E}\, \mathrm{d}V,$$

and

$$\delta \tilde{F}_{mt} = -\int \varepsilon_0 \mathbf{E}^* \cdot \delta \mathbf{E}^*\, \mathrm{d}V,$$

so that

$$\delta F^* = -S\,\delta T + \int (\mathbf{E} \cdot \delta \mathbf{D} - \varepsilon_0 \mathbf{E}^* \cdot \delta \mathbf{E}^*)\, \mathrm{d}V, \tag{15.54}$$

and

$$\delta \tilde{F}^* = -S\,\delta T - \int (\mathbf{D} \cdot \delta \mathbf{E} - \varepsilon_0 \mathbf{E}^* \cdot \delta \mathbf{E}^*)\, \mathrm{d}V. \tag{15.55}$$

We can use these expressions to obtain simpler forms for δF^* and $\delta \tilde{F}^*$ as follows. In δF^* we have

$$\mathbf{E} \cdot \delta \mathbf{D} - \varepsilon_0 \mathbf{E}^* \cdot \delta \mathbf{E}^* = \mathbf{E} \cdot (\delta \mathbf{D} - \varepsilon_0\, \delta \mathbf{E}^*) + (\varepsilon_0 \mathbf{E} - \mathbf{D}) \cdot \delta \mathbf{E}^* + (\mathbf{D} - \varepsilon_0 \mathbf{E}^*) \cdot \delta \mathbf{E}^*$$

and, if we let ϕ_k^* and q_k^* refer to the empty system and ϕ_k, q_k to the system with the body present, the volume integral in (15.54) gives

$$\delta F^* = -S\,\delta T + \sum_k \phi_k (\delta q_k - \delta q_k^*) + \int (\varepsilon_0 \mathbf{E} - \mathbf{D}) \cdot \delta \mathbf{E}^*\, \mathrm{d}V + \sum_k (q_k - q_k^*)\, \delta \phi_k^*.$$

But F^* is defined with $q_k = q_k^*$ and therefore also $\delta q_k = \delta q_k^*$, so that the two sums vanish. In the remaining integral $\varepsilon_0 \mathbf{E} - \mathbf{D} = -\mathbf{P}$ is non-zero only in the body and so

$$\delta F^* = -S\,\delta T - \int_{\text{body}} \mathbf{P} \cdot \delta \mathbf{E}^*\, \mathrm{d}V. \tag{15.56}$$

If the applied field $\mathbf{E}^*$ is uniform, or varies slowly over the region occupied (or to be occupied) by the body, which does not imply that $\mathbf{E}$, $\mathbf{P}$ or $\mathbf{D}$ is uniform, then we can take $\delta \mathbf{E}^*$

outside the integral and

$$\delta F^* = -S\,\delta T - \mathfrak{P}\,.\,\delta \mathbf{E}^* \tag{15.57}$$

where

$$\mathfrak{P} = \int_{\text{body}} \mathbf{P}\,\mathrm{d}V \tag{15.58}$$

is the total dipole moment of the body.

In $\delta\tilde{F}^*$ we encounter

$$\mathbf{D}\,.\,\delta\mathbf{E} - \varepsilon_0\mathbf{E}^*\,.\,\delta\mathbf{E}^* = \mathbf{D}\,.\,(\delta\mathbf{E}-\delta\mathbf{E}^*)+(\mathbf{D}-\varepsilon_0\mathbf{E})\,.\,\delta\mathbf{E}^*+(\mathbf{E}-\mathbf{E}^*)\,.\,\varepsilon_0\,\delta\mathbf{E}^*.$$

The volume integral of the first term $\sum_k q_k(\delta\phi_k - \delta\phi_k^*)$ vanishes because $\tilde{F}^*$ is defined with reference to electrodes whose potentials are the same with or without the body. The third term leads to $\sum_k(\phi_k - \phi_k^*)\delta q_k^*$ which vanishes for the same reason and so, once again, we have

$$\delta\tilde{F}^* = -S\,\delta T - \int_{\text{body}} \mathbf{P}\,.\,\delta\mathbf{E}^*\,\mathrm{d}V, \tag{15.59}$$

and, for a uniform applied field,

$$\delta\tilde{F}^* = -S\,\delta T - \mathfrak{P}\,.\,\delta\mathbf{E}^*. \tag{15.60}$$

Although the actual field $\mathbf{E}$ in the body will only be uniform if the body is a homogeneous ellipsoid and $\mathbf{E}^*$ is uniform, the use of these results only requires $\mathbf{E}^*$ to be uniform. This is much easier to achieve experimentally. For example, equation (15.57) or (15.60) is applicable to the system shown in Fig. 15.5.

FIG. 15.5. A dielectric object in an applied field.

If the body is inserted between two plates held at constant charge we have

$$\mathbf{P}\,.\,\mathbf{E}^* = (\mathbf{D}-\varepsilon_0\mathbf{E})\,.\,\mathbf{E}^* = (\mathbf{D}-\varepsilon_0\mathbf{E}^*)\,.\,\mathbf{E}^*+(\mathbf{E}^*-\mathbf{E})\,.\,\varepsilon_0\mathbf{E}^*.$$

The volume integral of the first term on the right gives $(q-q^*)\phi^*$ which is zero since $q = q^*$ and the second term gives $(\phi^* - \phi)q^*$. Thus

$$\mathfrak{P}\,.\,\mathbf{E}^* = (\phi^* - \phi)q^*. \tag{15.61}$$

For parallel plates a distance d apart $E^* = \phi^*/d$ and so the component $\mathfrak{P}_n$ of $\mathfrak{P}$ normal to the plates is

$$\mathfrak{P}_n = \left(1 - \frac{\phi}{\phi^*}\right)q^* d.$$

Thus we can obtain $\mathfrak{P}_n$ by electrical measurements of the plate charge and potential. Indeed since $q^* = (\varepsilon_0 A/d)\,\phi^*$, we have

$$\mathfrak{P}_n = \varepsilon_0 A(\phi^* - \phi).$$

Similarly if we insert the body between two plates held at constant potential we obtain

$$(q-q^*)\phi^* = \mathbf{E}^*\,.\,\mathfrak{P} \tag{15.62}$$

and for parallel plates $\mathfrak{P}_n = d(q-q^*)$.

We see then that the functions F^* and $\tilde{F}^*$ yield thermodynamic equations which relate an experimentally measurable quantity $\mathfrak{P}$ to an experimentally controllable field $\mathbf{E}^*$. It is no longer necessary to restrict our considerations to bodies of a special shape. On the other hand, of course, unless the body has a simple shape neither $\mathbf{P}$ nor $\mathbf{E}$ will be uniform in the body and we shall not be able to relate our results to the constitutive relation for the material of the body.

The functions F^* and $\tilde{F}^*$ are not only useful in practical calculations but also, as we shall be seeing later, are closely related to the free energy function used in statistical mechanical calculations of dielectric properties. This function, which we denote by F^s, turns out to be equal to F^* when the field E^* is either produced by electrodes at a great distance or fixed point charges rather than by charges on nearby macroscopic electrodes. It is also equal to $\tilde{F}^*$ when the electrodes are distant and, like both F^* and $F,^*$ it satisfies

$$\delta F^s = -S\,\delta T - \mathfrak{P}\,.\,\delta\mathbf{E}^*. \tag{15.63}$$

The difference between the three functions F^*, $\tilde{F}^*$ and F^s arises from the interaction between the body and mobile charges on the electrodes which produce the field. This interaction exists, i.e., there are image charges on the electrodes, even if the electrodes initially have either zero net charge or are held at zero potential. Thus in either case there may be a change in the free energy of the system as the body is brought near the electrodes, even if $\mathbf{E}^*$ is zero. This is easy to see if we consider bringing a body with a permanent moment from outside into the space between the plates of a condenser, but it may also occur even if the body has no net moment in zero field. In this case the change as the body is brought near the electrodes is due to the way the presence of the electrodes modifies the interactions between the individual dipoles within the body. These effects, which arise from the redistribution of charge over the surface of the electrodes, are absent if $\mathbf{E}^*$ is due to an array of fixed point charges but they also become vanishingly small if the electrodes are well removed from the body. In this case $F^* = \tilde{F}^* = F^s$, when the field $\mathbf{E}^*$ is the same in all three cases. Thus, although we should have to allow for these effects if we wished for example to calculate the force acting on a polarized body in the vicinity of a set of electrodes, we can almost always neglect this complication.

We now have an array of functions F, $\tilde{F}$, F^* and $\tilde{F}^*$ to which we might also add

$$F^\dagger = F - \tfrac{1}{2}\int \varepsilon_0 E^2\,\mathrm{d}V$$

and

$$\tilde{F}^\dagger = F + \tfrac{1}{2}\int \varepsilon_0 E^2\,\mathrm{d}V$$

all of which lead to useful thermodynamic relations, and are, in one way or another, related to the free energy of a system of dielectric bodies in a field. The function F with the identity

$$\delta F = -S\,\delta T + \int \mathbf{E}\,.\,\delta\mathbf{D}\,\mathrm{d}V$$

is the free energy of a system in which any batteries connected to electrodes are regarded as outside the system. The function $\tilde{F}$ with

$$\delta\tilde{F} = -S\,\delta T - \int \mathbf{D}\,.\,\delta\mathbf{E}\,\mathrm{d}V$$

is the free energy of a system in which the batteries are regarded as within the system and the same batteries remain permanently connected to the same electrodes. The functions F^* and $\tilde{F}^*$ with the thermodynamic identities (15.56) and (15.59) and the function F^s with the identity (15.63) are not free energy functions for any actual physical system. They refer both to the empty system and the system with the bodies present. It is, however, legitimate to regard each of these functions as the free energy of a dielectric body placed in a field $\mathbf{E}^*$ due to distant charges or electrodes. We could use F^*, $\tilde{F}^*$ or F^s to calculate any effect solely relating to the body but not, for example, to calculate the force on one of the electrodes which produces the field. The functions F^+ and $\tilde{F}^+$ have no direct physical interpretation. Whatever the status of the various functions they are all well-defined functions of state and legitimate targets for mathematical manipulation or the construction of Maxwell relations. This is also true of the various free energy densities. Even if they only apply directly to very special experimental situations, a relation such as equation (15.45) is of general validity, although in an actual experimental situation we might have great difficulty in measuring $\mathbf{P}$ or defining $\mathbf{E}$.

Magnetic Media

Electric results which refer to fields $\mathbf{E}$ and $\mathbf{D}$ can be transposed directly to magnetic results; for example, the equivalent of equation (15.45) is

$$\left(\frac{\partial \sigma}{\partial H_i}\right)_T = \mu_0\left(\frac{\partial M_i}{\partial T}\right)_H . \tag{15.64}$$

On the other hand, results which involve an impressed field $\mathbf{E}^*$ must be reconsidered, for the relations between electric and magnetic fields and their sources are quite different. In magnetism we generally deal with fields due to a specified set of constant currents or permanent magnets with a fixed magnetization.

We begin by proving a simple general result. In an isolated system in which $\mathbf{B} = \nabla \wedge \mathbf{A}$ and $\nabla \wedge \mathbf{H} = \mathbf{J}$ we have

$$\int(\mathbf{B}\,.\,\mathbf{H}-\mathbf{A}\,.\,\mathbf{J})\,\mathrm{d}V = \int\nabla\,.\,(\mathbf{A}\wedge\mathbf{H})\,.\,\mathrm{d}V = \oint(\mathbf{A}\wedge\mathbf{H})\,.\,\mathrm{d}\mathbf{S}$$

and, since the surface is at a distance from the sources of the fields, the surface integral is zero. Thus, for a complete system

$$\int\mathbf{B}\,.\,\mathbf{H}\,\mathrm{d}V = \int\mathbf{A}\,.\,\mathbf{J}\,\mathrm{d}V. \tag{15.65}$$

If we apply this to a system in which $\mathbf{J}=0$ and the fields are due to permanent magnets, and divide the system into an outer region 0 and the region m within the magnets we have

$$\int_0 \mathbf{B}\,.\,\mathbf{H}\,\mathrm{d}V = -\int_m \mathbf{B}\,.\,\mathbf{H}\,\mathrm{d}V.$$

In this system the impressed or controlled variable is $\mathbf{M}_0$ the magnetization of the magnet and, if we now define $\mathbf{b}$ as equal to $\mathbf{B}$ outside the magnet and as equal to $\mu_0\,\mathbf{H}$ in the magnet, we have

$$\int_{0+m} \mathbf{b}\,.\,\mathbf{H}\,\mathrm{d}V = -\mu_0\int_m \mathbf{M}_0\,.\,\mathbf{H}\,\mathrm{d}V,$$

and also

$$\int_{0+m} \mathbf{H}\,.\,\delta\mathbf{b}\,\mathrm{d}V = -\mu_0\int_m \mathbf{H}\,.\,\delta\mathbf{M}_0\,\mathrm{d}V.$$

For the system as a whole $\int\mathbf{H}\,.\,\delta\mathbf{B}\,\mathrm{d}V = 0$ and so, if we regard the thermodynamic system as including both the space inside the magnet and the space outside, but regard the work done to change $\mathbf{M}_0$ by $\delta\mathbf{M}_0$ as provided from outside we obtain

$$\delta F = -S\,\delta T - \mu_0\int_m \mathbf{H}\,.\,\delta\mathbf{M}_0\,\mathrm{d}V. \tag{15.66}$$

This is the magnetic analogue of the electric equation

$$\delta F = -S\,\delta T + \sum_k \phi_k\,\delta q_k.$$

It is more complicated because magnetism is a volume effect whereas charges can exist either independently or reside on surfaces. In the electric case we do not have to consider the interior of the electrodes. We could use (15.66) to define a function F^*, which would be the difference in the free energy which results from introducing a magnetizable body into the field of a permanent magnet. In practice this is of little use because magnets of precisely fixed, or experimentally variable, magnetization $\mathbf{M}_0$ do not exist. We therefore confine our attention to fields due to currents.

The function

$$\tilde{F} = F - \int \mathbf{H}\,.\,\mathbf{B}\,\mathrm{d}V \tag{15.67}$$

has the thermodynamic identity

$$\delta\tilde{F} = -S\,\delta T - \int \mathbf{B}\,.\,\delta\mathbf{H}\,\mathrm{d}V \tag{15.68}$$

which we can also express as

$$\delta\tilde{F} = -S\,\delta T - \int \mathbf{A}\,.\,\delta\mathbf{J}\,\mathrm{d}V. \tag{15.69}$$

For an empty system

$$\tilde{F}_{mt} = -\tfrac{1}{2}\mu_0 \int \mathbf{H}^*\,.\,\mathbf{H}^*\,\mathrm{d}V \tag{15.70}$$

and we define

$$\tilde{F}^* = \tilde{F} - \tilde{F}_{mt} \tag{15.71}$$

so that

$$\delta\tilde{F}^* = -S\,\delta T - \int (\mathbf{B}\,.\,\delta\mathbf{H} - \mu_0\mathbf{H}^*\,.\,\delta\mathbf{H}^*)\,\mathrm{d}V. \tag{15.72}$$

The integrand may be written as

$$\mathbf{B}\,.\,\delta\mathbf{H} - \mu_0\mathbf{H}^*\,.\,\delta\mathbf{H}^* = \mathbf{B}\,.\,(\delta\mathbf{H} - \delta\mathbf{H}^*) + (\mathbf{B} - \mu_0\mathbf{H})\,.\,\delta\mathbf{H}^* + (\mathbf{H} - \mathbf{H}^*)\,.\,\mu_0\,\delta\mathbf{H}^*.$$

Because the currents are the same with and without the body present, the first and last terms give vanishing volume integrals (see equation (15.65)) and so

$$\delta\tilde{F}^* = -S\,\delta T - \int (\mathbf{B} - \mu_0\mathbf{H})\,.\,\delta\mathbf{H}^*\,\mathrm{d}V,$$

or

$$\delta\tilde{F}^* = -S\,\delta T - \int_{\text{body}} \mu_0\mathbf{M}\,.\,\delta\mathbf{H}^*\,\mathrm{d}V. \tag{15.73}$$

Since $\mu_0\mathbf{H}^* = \mathbf{B}^*$ we can also write this as

$$\delta F^* = -S\,\delta T - \int_{\text{body}} \mathbf{M}\,.\,\delta\mathbf{B}^*\,\mathrm{d}V \tag{15.74}$$

which is perhaps more in keeping with our general attitude that **B**, not **H**, is the fundamental magnetic vector (but see problem 12). In a uniform impressed field

$$\delta\tilde{F}^* = -S\,\delta T - \mathfrak{M}\,.\,\delta\mathbf{B}^* \tag{15.75}$$

where

$$\mathfrak{M} = \int_{\text{body}} \mathbf{M}\,\mathrm{d}V \tag{15.76}$$

is the magnetic moment of the body.

If the field $\mathbf{B}^*$ is due to currents in wires, or filamentary conductors, no redistribution of current occurs when the body is introduced and so, in magnetism, we are not plagued with the image charge effects which occur with finite electrodes at a finite distance. Thus in this case $\tilde{F}^*$ can be used without qualification in discussing changes which affect the body alone (not the conductors which produce the field) and $\tilde{F}^*$ is equal to F^s the statistical free energy function.

The quantity $\mathbf{B}^*$ is an experimentally accessible and controllable parameter and we now show that $\mathfrak{M}$ is measurable. If a coil, characterized by a line element $\mathbf{dl}$, carrying a current i produces a field $\mathbf{b}^* = \mathbf{G}i$, over the region of the body, where $\mathbf{G}$ is a geometric factor we have, using (15.65),

$$\mathbf{b}^* \cdot \frac{\partial \mathfrak{M}}{\partial t} = i \int (\dot{\mathbf{A}} - \dot{\mathbf{A}}^*) \cdot \mathbf{dl} = -i \oint (\mathbf{E} - \mathbf{E}^*) \cdot \mathbf{dl}$$

and so the induced e.m.f. for zero current (when $\mathbf{E}^* = 0$) is

$$\phi = \oint \mathbf{E} \cdot \mathbf{dl} = -\mathbf{G} \cdot \frac{\partial \mathfrak{M}}{\partial t}. \tag{15.77}$$

If an auxiliary coil produces a small a.c. field $\mathbf{b}^*$, in addition to whatever other field $\mathbf{B}^*$ is present, the e.m.f. induced in a secondary coil, characterized by the geometric factor $\mathbf{G}$ with components G_i, due to the presence of the body is

$$\phi = -G_i \frac{\partial \mathfrak{M}_i}{\partial B_j^*} \frac{\partial b_j^*}{\partial t}. \tag{15.78}$$

This result is the basis of a method of obtaining $\partial \mathfrak{M}_i / \partial B_j^*$ by mutual inductance measurements. Depending on the experimental conditions we can determine either the isothermal magnetizability of the body or the isentropic value. If, for example, we have a sphere of volume V the quantity we obtain in this way will be

$$\chi^* = V \frac{\chi}{1 + \frac{1}{3}\chi} \tag{15.79}$$

where χ is the isothermal susceptibility χ_T or the isentropic susceptibility χ_S. Note, however, that χ is the value appropriate to a field $\mathbf{H} = \mathbf{H}^* - \frac{1}{3}\mathbf{M}$, not to a field $\mathbf{H} = 1/\mu_0 \mathbf{B}^* = \mathbf{H}^*$. Thus

$$\chi\left(H^* - \frac{M}{3}\right) = \frac{\chi^*(H^*)}{3V - \chi^*(H^*)}.$$

Equations (15.74) and (15.75) are the basic results used in discussing the thermodynamics of magnetic cooling experiments and in relating magnetic properties to the results of calculations in statistical mechanics. In a later section we shall give an example of the use of these results.

Superconductors

A superconductor is primarily a medium in which persistent currents can flow without the irreversible generation of heat. All known superconductors have, however, a further property which, for the purposes of this section, is equally important. A superconducting body exists in a definite, i.e., unique, equilibrium state which is solely determined by the temperature T and the applied magnetic field $\mathbf{B}^*$. Thus the condition of a superconducting body placed in a field $\mathbf{B}^*$ and then cooled down below the transition temperature is the same as its condition if it is cooled first and then subjected to the field. This property is manifest in the Meissner effect in which a body cooled through the transition temperature in a field below the critical field expels the field at the transition temperature.

In dealing with superconductors we have to deal with systems in which $\mathbf{B} = \mu_0\mathbf{H}$ and $\mathbf{M} = 0$. The elementary treatment which treats a superconductor as a medium with magnetic susceptibility $\chi = -1$ is both misleading and meaningless. If $\mathbf{j}$ is the current density the thermodynamic identity for the free energy is

$$\delta F = -S\,\delta T + \int\left(\frac{1}{\mu_0}\mathbf{B}\,.\,\delta\mathbf{B} - \mathbf{j}\,.\,\delta\mathbf{A}\right)\mathrm{d}V. \tag{15.80}$$

If the field is due to an external current of density $\mathbf{J}^*$ we have

$$\int\frac{1}{\mu_0}\mathbf{B}\,.\,\delta\mathbf{B}\,\mathrm{d}V = \int(\mathbf{J}^*+\mathbf{j})\,\delta\mathbf{A}\,\mathrm{d}V \tag{15.81}$$

and so

$$\delta F = -S\,\delta T + \int\mathbf{J}^*\,.\,\delta\mathbf{A}\,\mathrm{d}V. \tag{15.82}$$

We now define

$$\tilde{F} = F - \int\mathbf{J}^*\,.\,\mathbf{A}\,\mathrm{d}V \tag{15.83}$$

so that

$$\tilde{F}_{mt} = -\tfrac{1}{2}\int\mathbf{J}^*\,.\,\mathbf{A}^*\,\mathrm{d}V. \tag{15.84}$$

The function

$$\tilde{F}^* = \tilde{F} - \tilde{F}_{mt} \tag{15.85}$$

is the sum of the free energy of the body in zero field together with the work done to introduce it isothermally and reversibly into the field with $\mathbf{J}^*$ the *external* current held constant. We see that

$$\tilde{F}^* = \tilde{F} - \int(\mathbf{J}^*\,.\,\mathbf{A} - \tfrac{1}{2}\mathbf{J}^*\,.\,\mathbf{A}^*)\,\mathrm{d}V.$$

For a change $\delta\mathbf{J}^*$ in the external current

$$\delta\tilde{F}^* = -S\,\delta T + \int(\mathbf{J}^*\,.\,\delta\mathbf{A} - \delta(\mathbf{J}^*\,.\,\mathbf{A}) + \mathbf{A}^*\,.\,\delta\mathbf{J}^*)\,\mathrm{d}V,$$

which gives

$$\delta\tilde{F}^* = -S\,\delta T - \int(\mathbf{A}-\mathbf{A}^*)\,.\,\delta\mathbf{J}^*\,\mathrm{d}V.$$

We also have

$$\int(\mathbf{A}-\mathbf{A}^*)\,.\,\delta\mathbf{J}^*\,\mathrm{d}V = \int\frac{1}{\mu_0}(\mathbf{B}-\mathbf{B}^*)\,.\,\delta\mathbf{B}^*\,\mathrm{d}V = \int(\mathbf{J}-\mathbf{J}^*)\,.\,\delta\mathbf{A}^*\,\mathrm{d}V,$$

where $\mathbf{J} = \mathbf{j}+\mathbf{J}^* = 1/\mu_0\nabla \wedge \mathbf{B}$. However, $\mathbf{J} = \mathbf{J}^*$ except in the superconductor where $\mathbf{J} = \mathbf{j}$ and $\mathbf{J}^* = 0$ and so $\delta\tilde{F}^*$ can be expressed as

$$\delta\tilde{F}^* = -S\,\delta T - \int_{\text{body}}\mathbf{j}\,.\,\delta\mathbf{A}^*\,\mathrm{d}V. \tag{15.86}$$

This is the basic thermodynamic identity but we can also express it in a form which is more useful in practical calculations. Since $\nabla\,.\,\mathbf{j} = 0$ we can express $\mathbf{j}$ as $\mathbf{j} = \nabla\wedge\mathbf{m}$ where $\mathbf{m}$ vanishes outside the body, and therefore

$$\int\mathbf{j}\,.\,\delta\mathbf{A}^*\,\mathrm{d}V = \int(\nabla \wedge \mathbf{m})\,.\,\delta\mathbf{A}^*\,\mathrm{d}V = \int\mathbf{m}\,.\,(\nabla \wedge \delta\mathbf{A}^*)\,\mathrm{d}V + \oint(\mathbf{m} \wedge \delta\mathbf{A}^*)\,.\,\mathrm{d}\mathbf{S}.$$

The surface integral vanishes because $\mathbf{m} = 0$ on the surface, and with $\nabla \wedge \delta\mathbf{A}^* = \delta\mathbf{B}^*$ the result is

$$\delta\tilde{F}^* = -S\,\delta T - \int_{\text{body}}\mathbf{m}\,.\,\delta\mathbf{B}^*\,\mathrm{d}V. \tag{15.87}$$

Thermodynamics

If the *applied* field $\mathbf{B}^*$ is sensibly uniform over the region of the body we can take $\delta\mathbf{B}^*$ outside the integral, and we also have

$$\int \mathbf{m}\,\mathrm{d}V = \tfrac{1}{2}\int \mathbf{r} \wedge (\nabla \wedge \mathbf{m})\,\mathrm{d}V = \tfrac{1}{2}\int (\mathbf{r} \wedge \mathbf{j})\,\mathrm{d}V = \mathfrak{M}$$

where $\mathfrak{M}$ is the magnetic moment of the body, due to the circulating currents carried by mobile charges. Thus

$$\delta\tilde{F}^* = -S\,\delta T - \mathfrak{M}\,.\,\delta\mathbf{B}^*. \tag{15.88}$$

This looks exactly like the thermodynamic identity for a magnetic body but its physical significance is quite different. In a magnetic body $\mathfrak{M} = \int M\,\mathrm{d}V$ is the integral of a magnetization vector $\mathbf{M}$ which is finite and non-zero everywhere in the body and changes discontinuously at the surface. In a superconductor $\mathbf{m}$ is not the magnetization and in the body $\mathbf{H} \neq (1/\mu_0)\mathbf{B} - \mathbf{m}$ but is equal to $(1/\mu_0)\mathbf{B}$ and is zero. The effects due to the currents described by $\mathbf{m}$ arise only in a thin layer at the surface.

The application of these results is not simple and, in particular, requires careful discussion if the body is multiply connected. We refer the reader to Landau and Lifshitz (1960).

Thermodynamics and Statistical Mechanics

We shall be discussing the statistical mechanics of magnetic and dielectric bodies in the next chapter and here we are only concerned with the relation to statistical mechanics of the results we have so far derived.

In statistical mechanics the structure of a system, e.g., its chemical composition and crystalline form, determines an expression for the Hamiltonian energy function $\mathscr{H}$. This is a function of the microscopic variables of the system and a number of external parameters λ_n, which may, for example, refer to impressed fields or mechanical constraints such as the volume. A formal procedure then leads from $\mathscr{H}$ to an expression for the free energy function $F^s(\lambda_n, T)$ of the system in thermal equilibrium at a temperature T. This function is the free energy of the system, in the sense that if one set of values of the parameters $\lambda_n^{(1)}$ determines one equilibrium state at T and another set $\lambda_n^{(2)}$ determines a second equilibrium state, the reversible isothermal work done by external agencies to change the system from one state to the other is $F(\lambda_n^{(2)}, T) - F(\lambda_n^{(1)}, T)$. The structure of statistical mechanics and its relation to thermodynamics is then completed by the demonstration that, in an equilibrium state, the entropy is

$$S = -\left(\frac{\partial F^s}{\partial T}\right)_{\lambda_n}, \tag{15.89}$$

and the expectation value of the microscopic variable $\partial\mathscr{H}/\partial\lambda_n$ is

$$\left\langle \frac{\partial\mathscr{H}}{\partial\lambda_n} \right\rangle = \left(\frac{\partial F^s}{\partial\lambda_n}\right)_T. \tag{15.90}$$

In the theory of dielectrics the only variables which appear in the Hamiltonian $\mathscr{H}$ are those which refer to the body itself. In particular any fields which are explicitly included in $\mathscr{H}$ are regarded as impressed fields due to agencies outside the body. Thus the function F^s which results is not the free energy of the system which includes the body, nor is it necessarily the difference between the free energy of the system and the free energy of an identical system in the absence of the body. If $\mathscr{H}_\mathrm{b}$ is the Hamiltonian of the body outside the field,

$\mathscr{H}_{\text{mt}}$ is the Hamiltonian of the empty system and $\mathscr{H}_{\text{int}}$ describes the interaction between the body and the system which produces the field, then the free energy of the complete system is derived from $\mathscr{H}_{\text{b}}+\mathscr{H}_{\text{mt}}+\mathscr{H}_{\text{int}}$, and the free energy of the empty system from $\mathscr{H}_{\text{mt}}$. The thermodynamic function F^* or $\tilde{F}^*$ (the difference between these functions corresponds to a difference in the specification of the empty system) is related to $\mathscr{H}_{\text{b}}+\mathscr{H}_{\text{int}}$ and one or the other of these functions is equal to F^s if the term $\mathscr{H}_{\text{int}}$ is complete. Thus $\mathscr{H}_{\text{int}}$ must include not only terms which refer to impressed field $\mathbf{E}^*$ but also terms which refer, for example, to image charges on the electrodes which produce $\mathbf{E}^*$.

Now in general, in calculations in statistical mechanics these terms which refer both to the body and to the external environment are omitted, and the only interaction term included in the Hamiltonian is the impressed field term $-\mathbf{E}\,.\,\mathfrak{P}$, where $\mathfrak{P}$ is the microscopic variable or operator corresponding to the total dipole moment of the body. Equation (15.90) then yields

$$\langle\mathfrak{P}_i\rangle = -\left(\frac{\partial F^s}{\partial E_i^*}\right)_T \tag{15.91}$$

which coincides with the result obtained from F^* or $\tilde{F}^*$ even when F^s is not equal to these functions. However, as we remarked on page 171, the additional interaction terms vanish for distant electrodes and in this case $F^* = \tilde{F}^* = F^s$. These terms are always absent in the magnetic case, with fields due to filamentary conductors, and in magnetism $F^s = \tilde{F}^*$.

In some ways this is the most important result of this whole chapter. If F^s is the free energy function derived by a conventional calculation in statistical mechanics, the electric and magnetic dipole moments of the body are

$$\mathfrak{P}_i = -\left(\frac{\partial F^s}{\partial E_i^*}\right)_T \tag{15.92a}$$

$$\mathfrak{M}_i = -\left(\frac{\partial F^s}{\partial B_i^*}\right)_T. \tag{15.92b}$$

We have spent some time on the steps which lead to these results because these results are widely used and the literature on the subject though extensive is, with the notable exception of Landau and Lifshitz (1960), not always entirely free from confusion. The transition from a free energy function F satisfying

$$\delta F = -S\,\delta T+\int \mathbf{H}\,.\,\delta\mathbf{B}\,\mathrm{d}V \tag{15.93a}$$

to a function F^s satisfying

$$\delta F^s = -S\,\delta T-\int \mathbf{M}\,.\,\delta\mathbf{B}^*\,\mathrm{d}V \tag{15.93b}$$

is perhaps less puzzling once we realize that in (15.93a) the field $\mathbf{B}$ is the actual field in the system while in (15.93b) $\mathbf{B}^*$ is the impressed field, i.e., the field that would be present if the one part of the system under discussion were absent.

Paramagnetic Relaxation

In this section we discuss a specific application of the results arrived at in the section on Magnetic Media (pp. 172–4), which illustrates the power of thermodynamic arguments. We shall see that the entire thermodynamic behaviour of a paramagnetic body can be derived solely by magnetic measurements.

The heat capacity of a body at constant magnetization C_M is related to the capacity at constant impressed field C_H by

$$C_M = C_H + T\left(\frac{\partial S}{\partial H}\right)_T\left(\frac{\partial H}{\partial T}\right)_M,$$

where

$$C_M = T\left(\frac{\partial S}{\partial T}\right)_M$$

and

$$C_H = T\left(\frac{\partial S}{\partial T}\right)_H$$

(we use H rather than B* to conform to the conventional notation). Equation (15.75) yields the Maxwell relation

$$\left(\frac{\partial S}{\partial H}\right)_T = \mu_0\left(\frac{\partial \mathfrak{M}}{\partial T}\right)_H,$$

and so

$$C_M = C_H + \mu_0 T\left(\frac{\partial \mathfrak{M}}{\partial T}\right)_H\left(\frac{\partial H}{\partial T}\right)_M. \tag{15.94}$$

If we consider a reversible charge, in terms of $\mathfrak{M}$ and H as the independent variables,

$$T\,\delta S = T\left(\frac{\partial S}{\partial T}\right)_H\left(\frac{\partial T}{\partial \mathfrak{M}}\right)_H \delta\mathfrak{M} + T\left(\frac{S\partial}{\partial T}\right)_M\left(\frac{\partial T}{\partial H}\right)_M \delta H,$$

and so, if we define the isentropic magnetizability of the body as

$$\chi_S^* = \left(\frac{\partial \mathfrak{M}}{\partial H}\right)_S$$

we have

$$\chi_S^* = -\frac{C_M}{C_H}\left(\frac{\partial T}{\partial H}\right)_M\left(\frac{\partial \mathfrak{M}}{\partial T}\right)_H; \tag{15.95}$$

the relation

$$\left(\frac{\partial H}{\partial T}\right)_M\left(\frac{\partial T}{\partial \mathfrak{M}}\right)_H\left(\frac{\partial \mathfrak{M}}{\partial H}\right)_T = -1$$

then leads to

$$\chi_S^* = \frac{C_M}{C_H}\chi_T^* \tag{15.96}$$

where

$$\chi_T^* = \left(\frac{\partial \mathfrak{M}}{\partial H}\right)_T. \tag{15.97}$$

(Compare the relation between the isothermal and adiabatic bulk moduli in a gas.)

The same relation used in (15.94) yields

$$C_H - C_M = \mu_0 T\left(\frac{\partial \mathfrak{M}}{\partial T}\right)_H^2 (\chi_T^*)^{-1}$$

and so

$$C_H = \frac{\chi_T^*}{\chi_S^*}C_M = \frac{\mu_0 T}{\chi_T^* - \chi_S^*}\left(\frac{\partial \mathfrak{M}}{\partial T}\right)_H^2. \tag{15.98}$$

If $\mathfrak{M}$ is proportional to H this can be expressed as

$$C_H = \frac{\mu_0 T H^2}{\chi_T^* - \chi_S^*}\left(\frac{\partial \chi_T^*}{\partial T}\right)^2. \quad (15.99)$$

At liquid helium temperatures the spin–lattice relaxation time is generally of the order of a few milliseconds and so measurements of χ^* at radio frequencies yield χ_s^* and static measurements yield χ_T^*. By combining these two types of measurement over a range of fields and temperatures we can calculate C_H and C_M and construct all the thermodynamic functions for the spin system. Since, in addition, at these temperatures virtually all the entropy is associated with the spin system this is adequate information about the system as a whole. Many of the most important materials used in very low temperature work were first discovered by a study of their paramagnetic relaxation behaviour at helium temperatures. Although perhaps not, in itself, a central topic in the structure of physical science, techniques based on paramagnetic relaxation studies and magnetic cooling methods were used in the crucial experiments which demonstrated parity non-conservation in weak interactions and revolutionized our approach to many fundamental topics in physics.

Piezo-electricity

In a piezo-electric medium work done electrically on the medium is coupled to work done mechanically. If we restrict our attention to a homogeneous medium subject to stress S_{ij} with a strain u_{ij} the free energy density f satisfies

$$\delta f = -\sigma\, \delta T + E_i\, \delta D_i + S_{ij}\, \delta u_{ij}. \quad (15.100)$$

We define

$$\approx\!\!\!\!\! f = f - S_{ij} u_{ij} - E_i D_i \quad (15.101)$$

and then

$$\delta \approx\!\!\!\!\! f = -\sigma\, \delta T - D_i\, \delta E_i - u_{ij}\, \delta S_{ij}. \quad (15.102)$$

This yields the Maxwell relation

$$\left(\frac{\partial u_{ij}}{\partial E_k}\right)_{T,S} = \left(\frac{\partial D_k}{\partial S_{ij}}\right)_{E,T} = \left(\frac{\partial P_k}{\partial S_{ij}}\right)_{E,T} \quad (15.103)$$

Thus the isothermal piezo-electric coefficient for the direct effect defined by

$$P_i = d_{ijk} S_{jk} \quad (15.104a)$$

and that for the converse effect defined by

$$u_{ij} = c_{ijk} E_k \quad (15.104b)$$

satisfy

$$c_{ijk} = d_{kij}. \quad (15.105)$$

This result is clearly of importance in any practical discussion of piezo-electric effects. Media which display a strong direct effect display a strong converse effect and vice versa. If instead of $\approx\!\!\!\!\! f$ we use $\approx\!\!\!\!\! u = \approx\!\!\!\!\! f + TS$ we can derive similar relations for the isentropic coefficients needed at very high frequencies.

Conclusion

In this chapter we have been concerned more with the transformation of the fundamental equations, based on equations (15.12) and (15.13), into forms suitable for application to broad classes of problem and related to the results of statistical mechanics than with applications of the results. Applications can be found in variety elsewhere but neither texts on electromagnetism nor texts on thermodynamics normally give more than the most cursory attention to the fundamental principles.

The reader may perhaps wonder why we have emphasized the free energy function F rather than the total energy function U or the reversible work term W. There are three reasons. First of all, the use of F reminds us that the notions of an equilibrium state and a function of state are thermodynamic notions. Secondly, the function F is more closely related to the results of statistical mechanics, and finally most real systems undergo changes at constant temperature rather than constant entropy. We will, however, end by using the work term alone to derive an important result.

If a small body has a dipole moment $\mathbf{p}$ the expression $-\mathbf{p} \cdot \mathrm{d}\mathbf{E}^*$ is the external work required to move it from a position where the impressed field is $\mathbf{E}^*$ to a position where it is $\mathbf{E}^* + \mathrm{d}\mathbf{E}^*$. If the displacement is $\mathbf{dr}$ the force acting on the body $\mathbf{F}$ satisfies

$$\mathbf{F} \cdot \mathbf{dr} = \mathbf{p} \cdot \mathrm{d}\mathbf{E}^* \tag{15.106}$$

and this is valid whether or not $\mathbf{p}$ depends on $\mathbf{E}^*$. Equation (15.106) is therefore the general rule for calculating the force on a dipole. If $\mathbf{p} = \varepsilon_0 \alpha \mathbf{E}^*$ we have $\mathbf{F} = \frac{1}{2}\varepsilon_0 \alpha \nabla (E^*)^2$ but in general

$$F_i = p_j \frac{\partial E_j^*}{\partial r_i}. \tag{15.107}$$

If $\mathbf{E}^*$ satisfies $\nabla \wedge \mathbf{E}^* = 0$ so that $\partial E_j^*/\partial r_i = \partial E_i^*/\partial r_j$ we can express this as $F_i = p_j(\partial E_i^*/\partial r_j)$ or $\mathbf{F} = (\mathbf{p} \cdot \nabla)\mathbf{E}$.

Problems

1. A small charged body is attached to a rigid support by a spring. The electric field acting on the body changes slowly with time. Show that the mechanical energy stored in the spring changes at a rate which can be expressed as $\int \mathbf{E} \cdot \mathbf{J} \, \mathrm{d}V$.

2. The susceptibility of a dielectric medium obeys the Langevin–Debye law. Show that the application of an electric field leads to the evolution of heat. Under what circumstances does this lead to irreversible effects?

3. Use the result of problem 2 to discuss the radio-frequency loss in nitro-benzene.

4. A plane-parallel condenser of area A has an elastic dielectric spacer of Young's modulus Y, unstrained thickness x_0 and dielectric constant ε. Obtain an expression for the free energy function in terms of q, the charge on the plates. An alternating current $I_0 \cos \omega t$ flows through the condenser. Show that the potential difference has a component at a frequency 3ω.

5. A vacuum spaced capacitor of capacitance 1μ F is immersed in a liquid at $T = 300$ K whose dielectric constant is $\varepsilon = 3 + 3000/T$. Calculate the heat evolved as the potential across the plates is slowly raised to 1000 V.

6. A radio tuning capacitor has a capacitance expressed in terms of the angle θ of rotation by $C = C_0 + \gamma\theta$. Calculate the torque acting on the rotor when the plates are at a potential difference ϕ.

7. The capacitor in problem 6 is immersed in a liquid with $\varepsilon = 1+\tau/T$ with the plates set at $\theta = 0$. Calculate the heat evolved when θ is increased to π, keeping the potential difference ϕ constant.

8. A sphere of dielectric constant ε and radius a is inserted symmetrically between two plane-parallel electrodes of separation h with charge q^* and potential difference ϕ^*. Show that as $h/a \to \infty$ keeping q^* or $E^* = \phi^*/h$ constant, $\tilde{F}^* \to F^*$.

9. A dielectric sphere is inserted between the plates of a condenser with a fixed potential difference ϕ. The plates are then disconnected from the battery and connected to a battery of e.m.f. $\phi+\delta\phi$. The sphere is removed and the batteries interchanged again. Show (a) that the mechanical work done on the system in this process is $\delta\tilde{F}^*$ and (b) that the change in the total free energy of the system when the batteries are first changed over is negligible.

10. A Helmholtz coil produces a uniform field of 1 tesla and a sphere of linear susceptibility $\chi = 1$ and radius 10 cm is inserted into the field. Calculate the work required to remove the sphere from the field.

11. The magnetic moment of a spherical sample of paramagnetic salt is found to be given by $M = \tanh B^*/T$ in S.I. units. The thermal capacity C_H in zero field is $10^{-5}/T^2$ joule deg^{-1}. The sample is isothermally magnetized in a field $B^* = 1$ tesla at 1 K and then isentropically demagnetized to zero field. Find the final temperature.

12. An ellipsoidal body has a permanent, field independent, uniform magnetization $\mathbf{M}^\circ$ parallel to a principal axis. It is immersed in a paramagnetic fluid in which a set of coils produce a field $\mathbf{B}^* = \mu\mu_0\mathbf{H}^*$ in the absence of the body and which is uniform over the region occupied by the body. In the presence of the body the fields are $\mathbf{B}$ and $\mathbf{H}$ and $\tilde{F}^* = \tilde{F}-\tilde{F}_{\text{mt}}$ is the work required to introduce the body. Show that, for an isothermal change of the currents in the coils,

$$\delta\tilde{F}^* = -\int (\mathbf{B}-\mu\mu_0\mathbf{H}) \cdot \delta\mathbf{H}^* \, dV$$

and thus, that

$$\tilde{F}^* = -\mu_0(1+\gamma(\mu-1))\mathbf{H}^* \cdot \mathfrak{M}^\circ + \tfrac{1}{2}V_{\text{body}}(\mu-1)\mu_0\mathbf{H}^* \cdot d\mathbf{H}^*$$

where $\mathfrak{M}^\circ$ is the total moment of the body and γ the (positive) demagnetizing factor appropriate to the axis of magnetization of the ellipsoid. Use this result to show that the couple acting on a long thin bar magnet is $\mu_0\mathfrak{M}^\circ \wedge \mathbf{H}^*$ and calculate the couple acting on (a) spherical magnet and (b) a thin disc magnet.

13. A plane loop of wire of area A carrying a current I is immersed in the paramagnetic fluid of problem 12. What is the magnetic moment of the loop? What is the couple acting on the loop? Would the relation between moment and couple be the same for a long thin solenoid?

CHAPTER 16

Statistical Mechanics

Introduction

In classical mechanics the microscopic state of a system is completely specified when the coordinates $\mathbf{r}(n)$ and momenta $\mathbf{p}(n)$ of all the N particles of the system are known. In quantum mechanics the nearest we can get to an exact specification is knowledge of the pure quantum state of the system. In macroscopic physics we usually have to be content with much less information about a system. We may know its structure, composition and the dynamical laws it obeys. We may also be able to calculate, for each microstate, the values of all possible macroscopic observables connected with the system but all that we know about the microstate is that it is one of an enormous number of microstates compatible with the rather limited and imprecise information available to us about the macroscopic state of the system. The statistical task is to find, given a macroscopic description of the system, a probability distribution function for this ensemble of microstates. Then, if $X(n)$ is the value of the macroscopic observable X for the microstate (n), whose probability is $w(n)$, we have, for the expectation value of X,

$$X = \sum_n X(n)\, w(n),$$

where the sum is over all possible microstates. It makes for greater clarity if, instead of considering the ensemble of microstates for a single system, we regard the one system under consideration as a representative member of an ensemble of systems, each of which is in a particular microstate. The function $w(n)$ is then the probability that a system taken at random from the ensemble is in the microstate (n). The remarkable feature of statistical mechanics is that we can give a simple and general answer to the problem of calculating $w(n)$. The form of the theory is slightly different in the classical and quantum mechanical cases. We begin with the classical result.

If we have a system of N particles, each possible microstate can be represented as a point in a phase space of $6N$ dimensions, corresponding to the $3N$ components of the particle coordinates and the $3N$ components of the momenta. If we let $\mathrm{d}\Gamma$ be a volume element of phase space we can define the ensemble distribution function f so that the probability that the representative point of a system lies in $\mathrm{d}\Gamma$ is

$$\mathrm{d}w = f\,\mathrm{d}\Gamma.$$

The classical result is then that, for an ensemble representing systems in thermal equilibrium at a temperature T,

$$f = Z^{-1} \exp\left(-\frac{\mathscr{H}}{kT}\right), \tag{16.1}$$

where $\mathscr{H}$ is the Hamiltonian energy function for the system and Z is a normalizing factor.

In general $\mathscr{H}$ will be a function of the particle coordinates and momenta together with a number of macroscopic parameters λ_s, which for example might be the volume, the applied electric field or the magnetic field. The normalizing factor

$$Z = \int \exp\left(-\frac{\mathscr{H}}{kT}\right) d\Gamma \tag{16.2}$$

is therefore a function of T and the λ_s. It is known as the partition function. The expectation value of the energy is

$$U = \langle \mathscr{H} \rangle = \int \mathscr{H} f \, d\Gamma$$

and this is easily seen to be

$$U = -\frac{\partial}{\partial(1/kT)} \log Z = T\frac{\partial}{\partial T}(kT \log Z) - kT \log Z.$$

If we compare this with the thermodynamic relation

$$U = TS + F = -T\frac{\partial F}{\partial T} + F$$

we see that we can make the identification

$$F = -kT \log Z. \tag{16.3}$$

In quantum mechanics, if $|\rangle_k$ is the state function for the kth system of the ensemble and it is expanded in terms of some convenient set of basis states $|n\rangle$ as

$$|\rangle_k = \sum_n C_n(k)|n\rangle,$$

then the density matrix $\hat{\rho}$ is defined as the matrix whose elements are the ensemble averages

$$\rho_{nm} = \langle C_n(k) C_m^*(k) \rangle$$

over all the states k represented in the ensemble. The normalization condition is

$$Tr\hat{\rho} = \sum_n \rho_{nn} = 1.$$

If $\hat{\alpha}$ is the operator corresponding to a macroscopic observable α the expectation value of α in the state $|\rangle_k$ is

$${}_k\langle|\hat{\alpha}|\rangle_k = \sum_n \sum_m C_m^*(k)\langle m|\hat{\alpha}|n\rangle C_n(k)$$

and we see that the ensemble average is

$$\langle \alpha \rangle = \sum_n \sum_m \alpha_{mn}\rho_{nm} = \sum_m (\alpha\rho)_{mm} = Tr(\hat{\alpha}\hat{\rho}) = Tr(\hat{\rho}\hat{\alpha}). \tag{16.4}$$

If we change to a new representation with basic states

$$|\mu\rangle = \sum_n S_{n\mu}|n\rangle$$

the new density matrix has elements

$$\rho'_{\mu\nu} = (S^{-1})_{\mu n}\rho_{nm}S_{m\nu},$$

and, since this is the transformation law for a quantum mechanical operator, we can regard $\hat{\rho}$ as an operator. In particular, expectation values expressed as traces can be evaluated in any convenient representation, since

$$Tr(S^{-1}\hat{\rho}\hat{\alpha}S) = \sum_{nm\mu}(S^{-1})_{\mu n}(\hat{\rho}\hat{\alpha})_{nm}\, S_{m\mu} = \sum_{nm}(\hat{\rho}\hat{\alpha})_{nm}\delta_{nm} = \sum_{n}(\hat{\rho}\hat{\alpha})_{nn} = Tr(\hat{\rho}\hat{\alpha}).$$

If $\mathscr{H}$ is the Hamiltonian operator for the system the states satisfy the Schrödinger equation

$$i\hbar\frac{\partial}{\partial t}|\rangle_k = \mathscr{H}|\rangle_k$$

and this yields

$$i\hbar\frac{\partial\hat{\rho}}{\partial t} = \mathscr{H}\hat{\rho} - \hat{\rho}\mathscr{H}.$$

A stationary equilibrium ensemble in which the statistical distribution is constant must therefore have a density matrix which commutes with $\mathscr{H}$. It follows that $\hat{\rho}$ is diagonal in the energy representation and ρ_{nn} is the probability that a particular state of energy $E_n = \mathscr{H}_{nn}$ is represented in the ensemble. The quantum equivalent of equation (16.1) is then

$$\rho_{nn} = Z^{-1}\exp\left(-\frac{\mathscr{H}_{nn}}{kT}\right)$$

and so

$$Z = Tr\exp\left(-\frac{\mathscr{H}}{kT}\right). \tag{16.5}$$

Since the trace can be evaluated in any representation we can regard (16.5) as a general statement, independent of the representation, if we interpret the exponential function of an operator as

$$\exp\left(-\frac{\mathscr{H}}{kT}\right) = \sum_{\nu=0}^{\infty}\frac{1}{\nu!}\left(-\frac{\mathscr{H}}{kT}\right)^{\nu}. \tag{16.6}$$

The expectation value of the energy is

$$U = \langle\mathscr{H}\rangle = \frac{Tr\left\{\mathscr{H}\exp\left(-\frac{\mathscr{H}}{kT}\right)\right\}}{Tr\left\{\exp\left(-\frac{\mathscr{H}}{kT}\right)\right\}}$$

and this is easily seen to lead again to equation (16.3).

In a dielectric medium in a uniform applied field E_i^* the Hamiltonian, in either classical or quantum mechanics, is of the form

$$\mathscr{H} = \mathscr{H}_0 - E_i^*\mathfrak{P}_i,$$

where $\mathfrak{P}_i$ is the total dipole moment operator of the system or its classical equivalent. We see that $\mathfrak{P}_i = -(\partial\mathscr{H}/\partial E_i^*)$. Consider then the expectation value of $\partial\mathscr{H}/\partial\lambda$, where λ is a macroscopic parameter, for example, a field or a mechanical quantity such as the volume. In classical mechanics

$$\left\langle\frac{\partial\mathscr{H}}{\partial\lambda}\right\rangle = \frac{\int\frac{\partial\mathscr{H}}{\partial\lambda}\exp\left(-\frac{\mathscr{H}}{kT}\right)d\Gamma}{\int\exp\left(-\frac{\mathscr{H}}{kT}\right)d\Gamma}$$

which immediately yields

$$\left\langle \frac{\partial \mathscr{H}}{\partial \lambda} \right\rangle = -kT \frac{\partial}{\partial \lambda} \log Z = \left(\frac{\partial F}{\partial \lambda} \right)_T. \tag{16.7}$$

The quantum mechanical result is easily seen to be identical.

We have now collected together the general results that we shall need in the remainder of our discussion. The reader who requires a more thorough treatment is referred to any of the numerous excellent texts on statistical mechanics. In our discussion we shall, apparently, use classical and quantum concepts quite indiscriminately, choosing whichever leads to the most compact formulation. In reality we are almost always implying a quantum mechanical treatment, even if the calculation is phrased in classical terms. The very assumption that we are dealing with atoms, ions, or molecules, characterized by definite moments or moment operators, implies that there are some aspects of the structure of the system which have already been quantized. In this connection the reader will find the discussion of Miss van Leuwen's theorem in van Vleck (1932) instructive. This theorem demonstrates that in a purely classical theory all magnetic effects vanish identically.

In statistical mechanics the fields are regarded as impressed external parameters and so the theory leads to the free energy function F^s discussed in chapter 15. We shall, for brevity, use the unadorned letter F, but the reader should keep in mind the fact that F is now no longer the complete free energy of a complete thermodynamic system.

From this point onwards the sole problem in statistical mechanics is the evaluation of the partition function Z from the Hamiltonian $\mathscr{H}$ of the system. This problem is rarely trivial and, if there are appreciable interactions between the atoms of the system, is often thoroughly intractable even by a process of approximation. Before we embark on a study of real physical systems it will be useful to consider some of the aspects of the relations between F, Z and $\mathscr{H}$ and some of the cases where major simplifications can be effected.

In some cases $\mathscr{H}$ can be separated with a sum of terms $\mathscr{H}_\mu$ which are independent, in the sense that microscopic variables which appear in one term do not appear in any of the other terms. We then have

$$\mathscr{H} = \sum_\mu \mathscr{H}_\mu, \tag{16.8a}$$

and the partition function is the continued product

$$Z = \Pi_\mu Z_\mu, \tag{16.8b}$$

with

$$Z_\mu = \int \exp\left(-\frac{\mathscr{H}_\mu}{kT} \right) d\Gamma_\mu, \tag{16.8c}$$

where $d\Gamma_\mu$ corresponds only to the variables in $\mathscr{H}_\mu$. Further, since $F = -kT \log Z$ we have

$$F = \sum_\mu F_\mu. \tag{16.8d}$$

In this case, therefore, the whole problem can be separated into problems of, we hope, lesser complexity.

One obvious case of this procedure is the separation of the terms in $\mathscr{H}$ which refer to internal motions in atoms from the terms which refer to the centre of mass motion of the atoms. A second case occurs when we have a system of N non-interacting atoms, which, for simplicity, we assume to be identical, and each to have an individual Hamiltonian $\mathscr{H}_a$, so that

$$\mathscr{H} = \sum_a \mathscr{H}_a, \tag{16.9a}$$

$$Z = (Z_a)^N, \tag{16.9b}$$

$$F = -NkT \log Z_a. \tag{16.9c}$$

We now consider some special cases for a system of this type.

If the energy levels of the atoms are W_0, W_1, W_2, etc., with degeneracies g_0, g_1, g_2, etc., we have

$$Z_a = \sum_{\text{states}} \exp\left(-\frac{W_v}{kT}\right) = \sum_{\text{levels}} g_v \exp\left(-\frac{W_v}{kT}\right), \tag{16.10}$$

In some cases, e.g., in dealing with electronic polarization or diamagnetism, we know that $W_1 - W_0 \gg kT$ and only the lowest level contributes to Z_a so that

$$Z_a = g_0 \exp\left(-\frac{W_0}{kT}\right), \tag{16.11a}$$

$$F = N(W_0 - kT \log g_0), \tag{16.11b}$$

$$S = Nk \log g_0, \tag{16.11c}$$

$$U = NW_0. \tag{16.11d}$$

The electric dipole moment of the system, for example is given by

$$\mathfrak{P} = -\left(\frac{\partial F}{\partial E^*}\right)_T = -N\frac{\partial W_0}{\partial E^*}$$

and, in order to obtain $\mathfrak{P}$, we need only engage in an atomic calculation of the ground state energy W_0 as a function of E^*.

A second case, which occurs frequently, corresponds to atoms, molecules, etc., with a small number of low-lying levels W_0 - - - W_l and then a gap $W_{l+1} - W_l \gg kT$ to the next level. A familiar example is, of course, an atom or ion with a net angular momentum $J\hbar$ in the lowest level. In a magnetic field this splits into $2J+1$ levels. In many cases of this type it is possible to devise an effective Hamiltonian $\mathscr{H}_{al}$ (spin Hamiltonian) which makes no explicit reference to levels above W_l. We then have to evaluate

$$Z_a = Tr \exp\left(-\frac{\mathscr{H}_{al}}{kT}\right). \tag{16.12a}$$

Because the multiplicity of the system is low the evaluation of Z_a is at least a finite task. In some cases it can be done exactly. In other cases we can use the power series expansion

$$Z_a = Tr\, 1 + Tr\frac{-\mathscr{H}_{al}}{kT} + \frac{1}{2!}\, Tr\left(\frac{-\mathscr{H}_{al}}{kT}\right)^2 + \text{etc.} \tag{16.12b}$$

which always converges. If, in addition, the energy difference $W_l - W_0$ is small compared with kT the convergence is very rapid and often only the first few terms are required. There are a number of powerful techniques for evaluating sums such as (16.12b) and this is often a very useful approach to problems in paramagnetism (see, e.g., Simon *et al.*, 1951).

In general, if we have a system of interacting atoms, the evaluation of Z is difficult, if not impossible, with present techniques. There is, however, one simple case which was first discussed by van Vleck (1937). If we have N atoms, each of which has a small group of low-lying levels, and then a gap, large compared with kT to the next level, we can define a Hamiltonian $\mathscr{H}_l$ for the whole system which refers only to states built up from these low-lying atomic levels. We then have

$$Z = Tr\,1 - Tr\left(\frac{\mathscr{H}_l}{kT}\right) + \frac{1}{2!}\,Tr\left(\frac{\mathscr{H}_l}{kT}\right)^2 + \text{etc.}$$

Even if the spacings of the individual atomic levels and the interaction energies per atom are small compared with kT this converges very slowly if N is large but, nevertheless, yields an expression for F of the form

$$F = N\{f_0 + f_1T^{-1} + f_2T^{-2} + \text{etc.}\}$$

which converges rapidly. This has been used by van Vleck (1937) to give a treatment of dipole–dipole and exchange interaction in paramagnetic media which, if not exact, can at least be improved by the tedious but determinate process of calculating more terms in the expansion. Unfortunately, the convergence is slow if the strength of the interaction is comparable with kT and experimentally this is often the situation of greatest interest.

Non-interacting Systems

We now use these results to discuss a few special cases, beginning with systems of non-interacting atoms. The simplest case of all occurs when the atoms have a singlet ground state and the next level is at an energy large compared with kT. We then have

$$F = U = NW_0$$

and the entropy is zero. If W_{00} is the ground state energy in zero field an atomic calculation will yield either

$$W_0 = W_{00} + a_1E^* + a_2E^{*2} + \text{etc.}$$

or

$$W_0 = W_{00} + b_1H^* + b_2H^{*2} + \text{etc.}$$

(we use H^* rather than B^* to conform to the conventional notation used in this subject). If the system has a centre of symmetry $a_1 = a_3 = a_5 = \text{etc.} = 0$ while if it has zero angular momentum $b_1 = b_3 = \text{etc.} = 0$. The electric case corresponds to induced polarization and the magnetic case to diamagnetism. A slightly more complicated case corresponds to a degenerate ground state, whose degeneracy is not lifted by the field. This will be the case for atomic hydrogen in an electric field but not in a magnetic field (which lifts the two-fold spin degeneracy).

The next simplest case is represented by a paramagnetic atom or ion whose isolated low-lying level in zero field has angular momentum $J\hbar$ and degeneracy $2J+1$. The effective Hamiltonian in a magnetic field $\mathbf{H}^*$ is

$$\mathscr{H}_{al} = -\mu_0 g\beta\mathbf{H}^*\,.\,\mathbf{J}, \tag{16.13}$$

and

$$Z_a = \sum_{m=-J}^{J} \exp\left(\frac{\mu_0 g\beta H^*}{kT}\,m\right). \tag{16.14}$$

This can be evaluated exactly but, since the calculation is so familiar, we shall not repeat it. If $x = (\mu_0 g\beta H^*/kT) \ll 1$ we can also evaluate it as

$$Z_a = Tr\,\mathbf{1} + xTr\hat{J}_z + \tfrac{1}{2}x^2 TrJ_z^2 + \text{ etc.},$$

where we have taken H^* to define the z axis. But $Tr\,\mathbf{1} = 2J+1$, $TrJ_z = 0$ and $TrJ_z^2 = \frac{1}{3}J(J+1)(2J+1)$ so that

$$Z_a = (2J+1)\{1+\tfrac{1}{6}J(J+1)x^2 + \text{ etc.}\}, \tag{16.15a}$$

$$F \approx -NkT\log(2J+1) - \tfrac{1}{6}J(J+1)NkTx^2 + \text{ etc.}, \tag{16.15b}$$

$$\mathfrak{M} = -\frac{1}{\mu_0}\frac{\partial F}{\partial H^*} \approx \frac{J(J+1)\mu_0 g^2\beta^2 N}{3kT} H^*, \tag{16.15c}$$

$$S = -\frac{\partial F}{\partial T} \approx Nk\{\log(2J+1) - \tfrac{1}{6}J(J+1)x^2\}, \tag{16.15d}$$

and

$$\chi_T = \frac{1}{V}\left(\frac{\partial \mathfrak{M}}{\partial H^*}\right) = \frac{J(J+1)\mu_0 g^2\beta^2 N}{3kT\,V}. \tag{16.15e}$$

In many paramagnetic media the $2J+1$ fold level is split by crystal fields and this simple treatment is inadequate. We may distinguish two extreme cases in which the splitting is either large compared with kT or small compared with kT. The Co^{2+} ion, for example, has $J = 9/2$, but, in most cobalt salts at low temperatures, all, except a doubly degenerate ground state level, lie at high energy. In Mn^{++}, on the other hand, $J = 5/2$, and all six states lie close together in manganous salts. The effective Hamiltonian for cobalt, neglecting hyperfine structure, can be expressed in terms of a fictitious spin $S = \frac{1}{2}$ (since $2S+1=2$ is the degeneracy of the ground state in zero field). We have

$$\mathscr{H}_{al} = -\mu_0 H\beta_i^* g_{ij} S_j, \tag{16.16a}$$

where the axes of the g tensor are specified relative to axes in the crystal. In manganous salts the typical effective Hamiltonian is

$$\mathscr{H}_{al} = -\mu_0\beta g\mathbf{H}^*\cdot\mathbf{J} + DJ_z^2 + A\mathbf{J}\cdot\mathbf{I}, \tag{16.16b}$$

where the z axis is defined by the crystal and we have included hyperfine structure with a nuclear spin I.

In a principal axis system the Hamiltonian $\mathscr{H}_{al}$ of (16.16a) leads to

$$\chi_{ii} = \frac{S(S+1)\mu_0\beta^2 N}{3kT\,V} g_{ii}^2 \tag{16.17}$$

and the partition function is

$$Z_a = 2\cosh\left[\frac{\{(g_{11}H_{11}^*)^2 + (g_{22}H_{22}^*)^2 + (g_{33}H_{33}^*)^2\}^{\frac{1}{2}}\mu_0\beta}{2kT}\right]. \tag{16.18}$$

As a result F and S depend on the direction of the field relative to the crystal axes. In some rare earth salts the anisotropy is very pronounced with $g_{\min}/g_{\max} \to 0$. Rotation of a thermally isolated crystal in a field, so that S is constant, leads to a considerable change in temperature. This effect was used in the first demonstration of parity non-conservation (Ambler *et al.*, 1957), and is also the basis of the spin refrigerator (Jeffries, 1963; Abragam, 1961, p. 42; Robinson, 1963).

The last topic in non-interacting systems that we shall discuss is the response of molecules with a permanent electric dipole moment to an electric field. We have already considered this topic in chapter 13; the discussion here emphasizes a different aspect of the problem. This is only superficially similar to paramagnetism. We follow the treatment given by van Vleck (1932) and consider a system of molecules which can be treated as rigid linear dumbbells with a fixed electric dipole moment π parallel to the axis of the dumbbell. If the moment of inertia of a molecule about an axis normal to the length of the dumbbell is I, the energy levels in zero field are

$$W_0(J) = \hbar^2 \frac{J(J+1)}{2I},$$

where $J\hbar$ is the angular momentum. Each level is $2J+1$ fold degenerate. If $m\hbar$ is the projection of $\mathbf{J}\hbar$ on the direction of the field E^*, the perturbed levels, to second order in E^*, are given by

$$W(J, m) = W_0(J) + \frac{I\pi^2 E^{*2}}{\hbar^2} \frac{J(J+1)-3m^2}{J(J+1)(2J-1)(2J+3)}.$$

Notice that, for $J > 0$, a level corresponding to rotation about an axis perpendicular to E^*, i.e. with $m = 0$, has an increased energy and so contributes a negative polarization.

In practice we normally have $\pi E^* \ll kT$ and so, if we expand Z_a, we need only keep the first non-vanishing term in E^*. We have

$$Z_a = \sum_J \sum_m \exp\left(-\frac{W_0(J)}{kT}\right)\left\{1 - \frac{I\pi^2 E^{*2}}{\hbar^2 kT} \frac{J(J+1)-3m^2}{J(J+1)(2J-1)(2J+3)}\right\}.$$

In the sum over m, the terms quadratic in E^* vanish except when $J = 0$ so that

$$Z_a(E^*) = Z_a(0) + \frac{I\pi^2 E^{*2}}{3\hbar^2 kT},$$

where

$$Z_a(0) = \sum_{J=0}^{\infty} (2J+1) \exp\left(-\frac{\hbar^2 J(J+1)}{2IkT}\right).$$

If we also have $kT \gg \hbar^2/I$, so that several rotational levels are excited,

$$Z_a(0) \approx \frac{2I}{\hbar^2} kT,$$

and thus

$$Z_a(E^*) \approx \frac{2IkT}{\hbar^2}\left\{1 + \frac{1}{6}\frac{\pi^2 E^{*2}}{k^2T^2}\right\}.$$

This yields

$$F = F(0) - \frac{N}{6}\frac{\pi^2 E^{*2}}{kT}$$

and

$$\mathfrak{P} = -\frac{\partial F}{\partial E^*} = \frac{N}{3}\frac{\pi^2}{kT} E^*,$$

or

$$\chi = \frac{\pi^2}{3kT}\frac{N}{\varepsilon_0 V},$$

which is the Langevin–Debye result. We may remark that the usual elementary classical treatment of both the Curie and Langevin–Debye laws relies on an *ad hoc* assumption about the appropriate volume element in phase space. It also obscures the very real difference between the nature of the two phenomena.

Interacting Systems

In solids and liquids and even to some extent in gases we cannot neglect the interactions between the atoms and molecules of the system. Indeed, these interactions are largely responsible for the structure of the system. If we have a system of N units, not necessarily all the same, the total Hamiltonian of the system can be written as $\mathscr{H}_0+\mathscr{H}_{\text{int}}$ where $\mathscr{H}_0$ is the sum of the individual atomic Hamiltonians and $\mathscr{H}_{\text{int}}$ is the interaction term. This term will contain contributions from all sorts of interactions but amongst others it will contain a term representing electric or magnetic dipolar interaction and, if we propose to use statistical mechanics to discuss the polarizability of the medium or its magnetic properties, this term is of crucial interest to us. All the other terms we can regard as leading to interesting and possibly insoluble problems in atomic and solid state physics but the dipolar term is our particular concern. It is obviously closely connected with the local field problem.

We take the attitude that the other terms in $\mathscr{H}_{\text{int}}$ have been dealt with and that, as a result, the system is classified as a set of identical units which may be atoms, molecules or unit cells, each of which is characterized by an individual term in the Hamiltonian. The interaction term then involves only the dipolar interaction between these units. (In the magnetic case it is not difficult to extend this to include exchange interaction.) We will initially express our results in electrical terms and only later consider magnetic effects.

If we let $\hat{\Pi}_i(n)$ be the electric dipole moment operator for the nth unit located at $\hat{\mathbf{r}}(n)$ and write $\hat{\mathbf{r}}(n,m) = \hat{\mathbf{r}}(n)-\hat{\mathbf{r}}(m)$ the electric dipole interaction energy between units (n) and (m) can be expressed as $-\hat{\Pi}_i(n)\hat{\Gamma}_{ij}(n, m)\hat{\Pi}_j(m)$ where

$$\hat{\Gamma}_{ij}(n, m) = \frac{3\hat{r}_i(n, m)\hat{r}_j(n, m)-\hat{r}^2(n, m)\delta_{ij}}{4\pi\varepsilon_0\hat{r}^5(n, m)}. \tag{16.19}$$

Notice that $\hat{\Gamma}_{ij}(n, m)$ is symmetric in both the coordinate suffixes (i, j) and the atomic labels (n, m). If $\mathscr{H}_0(n)$ is the Hamiltonian of the nth unit the Hamiltonian of the whole system in a field E_i^* due to external charges is

$$\mathscr{H} = \sum_n\{\mathscr{H}_0(n)-E_i^*(r(n)\hat{\Pi}_i(n)\}-\tfrac{1}{2}\sum_{m\neq n}\sum \hat{\Pi}_i(n)\hat{\Gamma}_{ij}(n, m)\hat{\Pi}_j(m). \tag{16.20}$$

The tensor coefficients $\hat{\Gamma}_{ij}(n, m)$ are functions of the microscopic coordinates $\hat{\mathbf{r}}(n)$ and $\hat{\mathbf{r}}(m)$ and are, therefore, strictly speaking, microscopic variables. This is important in some problems but we shall arbitrarily assume, either that the $\hat{\mathbf{r}}(n)$ are fixed, or that some form of average has already been taken. Thus we shall treat the $\Gamma_{ij}(n, m)$ as parameters and, to indicate this, we drop the "hat". We shall also assume that the impressed field is uniform so that we have only to consider

$$\mathscr{H} = \sum_n \mathscr{H}_0(n)-E_i^*\sum_n \hat{\Pi}_i(n)-\tfrac{1}{2}\sum_{n\neq m}\sum \hat{\Pi}_i(n)\Gamma_{ij}(n, m)\hat{\Pi}_j(m). \tag{16.21}$$

We suppose, in addition, that the problem for non-interacting atoms is soluble so that we can concentrate our attention on the effects of the interaction. One approach is that used by van Vleck (1937) in which the partition function is expanded in powers of $1/kT$. This is relatively successful in paramagnetism where there is a useful range of temperatures in which the series converges rapidly. It is less useful in dielectrics. The second approach, which we adopt, is to introduce a local field E_i^{loc} to replace E_i^*. In Onsager's treatment $E_{(n)}^{loc}$ is set equal to the expectation value of the field at the site (n) when the dipole at this site is held fixed, and all other dipoles have their expectation values under the combined influence of the impressed field $\mathbf{E}^*$, and the fields due to each other, but not the field of the one fixed dipole. This makes $E^{loc}(n)-E^*$ a function of the moment operator at site (n) (see chapter 13). We shall consider the simpler form of the theory in which $E_{(n)}^{loc}-E^*$ is set equal to the field due to all the other dipoles when they have their final expectation values. Thus if $\langle\Pi_j(m)\rangle$ is the expectation value of the jth component of the moment of unit (m)

$$E_i^{loc}(n) = E_i^* + \sum_{m\neq n} \Gamma_{ij}(n, m)\langle\Pi_j(m)\rangle. \tag{16.22}$$

We have now to discuss the evaluation of the sum in this expression. If we assume that the final polarization of the medium is to be uniform, which will certainly mean that the medium must be homogeneous and of some special shape, e.g., an ellipsoid, then $\langle\Pi_j(m)\rangle$ is independent of (m) and we have

$$E_i^{loc}(n) = E_i^* + \langle\Pi_j\rangle \sum_{m\neq n} \Gamma_{ij}(n, m). \tag{16.23}$$

We take a closed surface σ surrounding the point (n) and let

$$S_{ij}(\sigma) = \sum_{m\neq n}^{\sigma} \Gamma_{ij}(m, n)$$

be the sum over all sites within σ except (n). Because Γ_{ij} is symmetrical we can choose a principal axis system in which S_{ij} is diagonal, and

$$S_{ii}(\sigma) = \sum_{m\neq n}^{\sigma} \frac{3r_i^2-r^2}{4\pi\varepsilon_0 r^5}.$$

We can obviously find an ellipsoidal surface σ for which each of the three diagonal elements S_{11}, S_{22} and S_{33} is zero. Thus it is possible to choose σ so that $S_{ij} = 0$. If we choose a large enough surface the medium outside σ can be treated as a continuum with a polarization $P_j = N/V\langle\Pi_j\rangle$ and so we can express $E^{loc}-E^*$ in terms of the shape factor γ_{ij}^{σ} for the surface σ and the corresponding factor γ_{ij}^{S} for the external surface of the body. Thus

$$E_i^{loc} = E_i + \frac{1}{\varepsilon_0}\gamma_{ij}^{\sigma} P_j = E_i^* + \frac{1}{\varepsilon_0}\gamma_{ij} P_j \tag{16.24}$$

where

$$\gamma_{ij} = \gamma_{ij}^{\sigma} + \gamma_{ij}^{S}. \tag{16.25}$$

We also obviously have

$$\gamma_{ij} = \frac{V}{N}\varepsilon_0 \sum_{m\neq n} \Gamma_{ij}(m, n). \tag{16.26}$$

We now introduce the microscopic variable

$$\Delta\hat{\Pi}_j(m) = \hat{\Pi}_j(m) - \langle\Pi_j\rangle \tag{16.27}$$

and write the Hamiltonian as

$$\mathscr{H} = \sum_n \{\mathscr{H}_0(n) - E_i^{\text{loc}}\hat{\Pi}_i(n)\} - \tfrac{1}{2} \sum_{m \neq n}\sum \Delta\hat{\Pi}_i(n)\Gamma_{ij}(n, m)\Delta\hat{\Pi}_j(m)$$
$$+\tfrac{1}{2} \sum_{m \neq n}\sum \langle \Pi_i(n)\rangle \Gamma_{ij}(n, m)\langle \Pi_j(m)\rangle. \tag{16.28}$$

This, within the limitations of our original dipole approximation, is still exact. Using (16.26) we can write the last sum in (16.28) as

$$\tfrac{1}{2} \sum_{m \neq n}\sum \langle \Pi_i(n)\rangle \Gamma_{ij}(n, m)\langle \Pi_j(m)\rangle = \tfrac{1}{2}\frac{V}{\varepsilon_0} P_i\gamma_{ij}P_j = \frac{1}{2\varepsilon_0 V}\mathfrak{P}_i\gamma_{ij}\mathfrak{P}_j \tag{16.29}$$

in terms of the macroscopic quantities P_i, the polarization, or $\mathfrak{P}_i$, the total dipole moment. This will appear as an additive term in F and so we need only consider the free energy F' derived from

$$\mathscr{H} = \sum_n (\mathscr{H}_0(n) - E_i^{\text{loc}}\hat{\Pi}_i(n)) - \tfrac{1}{2} \sum_{m \neq n}\sum \Delta\hat{\Pi}_i(n)\Gamma_{ij}(n, m)\Delta\hat{\Pi}_j(m). \tag{16.30}$$

In the final stages of the calculation we should have to replace F' by

$$F = F' + \tfrac{1}{2}\frac{V}{\varepsilon_0}\gamma_{ij}P_iP_j. \tag{16.31}$$

The whole point of the local field approach is that the one troublesome term in $\mathscr{H}$, i.e. the double sum, now only involves operators $\Delta\hat{\Pi}$ whose expectation value is zero. We therefore hope that its effect will be small, or at least that the term will exert a small effect on the field dependence of F. We shall examine this assumption later, but for the moment we assume that it is valid. We are then left with

$$\mathscr{H}' = \sum_n (\mathscr{H}_0(n) - E_i^{\text{loc}}\hat{\Pi}_i(n)) \tag{16.32}$$

which consists of a sum of individual non-interacting terms. We have assumed that this is a soluble problem. It leads to a free energy $F^{\text{loc}}(E^{\text{loc}}, T)$ identical in form with the free energy $F(E^*, T)$ obtained by neglecting interactions. Thus, if this is known to lead to $P = \varepsilon_0\chi^{\text{ni}}E^*$, we have

$$P = \varepsilon_0\chi^{\text{ni}}\left(E^* + \frac{1}{\varepsilon_0}\gamma P\right) = \varepsilon_0\chi^{\text{ni}}\left(E + \frac{1}{\varepsilon_0}\gamma^\sigma P\right)$$

and

$$P = \varepsilon_0 \frac{\chi^{\text{ni}}}{1 - \gamma\chi^{\text{ni}}} E^*,$$

or

$$P = \varepsilon_0 \frac{\chi^{\text{ni}}}{1 - \gamma^\sigma\chi^{\text{ni}}} E.$$

This gives

$$\chi = \frac{\chi^{\text{ni}}}{1 - \gamma^\sigma\chi^{\text{ni}}}.$$

In an isotropic medium $\gamma^\sigma = \frac{1}{3}$ and so this argument leads to the Lorentz local field correction. The objections to the Lorentz treatment, other than those which are general objections to the dipole approximation, must therefore be connected with dropping the terms in $\Delta\hat{\Pi}(n)$. This is rather obvious in the case of a medium consisting of molecules with permanent moments, for the dipolar coupling between neighbouring molecules correlates fluctuations $\Delta\hat{\Pi}(m)$ and $\Delta\hat{\Pi}(n)$ on nearby sites. It is, however, usually assumed that this does not occur when we are dealing solely with induced electronic moments. We now investigate this assumption.

We begin by remarking that the neglected term

$$-\tfrac{1}{2}\sum_{m\neq n}\sum \Delta\hat{\Pi}_i(m)\Gamma_{ij}(n,\ m)\Delta\hat{\Pi}_j(n)$$

is, in fact, responsible for van der Waals forces. These are not necessarily small, even if in many media they are small compared with other interactions. The approximate van der Waals contribution to the binding energy per atom in a solid is of the order of $u_0(N\alpha/4\pi V)^2$ where α is the atomic polarizability and u_0 the atomic ionization energy, e.g., about 10 eV. Since $N\alpha/V \approx \chi_{\text{electronic}} \approx 1$ we see that the van der Waals energy is an appreciable fraction of an electron-volt. Thus, the neglected term is itself large. It is still, however, possible that its effect on the *field dependence* of F is small. To investigate this we write

$$\mathscr{H}' = \mathscr{H}_0 - E^{\text{loc}}\hat{\mathfrak{P}} - \lambda\hat{W}$$

where $\hat{W}$ is the neglected van der Waals operator, and λ is a dummy parameter eventually to be set equal to unity. We obtain from this a free energy which yields a Maxwell relation

$$\frac{\partial\mathfrak{P}}{\partial\lambda} = \frac{\partial W}{\partial E^{\text{loc}}}.$$

Now the only way in which W can depend on E^{loc} is through the dependence of either the atomic polarizability α or the ionization potential u_0 on E^{loc}. In fact for any atomic system we find that both u_0 and α depend quadratically on the electric field, although the effect is small. We thus obtain

$$\frac{\partial\mathfrak{P}}{\partial\lambda} = \varepsilon_0\theta V E^{\text{loc}}, \tag{16.33}$$

where θ is a small numerical constant. The effect of the van der Waals term is therefore a small (usually positive) change in the susceptibility. It is not feasible to make a detailed calculation of the magnitude of this increase but a crude approximation indicates that it is unlikely ever to exceed 1%. Thus, although at first sight there is no justification for omitting the van der Waals term in dealing with electronic polarization, in practice it leads to errors which are insignificant in comparison with those inherent in the use of the dipole approximation or, for that matter, in our assumption that the problem of calculating the response of the non-interacting units is trivial.

The local field correction is usually presented as a method of relating the observed experimental susceptibility χ to an intrinsic susceptibility $\chi^{\text{ni}} = N\alpha/V$, with the implication either that α can be obtained from independent measurements, or that α can be calculated, with some exactness, theoretically. In magnetism this attitude is often justified but in dielectrics there is almost no way in which we can arrive at a reliable independent estimate of α. The strongest evidence for the general correctness of the local field correction in media with well-defined non-overlapping structural units comes from the additivity rules, discussed in chapter 13.

In magnetism we are primarily concerned with media containing ions, atoms or molecules with a permanent magnetic moment. However, the dipolar interactions are now much smaller. For a given interatomic spacing the ratio is of the order of $\varepsilon_0/\mu_0\ \beta^2/\delta^2$ where β is the Bohr magneton and δ the Debye. This ratio is approximately 10^{-4}. Further, whereas all atoms have appreciable electronic polarizabilities, only free radicals and transition group ions or atoms have permanent magnetic moments; thus, in a material such as $KCr(SO_4)_2 \cdot 12H_2O$ in which only the Cr^{3+} ions are magnetic, the spacing between magnetic ions is large. Indeed in this material the dipole–dipole term is small compared with kT above about 3×10^{-2} degrees K. For this reason van Vleck's method of expanding the partition function in powers of $1/kT$ is often very successful in dealing with paramagnetic interactions.

In zero field we have

$$\hat{\mathscr{H}} = \sum_n \hat{\mathscr{H}}_0(n) - \tfrac{1}{2} \sum_{n \neq m} \sum \hat{\mu}_i(n) \Gamma_{ij}(n, m) \hat{\mu}_j(m) \qquad (16.34)$$

and there is now no distinction between the magnetic moment operator $\hat{\mu}_i(n)$ and $\Delta\hat{\mu}_i(n)$ since $\langle \mu_i \rangle = 0$. If we omit the interaction term, the zero field atomic Hamiltonian $\hat{\mathscr{H}}_0(n)$ leads to a degenerate ground state level and at low temperatures all other levels will be at a high enough energy to be disregarded. As a result $F = NW_0 - NkT \log g_0$, $S = Nk \log g_0$ and the specific heat is zero. The inclusion of the interaction term (which would not appear if we took the simple local field correction) alters this situation and lifts some of the g_0^N fold degeneracy of the ground state. Van Vleck's treatment leads to a term which gives a thermal capacity per gram molecule of the form

$$C = R\frac{\theta^2}{T^2}. \qquad (16.35)$$

This is most easily observed experimentally using paramagnetic relaxation techniques (see chapter 15).

Magnetism, although usually much simpler than dielectric phenomena, and more amenable to calculation, largely because of the wealth of paramagnetic resonance data, is further complicated by exchange interaction. In some media this is so strong that it is quite pointless to attempt to treat the system in terms of individual dipoles. We shall not attempt to deal with this case since the exceedingly difficult problems it presents are less connected with magnetism than with the application of statistical mechanics to solid state physics.

So far we have been discussing essentially equilibrium properties in static fields, or fields whose periods were long compared with any natural period of the system. In some cases, e.g. optical phenomena in ionic crystals, we intuitively assume that the electronic polarization can be treated in isolation and this is by and large true, but uninteresting since neither thermodynamics nor statistical mechanics (unless we consider thermal radiation) adds much to the discussion. There are, however, some systems, e.g., paramagnetic materials, where we have two sub-systems each of which is separately characterized by short relaxation times, but whose interaction is characterized by a much longer relaxation time. It is then quite often useful to apply statistical mechanics to each sub-system separately. Thus, in paramagnetic media we can often consider the magnetic spin system as a distinct entity, and amenable to statistical calculation quite separately from the lattice system.

Although we originally introduced the quantum mechanical density matrix $\hat{\rho}$ or ρ_{nm} as the analogue of the classical ensemble distribution function, its use and properties are much more general. There is an excellent introduction to the use of the density matrix as a general tool in time-dependent electromagnetic problems by Bloembergen (1965); its application to magnetic resonance phenomena is treated by Abragam (1961).

Summary

We conclude by summarizing our results. Statistical mechanics furnishes a formal procedure for calculating macroscopic properties from the structure of a medium summarized in the expression for its Hamiltonian function $\mathscr{H}$. The primary result of a statistical calculation is an expression for the free energy of a body in an impressed field E^* or H^* as a function of the field and the temperature. In paramagnetism enough data are usually available from paramagnetic resonance experiments for us to construct an accurate effective Hamiltonian neglecting interactions, and we can often calculate macroscopic quantities such as χ or the entropy S more accurately and easily than we can measure them. Paramagnetic relaxation data also furnish approximate information about the effect of interactions. In discussing dielectric phenomena in any medium appreciably more complicated than a gas at atmospheric pressure, we rarely have enough *a priori* information to construct a worthwhile Hamiltonian; in addition the interaction energies are sufficiently large for us to be in some doubt about their effect. In this case statistical mechanics is of more use as descriptive theory than as a means of calculating useful information. It is easier to measure a refractive index to five-figure accuracy than to calculate it to two significant figures.

Problems

1. The effective Hamiltonian for the Ni^{2+} ion in an applied field $\mathbf{B}^*$ is known from paramagnetic resonance data to be of the form $\mathscr{H} = g\beta\mathbf{B}^* . \mathbf{J} + DJ_z^2$ where $J = 1$. Show that, to order $1/T$, the term DJ_z^2 has no effect on the paramagnetic susceptibility when $\mathbf{B}^*$ is parallel to the z axis.

2. The salt $Ce_2Mg_3(NO_3)_{12}.24H_2O$ has an effective Hamiltonian for the Ce^{3+} ions given by $\mathscr{H} = g\beta(S_xB_x^* + S_yB_y^*)$ with an effective spin $S = \frac{1}{2}$. If a single crystal is isothermally magnetized at 1 K in a field B_x^* such that $g\beta B^* \gg kT$ and then rotated so that $\mathbf{B}^*$ is parallel to the z axis it cools to 0·01 K. The protons in the water of crystallization have a nuclear moment such that $g_n\beta B^*$ is comparable with kT at 0·01 K in practicable fields. It has been proposed that this system might be used to produce a substantial degree of proton polarization. Comment on this proposal. (Hint: consider the entropy of the system.) In the light of your answer what can be said about the thermal equilibrium between the electronic and proton spin systems?

3. A gas consists of N molecules of permanent dipole moment π and moment of inertia I in a volume V. Calculate the dielectric susceptibility at a temperature low enough that no rotational levels with $J > 0$ are excited.

CHAPTER 17

Thermal Radiation

Thermal Energy in the Electromagnetic Field

In the last two chapters we have regarded the electromagnetic field as storing energy which is completely recoverable as work. The phenomena associated with radiant heat transfer, however, make it quite obvious that the field can also store energy as heat, because if we supply the energy to the field as heat it can never be completely recoverable as work. Since energy stored in static fields is clearly recoverable as work, thermal electromagnetic energy must be exclusively associated with oscillatory or wave fields. Not all wave energy, of course, is in the form of heat. The distinction between heat and work is a matter of the coherence of the field. It will be simplest and certainly involve us in fewer ill-defined terms if we discuss this notion in terms of the mechanical vibrations of a crystal.

Suppose that we have a perfect crystal of some simple and well-defined shape, e.g. a cube, of rock-salt and we excite one of its normal modes of resonance.This will require the expenditure of work. When the external excitation is removed the crystal will continue to vibrate with a well-defined amplitude and phase in the same normal mode. The stored energy remains available as work.

Now the concept of a normal mode depends on the assumption that the forces acting on the particles of the medium are harmonic so that the Hamiltonian of the system is a quadratic form in the $3N$ momentum components p_n and $3N$ coordinate components q_n of the N particles of the system, i.e., $\mathscr{H} = \sum_n \sum_m (\mu_{nm} p_n p_m + k_{nm} q_n q_m)$. In a real crystal this is of course not true and the actual Hamiltonian will contain further terms, e.g., $\sum_l \sum_n \sum_m \gamma_{lnm} q_l q_n q_m$ etc. which are anharmonic. The description in terms of normal modes depends on the assumption that the effect of these terms is small, so that, if we excite the system in one of the approximate normal modes, the excitation will persist in a form, recognizably associated with this mode, for at least a reasonable number of periods of the mode. After a long time, however, the excitation, due to the effect of the anharmonic terms, will take on a quite different form and, if we were to attempt to describe it in terms of a superposition of approximate normal modes, we would find that we required an exceedingly large number of terms, which, after an indefinitely long period, would approach $3N$, the total number of independent normal modes for the crystal. The detailed behaviour of the excitation has now become so complex that, not only are we unable to calculate it, but also we could not comprehend the result and its details would depend critically on the initial conditions, which we only know with macroscopic exactitude. Thus, we do not attempt to describe the ultimate excitation in determinate, but only in statistical, language and, in macroscopic terms, we describe the excitation as an incoherent thermal excitation. Clearly the process we have described above is normally referred to as dissipation. As dissipation proceeds we pass from a description of the excitation in terms of a single, approximate, normal mode, with a definite phase, amplitude and energy, to a description in which we are content to know the probability that each of the $3N$ modes is excited with a particular energy and we give up all hope of assigning phases to the excitations. This means that we cannot, using any device specified in macroscopic terms, hope to recover all the stored energy as work.

Clearly we can extend these ideas to a discussion of the excitation of the electromagnetic field although we must now remember that the field itself is a linear system with exact normal modes which persist indefinitely and that the non-linear interactions which lead to dissipation and the approach to thermal equilibrium are associated with the coupling between the field and matter. In addition we have also to remember that the modes of the field have an inherently continuous distribution in frequency and that the number of degrees of freedom of the field is infinite.

Thermal radiation in vacuum is primarily of interest in connection with the transport of heat but electromagnetic excitations of thermal origin also occur in one dimensional systems or circuits where they are known as Johnson noise. These incoherent fluctuations set an ultimate limit to the sensitivity of all types of electronic apparatus. Similar fluctuations occur in every form of apparatus but in practice only electronic systems are inherently so free from spurious fluctuations that the fundamental limit is of practical significance. We shall discuss noise and fluctuations in the next chapter but some of the examples at the end of this chapter are designed to show the link with the material of this chapter.

Thermodynamics

To discuss the thermodynamic properties of any system we need to know not only that it can store energy as heat, but also that it can exchange energy reversibly with other systems and that in this process heat can be converted to work and *vice versa.* We also need to be able to describe at least one such process in quantitative terms. In the case of a gas the key phenomenon is the work done by gas pressure in an expansion. In the case of radiation also, the simplest reversible processes involving work and heat are those associated with radiation pressure. Before we consider this phenomenon we have first to establish some of the fundamental properties of thermal radiation in equilibrium and to explain how we assign a temperature to a radiation field.

We imagine, first of all, that we have a large cavity with perfectly reflecting walls and that we excite, coherently, a number of its normal modes and then introduce into the cavity a small lossy body. In the fullness of time the coherent excitation will disappear and the body will reach a temperature T. It will then be radiating heat into the cavity at a rate which just balances the energy it absorbs. If we now introduce a second small body, also at T, into the cavity the second law ensures that there can be no net heat transfer between the bodies. Thus the radiation field in the cavity is in equilibrium with any body at the same temperature T. It can therefore be assigned the temperature T.

If two cavities both contain bodies at T and communicate via a small hole, the net energy transfer through the hole must be zero. The energy density in both cavities must therefore be the same and so the energy density u of radiation characterized by a temperature T is a function only of T.

By considering two bodies of different shape in the same cavity we conclude also that the radiation is isotropic, i.e., the energy flux in any direction is independent of that direction. Finally, by considering two bodies with different spectral responses to radiation, we may also conclude that the contribution $\mathrm{d}u$ to u associated with frequencies in $\mathrm{d}\nu$ at ν is a function only of ν and T. Thus

$$\mathrm{d}u = f(\nu, T)\,\mathrm{d}\nu. \tag{17.1}$$

Thermal Radiation

The stress tensor of the electromagnetic field is (see chapter 10)

$$T_{ij} = \varepsilon_0(E_iE_j - \tfrac{1}{2}E^2\,\delta_{ij}) + \frac{1}{\mu_0}(B_iB_j - \tfrac{1}{2}B^2\,\delta_{ij})$$

and the normal component of the force per unit area, i.e., the pressure, acting on a perfectly reflecting surface is $-T_{nn}$ where n is the normal into the surface. Thus

$$p = -\varepsilon_0(E_n^2 - \tfrac{1}{2}E^2) - \mu_0(B_n^2 - \tfrac{1}{2}B^2).$$

Since radiation is isotropic $E_n^2 = \tfrac{1}{3}E^2$ and $B_n^2 = \tfrac{1}{3}B^2$, thus

$$p = \tfrac{1}{3}\left\{\tfrac{1}{2}\varepsilon_0 E^2 + \frac{1}{2\mu_0}B^2\right\} = \tfrac{1}{3}u. \tag{17.2}$$

We now consider a hypothetical heat engine consisting of a cylinder whose partially reflecting walls can either be isolated, or exchange heat with a reservoir at T. The cylinder is closed at one end by a fixed wall and at the other end by a perfectly reflecting piston. The volume of the cylinder is V and the energy of the radiation is $U = uV$. For a general reversible change

$$T\,\mathrm{d}S = \mathrm{d}U + p\,\mathrm{d}V = \mathrm{d}(uV) + p\,\mathrm{d}V = 3\,\mathrm{d}(pV) + p\,\mathrm{d}V,$$

so that

$$\mathrm{d}S = \frac{3}{T}V\,\mathrm{d}p + \frac{4}{T}p\,\mathrm{d}V. \tag{17.3}$$

Since p and u are functions only of T this leads to

$$\mathrm{d}S = \frac{3V}{T}\frac{\mathrm{d}p}{\mathrm{d}T}\mathrm{d}T + 4\frac{p}{T}\mathrm{d}V,$$

and this then yields the Maxwell relation

$$\frac{\partial}{\partial V}\left(\frac{3V}{T}\frac{\mathrm{d}p}{\mathrm{d}T}\right) = \frac{\partial}{\partial T}\left(4\frac{p}{T}\right).$$

From this we obtain

$$\frac{\mathrm{d}p}{\mathrm{d}T} = 4\frac{p}{T}$$

and so

$$p = \tfrac{1}{3}aT^4, \tag{17.4a}$$

and

$$u = aT^4, \tag{17.4b}$$

where a is a universal constant.

The energy incident on unit area of the walls per unit time is

$$P = \frac{1}{4\pi}\int_0^{2\pi}\int_0^{\pi/2} uc\cos\theta\sin\theta\,\mathrm{d}\theta\,\mathrm{d}\phi = \tfrac{1}{4}uc,$$

so that

$$P = \tfrac{1}{4}acT^4 = \sigma T^4. \tag{17.5}$$

This must also be the rate at which energy is radiated per unit area by a perfectly absorbing black body at T. This result is known as Stefan's law. The constant σ, Stefan's constant, cannot be obtained from thermodynamic arguments. Its experimental value is approximately 6×10^{-8} watts m^{-2} deg^{-4} which is most easily remembered as 6 watts cm^{-2} at 1000 K. A sphere of radius 1 m at 300 K loses heat by radiation at a rate of 1 kilowatt. The large value of σ should not lead us, however, to believe that the density of radiant energy $(4\sigma/c)\, T^4$ is large. At 300 K it is $6{\cdot}5 \times 10^{-6}$ joule m^{-3} which may be compared with the thermal energy of a typical solid at the same temperature which is of the order of 10^8 joule m^{-3}.

The Doppler shift in the frequency of radiation reflected from a moving piston gives us a means of investigating the spectral composition of thermal radiation. If, in our engine, the piston, of area A, moves slowly outwards with a velocity v, then $\delta V = Av\,\delta t$. When no heat enters the system we can discuss the change in the spectral distribution during this isentropic change as follows. Radiation reflected into a solid angle $d\Omega$ at an angle θ to the normal, in a frequency interval $d\nu$ at ν, originally came from radiation incident at θ in $d\nu'$ at $\nu' = \nu(1+2v/c\cos\theta)$. The energy incident during δt is $cf(\nu')\,d\nu'\cos\theta\,(d\Omega/4\pi)A\,\delta t$. This exerts a force $2f(\nu')\,d\nu'\cos^2\theta\,A(d\Omega/4\pi)$ on the piston and so, in δt, does work $2f(\nu')\,d\nu'\cos^2\theta\,Av\,\delta t(d\Omega/4\pi.)$ The energy reflected is therefore $cf(\nu')\,d\nu'\cos\theta\,A\,\delta t\,(1-2v/c\cos\theta)\,(d\Omega/4\pi)$. Thus, during δt, the total energy in $d\nu$, i.e. $Vf(\nu)\,d\nu$, is changed by

$$\delta(Vf(\nu)\,d\nu) = \int\limits_{\Omega}\left(cf(\nu')\,d\nu'\left\{1-\frac{2v}{c}\cos\theta\right\}-cf(\nu)\,d\nu\right)A\delta t\,\frac{\cos\theta\,d\Omega}{4\pi}.$$

If we now put

$$f(\nu') = f(\nu)+\frac{\partial f}{\partial\nu}(\nu'-\nu)+\text{etc.}$$

we have

$$f(\nu') = f(\nu)+\frac{2v\cos\theta}{c}\nu\,\frac{\partial f}{\partial\nu}+\text{etc.}$$

and, since

$$\left(1-\frac{2v}{c}\cos\theta\right)d\nu' = \left\{1-\left(\frac{2v\cos\theta}{c}\right)\right\}^2 d\nu \approx d\nu,$$

we obtain

$$\delta\{Vf(\nu)\} = \int\limits_{\Omega} 2\nu\frac{\partial f}{\partial\nu}Av\,\delta t\,\frac{\cos^2\theta\,d\Omega}{4\pi}.$$

Since we also have $Av\,\delta t = \delta V$, this yields

$$\delta\{Vf(\nu)\} = \tfrac{1}{3}\nu\,\frac{\partial f}{\partial\nu}\,\delta V.$$

If we write this as

$$\frac{\partial}{\partial V}\left\{Vf(\nu)\right\} = \tfrac{1}{3}\nu\,\frac{\partial f}{\partial\nu}$$

the general solution is easily seen to be

$$f = \nu^3\psi(\nu^3 V), \tag{17.6}$$

where ψ is an arbitrary function of the argument $\nu^3 V$. Since the change was reversible and isentropic, we can use the relation

$$T\left(\frac{\partial S}{\partial T}\right)_V = V\frac{\partial u}{\partial T} = V\frac{du}{dT} = 4aVT^3$$

which gives

$$S = \frac{4}{3}aVT^3,$$

to express (17.6) as

$$f(\nu) = \nu^3\phi\left(\frac{\nu}{T}\right), \tag{17.7}$$

where again ϕ is an arbitrary function of ν/T. We have, therefore,

$$\frac{\partial f}{\partial \nu} = 3\nu^2\phi\left(\frac{\nu}{T}\right)+\frac{\nu^3}{T}\frac{\partial \phi}{\partial(\nu/T)}.$$

We see that the maximum intensity occurs at a frequency $\nu_{\max}$ for which

$$3\phi\left(\frac{\nu}{T}\right)+\frac{\nu}{T}\frac{\partial}{\partial(\nu/T)}\phi(\nu/T) = 0.$$

Thus $\nu_{\max}$ is proportional to T which is Wien's displacement law. This is about as far as we can go in analysing thermal radiation by purely thermodynamic arguments. But it is pertinent to remark that, had Planck stopped at this point, he would not have discovered the quantum theory.

Statistical Mechanics

We next turn to statistical mechanics, and now we regard the system of radiation, in a cavity exchanging heat through the walls with a reservoir at T, as a representative member of a canonical ensemble. If we assume that the coupling to the reservoir due to loss in the walls is small, the Hamiltonian of the radiation field can be expressed as a sum of independent harmonic oscillator terms, one for each normal of the cavity. We label the frequencies of these modes as ν_m so that the energy levels of each mode are $(n+\frac{1}{2})h\nu_m$, where n is integral, and we have

$$Z = \prod_m \sum_{n=0}^{\infty} \exp\left(-\frac{(n+\frac{1}{2})h\nu_m}{kT}\right). \tag{17.8}$$

From this we obtain the thermodynamic functions. In particular

$$U = \sum_m \left\{\tfrac{1}{2}h\nu_m+\frac{h\nu_m}{\exp(h\nu_m/kT)-1}\right\}. \tag{17.9}$$

We have now only to calculate the distribution of the normal modes. We begin with the usual elementary discussion and consider a large cubic cavity of side L. The frequencies of the normal modes are then given by

$$\left(\frac{2\nu L}{c}\right)^2 = p^2+q^2+r^2$$

where p, q and r are positive integers. The number of modes up to ν is equal to the number of points with integral coordinates in the positive octant of a sphere of radius $2\nu L/c$ and this is approximately

$$N(\nu) = \frac{1}{8}\frac{4\pi}{3}\cdot\left(\frac{2\nu L}{c}\right)^3.$$

A further factor 2 occurs because the electromagnetic field is a transverse field with two independent polarizations and so we obtain

$$N(\nu) = \frac{8\pi\nu^3}{3c^3}L^3 = \frac{8\pi\nu^3}{3c^3}V. \tag{17.10}$$

The number of modes in $d\nu$ at ν is therefore

$$dN(\nu) = \frac{8\pi\nu^2}{c^3}V\,d\nu. \tag{17.11}$$

It is possible to show that (17.10) and (17.11) are asymptotically correct (Weyl, 1911) for a cavity of any shape whenever $\nu R/c \gg 1$, where R is the least characteristic length or radius of curvature of the cavity. If then we neglect any effects due to possible departures from (17.10) and (17.11) at low frequencies (we consider this point at a later stage), we have, for the energy dU associated with frequencies in $d\nu$,

$$dU = V\left\{\tfrac{1}{2}h\nu + \frac{h\nu}{\exp(h\nu/kT)-1}\right\}\frac{8\pi\nu^2\,d\nu}{c^3}. \tag{17.12}$$

The energy density is

$$du = \left\{\tfrac{1}{2}h\nu + \frac{h\nu}{\exp(h\nu/kT)-1}\right\}\frac{8\pi\nu^2\,d\nu}{c^3}. \tag{17.13}$$

When we express this as

$$\frac{du}{d\nu} = \nu^3\left\{\frac{8\pi h}{c^3}\right\}\left\{\tfrac{1}{2} + \frac{1}{\exp(h\nu/kT)-1}\right\},$$

we see that it is in agreement with the thermodynamic result (17.7).

If, in (17.13), we take the limit as $h \to 0$ we have

$$\tfrac{1}{2}h\nu + \frac{h\nu}{\frac{h\nu}{kT}+\frac{1}{2}\frac{h^2\nu^2}{k^2T^2}+\ldots} = \tfrac{1}{2}h\nu + kT\left(1-\frac{1}{2}\frac{h\nu}{kT}\right) = kT,$$

and so we obtain

$$du = \frac{8\pi\nu^2}{c^3}kT\,d\nu. \tag{17.14}$$

This is the Rayleigh–Jeans law. The total energy density obtained from this law is

$$u = \int_0^\infty \frac{8\pi\nu^2}{c^3}kT\,d\nu \tag{17.15}$$

which diverges. Thus although (17.14) is an excellent approximation to the observed spectral distribution at low frequencies it must be inadequate at high frequencies.

Thermal Radiation

The total energy density obtained from the field expression (17.13) is

$$u = \int_0^\infty \left\{\tfrac{1}{2}h\nu + \frac{h\nu}{\exp(h\nu/kT)}\right\}\frac{8\pi\nu^2}{c^3}\,d\nu, \tag{17.16}$$

and this also diverges because of the zero-point term $\frac{1}{2}h\nu$. The divergence is, however, of a different nature, for the magnitude of the divergent part of the integral is independent of T. Thus, whereas the classical Rayleigh–Jeans formula leads to a divergent specific heat and entropy, the Planck formula does not. This is, of course, not entirely satisfactory but it is at least an improvement. For the moment we shall neglect the zero-point term and consider the remaining part of (17.16), i.e.,

$$u = \int_0^\infty \frac{8\pi h\nu^3\,d\nu}{(\exp(h\nu/kT)-1)c^3} = \frac{8\pi k^4}{h^3c^3}\int_0^\infty \frac{x^3\,dx}{\exp(x)-1} = \frac{8\pi^5k^4}{15h^3c^3}T^4. \tag{17.17}$$

Needless to say, $8\pi^5k^4/15h^3c^3 = 4\sigma/c$ and this agrees with the experimental value of Stefan's constant.

Zero-point Energy

We now return to the question of the zero-point energy term which we can isolate by the simple expedient of letting $T = 0$. We now have

$$u = \frac{4\pi h}{c^3}\int_0^\infty \nu^3\,d\nu, \tag{17.18}$$

but, of course, when we consider a real physical cavity of volume V with real material walls, there will certainly be some frequency ν_0 above which the walls fail to reflect radiation. Thus, if we evaluate either the pressure acting on the walls or the energy U associated with modes whose frequency distribution depends on the presence of the walls we shall have to cut off the frequency spectrum at high frequencies. Let us suppose that we have a large cavity of volume V_1 contained within a second cavity of similar construction of volume V_2 and that we cut off the spectrum by introducing some sort of arbitrary function $\Omega(\nu)$ such as $\exp(-\nu/\nu_0)$ into the spectral density. The energy of the system is now

$$U = V_1\int_0^\infty \frac{4\pi h\nu^3}{c^3}\,\Omega(\nu)\,d\nu + (V_2-V_1)\int_0^\infty \frac{4\pi h\nu^3}{c^3}\,\Omega(\nu)\,d\nu \tag{17.19}$$

which is independent of V_1. Thus, there is no net pressure acting on the walls of the inner cavity. The question of whether we can do this in a completely consistent way lies outside the scope of macroscopic electromagnetism and can only be answered in the context of quantum electrodynamics. If we take the answer to be in the affirmative the question then arises whether there are any observable macroscopic consequences of the zero-point energy term, for we have shown that it does not lead to either a specific heat or a macroscopic pressure in large systems. Now equation (17.19) relies on the validity of the asymptotic expression (17.11) for the spectral density of the modes of a cavity and, in particular, it relies on the assumption that this expression is adequate for all frequencies below ν_0, the cut-off frequency. This, of course, cannot be true at frequencies $\nu < c/R$ where R is one of the linear dimensions of the cavity. Consider, for example, the cubic cavity shown in Fig. 17.1 of side L with a partition at $l \ll L$ from one end. Up to frequencies of the order of $\nu_1 = c/2l$ the small flat cavity is essentially a two dimensional resonator and the spectral density of its modes, instead of being $(8\pi\nu^2/c^3)L^2l$, is nearer to $(2\pi\nu/c^2)L^2$. Thus, the zero-point contribution to the energy of this system from low frequency modes depends on l.

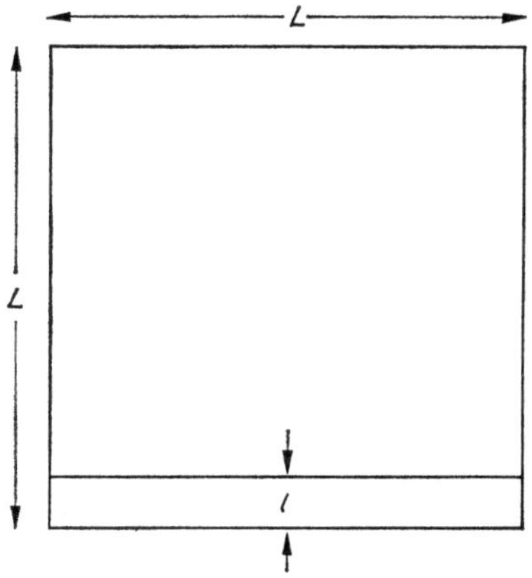

FIG. 17.1. Cross-section of a cubic cavity with a partition.

If U_0 is the zero-point energy in the absence of the partition

$$U \approx U_0 + \int_0^{c/2l} \tfrac{1}{2}h\nu \frac{2\pi L^2}{c^2}\nu \, d\nu - \int_0^{c/2l} \tfrac{1}{2}h\nu \frac{8\pi L^2 l}{c^3}\nu^2 \, d\nu$$

which gives

$$U = U_0 - \frac{\pi}{48} L^2 \frac{hc}{l^3}. \qquad (17.20)$$

The pressure acting on the partition in a direction tending to reduce l is therefore

$$p = \frac{1}{L^2}\frac{\partial U}{\partial l} = \frac{\pi}{16}\frac{hc}{l^4}. \qquad (17.21)$$

A more accurate calculation which leads to a slightly different numerical factor is given by Fierz (1960). The effect clearly leads to a force of attraction between parallel conducting plates at small separations. It was first predicted by Casimir (1948) and has been studied experimentally by Spaarnay (1958). The effect should not be confused with the van der Waals attraction between any two solid objects which obeys a different power law, although the two effects are connected. It is of some interest as an observable macroscopic consequence of the zero-point term even though it is exceedingly small. Thus, in c.g.s. units the pressure is about $5 \times 10^{-17} l^{-4}$ dyne cm^{-2}. Even if $l = 10^{-4}$ cm (1μm) this is only $\frac{1}{2}$ dyne cm^{-2} or 5×10^{-7} atmospheres.

If we discount the zero-point term the state of the thermal radiation field depends on its temperature and this in turn is defined to be equal to the temperature of those material bodies with which the field is in equilibrium. We have, however, discussed the properties of the field without any reference to the mechanism by which it interacts with matter. This has been possibly only through the introduction of such vague notions as a lossy body or an almost perfectly reflecting cavity wall. These are concepts which require more detailed specification and ultimately involve quantum electrodynamics as well as statistical mechanics. It is, however, possible to go some way in this direction without entirely leaving the context of macroscopic physics. Thus, H. A. Lorentz in *The Theory of Electrons* was able to derive the Rayleigh–Jeans law by considering the classical fluctuations in the fields due to electrons in the conducting walls of a cavity. The one dimensional aspect of this problem, which corresponds to Johnson, or thermal, noise in conductors, is treated in the next chapter.

Problems

1. Find the numerical value of $2\pi^5 k^4/15h^3c^2$ in S.I. units.

2. Find the wavelength at which the peak spectral intensity of black-body radiation occurs when the temperature is (a) 3 K, (b) 300 K, (c) 6000 K.

3. A perfectly reflecting cavity contains a body at 6000 K. What is the radiation pressure on the walls expressed in atmospheres?

4. Sodium has a single isotope of mass 23 and nuclear spin $\frac{3}{2}\hbar$. The Fermi temperature of the conduction electrons is approximately 10^4 K and the Debye temperature 10^2 K while the density is about 1 gm cm^{-3}. Compare the entropies per unit volume at 1 K due to (a) nuclear spin, (b) lattice vibrations, and (c) conduction electrons with the entropy per unit volume of black-body radiation at 1 K.

5. A resistor in the form of a thin cylinder is placed in a large cavity containing a body at a temperature T. Show that in equilibrium there must be a fluctuating e.m.f. across the ends of the resistor.

6. A long, slightly lossy transmission line has a fixed short at one end and a sliding short at the other. Relate the force acting on the sliding short to the electromagnetic energy per unit length u stored in the line. Use this result to show that u is proportional to T^2.

7. Show that the spectral density of the energy, in the line in problem 6, is described by the law

$$du(\nu) = \nu\psi\left(\frac{\nu}{T}\right)d\nu$$

where ψ is a universal function of ν/T. (Consider the Doppler shift as the short moves.)

8. By considering a long lossy line connected to a dipole in a large cavity containing a body at T show that $\psi(\nu/T)$ is, to within a multiplicative constant, the same as the function $\phi(\nu/T)$ in equation (17.7). (Hint: consider equilibrium at two different temperatures.)

9. What is the r.m.s. value of the electric field near the surface of a black body at 300 K (27° C)?

CHAPTER 18

Noise and Fluctuations

Introduction

If we observe a small mirror suspended in a carefully shielded vessel we notice that it executes small, but macroscopically observable, fluctuations about its mean position. The mean energy associated with these fluctuations is kT where T is the temperature of the system and they are due to slightly unbalanced torques acting on the mirror as molecules of the gas make random collisions with the mirror. This is probably the simplest mechanical demonstration of fluctuations of thermal origin yet it is still a difficult experiment to perform. By contrast it is extremely easy to build a high-gain electronic amplifier which displays a macroscopically fluctuating output arising from fluctuations of thermal origin in components in the input stage. For this reason the study of fluctuations, except in a formal context, finds its main application in connection with electronic systems and instruments. With few exceptions only electronic apparatus realizes the ultimate sensitivity allowed by the laws of thermodynamics and statistical mechanics.

The existence of macroscopically observable fluctuations arises because, especially in electronics, we can design systems whose macroscopic resolution is fine enough to distinguish between different states corresponding to the same thermodynamic or statistical ensemble. Thus, if we prepare a system using a macroscopic specification, e.g., it is a 1 Megohm resistance at 300 K, this defines such a coarse statistical ensemble of states that we may be able to detect the system changing from one group of states to another within the ensemble as it interacts with the reservoir which maintains its temperature.

We can express this notion in its simplest form as follows. If we take the 1 Megohm resistance and instantaneously measure the voltage across its ends and find a value V_1 then, at this instant, we know that the mobile carriers were in one of the states of the statistical ensemble in which there was an excess of carriers towards one end of the resistance. At this instant we know that it is in a sub-ensemble of the main ensemble. The interaction between the carriers and their environment is, however, so strong that very quickly they will be removed from this sub-ensemble to a new sub-ensemble. At a later instant we should measure $V_2 \neq V_1$. In practice we observe V over a finite interval of time with an apparatus which either has a definite resolving time or a definite spectral response or bandwidth. As a result the practical discussion of fluctuations and noise is much concerned with the spectral distribution of the fluctuations. The theory of noise and fluctuations is not in itself difficult, but very often the system in which the fluctuations are manifest is exceedingly complicated. Thus, for example, although we can give a relatively straightforward account of the fluctuations in the current emitted by a hot cathode, the effect that these fluctuations have on the behaviour of a space-charge-limited thermionic diode at high frequencies presents a problem of quite formidable difficulty.

Noise and Fluctuations

Johnson Noise

J. B. Johnson (1928) observed that when a sensitive amplifier was connected to a resistance R at a temperature T the output of the amplifier displayed fluctuations whose mean square value was proportional to the product RT and independent of the nature of the resistance. The theoretical interpretation of these results was given by Nyquist (1928) in a companion paper and we now reproduce the essential features of Nyquist's argument.

If two resistances R_1 and R_2 at the same temperature T are connected in parallel, and $\overline{V_1^2}$ and $\overline{V_2^2}$ are the mean-square open-circuit fluctuating voltages generated by R_1 and R_2, the power delivered to R_2 by R_1 is $\overline{V_1^2}\,(R_1/(R_1+R_2)^2)$, and the power delivered to R_1 by R_2 is $\overline{V_2^2}\,(R_2/(R_1+R_2)^2)$. These average powers must be equal, to avoid a violation of the second law of thermodynamics, and so $\overline{V_1^2}/R_1 = \overline{V_2^2}/R_2$. This equality must also hold for the contributions $\mathrm{d}\overline{V_1^2}$ to $\overline{V_1^2}$ and $\mathrm{d}\overline{V_2^2}$ to $\overline{V_2^2}$ from frequency components in any interval $\mathrm{d}\nu$. Thus, $\mathrm{d}\overline{V_1^2} = f(\nu,T)R_1\mathrm{d}\nu$ where $f(\nu,T)$ is a universal function of ν and T.

If two equal resistances R, both at T, are connected by a long lossless transmission line of characteristic impedance $Z_0 = R$, group velocity v_g and length l, the energy stored in the line when equilibrium has been attained is $2(\mathrm{d}V^2/4R)\,(l/v_g)$ for each frequency interval $\mathrm{d}\nu$. This is the thermal equilibrium energy for the line and, if the resistors are shorted (so that the line has resonances whenever $kl = \frac{1}{2}n$ where n is integral and k the wave number), it must be equal to the energy obtained from a statistical-mechanical calculation. This is

$$\mathrm{d}U = \frac{h\nu}{\exp(h\nu/kT)-1}\frac{\mathrm{d}n}{\mathrm{d}\nu}\mathrm{d}\nu$$

where $\mathrm{d}n/\mathrm{d}\nu$ is the number of modes per unit frequency interval. We have

$$n = 2kl$$

and so

$$\frac{\mathrm{d}n}{\mathrm{d}\nu} = 2l\frac{\mathrm{d}k}{\mathrm{d}\nu} = \frac{2l}{v_g}.$$

Thus

$$2\frac{\mathrm{d}\overline{V^2}}{4R}\frac{l}{v_g} = \frac{h\nu\,\mathrm{d}}{\exp(h\nu/kT)-1}\frac{2l}{v_g}\mathrm{d}\nu$$

and

$$\mathrm{d}\overline{V^2} = 4R\frac{h\nu\,\mathrm{d}\nu}{\exp(h\nu/kT)-1}. \tag{18.1}$$

In the usual case where $h\nu \ll kT$, i.e., in the Rayleigh–Jeans limit, this reduces to

$$\mathrm{d}\overline{V^2} = 4RkT\,\mathrm{d}\nu \tag{18.2}$$

which is the Johnson noise formula. If the fluctuations across R are observed using an amplifier of bandwidth $\delta\nu$ and voltage gain G, the output will display fluctuations of mean square amplitude $4RkTG^2\,\delta\nu$.

The available noise power in $\mathrm{d}\nu$ is $\mathrm{d}\overline{V^2}/4R$, i.e.,

$$\mathrm{d}P_n = kT\,\mathrm{d}\nu \tag{18.3}$$

or, if $h\nu$ is not negligible compared with kT,

$$\mathrm{d}P_n = \frac{h\nu\,\mathrm{d}\nu}{\exp(h\nu/kT)-1} \tag{18.4}$$

to which we might possibly add the zero-point fluctuations of spectral density $\frac{1}{2}h\nu\, d\nu$. It is possible to give an alternative demonstration of the validity of equation (18.4), or

$$dP_n = \frac{h\nu\, d\nu}{\exp(h\nu/kT) - 1} + \tfrac{1}{2}h\nu\, d\nu, \tag{18.5}$$

by treating a long transmission line, in which attenuation is described in language compatible with quantum mechanics as the transfer of energy from the normal modes of the line to the normal modes of an indefinitely large heat sink (Robinson, 1965). These results do not, therefore, depend on the concept of resistance, which is not normally regarded as a component of the theoretical structure of quantum electrodynamics.

If we restrict our attention to (18.3), then simple arguments based on the second law of thermodynamics lead to the conclusion that (18.3) applies to the fluctuating output of any two terminal network in thermal equilibrium at T. Thus, for example, an aerial in an enclosure at T produces an available noise power kT per unit frequency range at its terminals. By considering the noise power delivered and received by an aerial matched to a load resistance R at T, we can derive a relation between the transmitting and receiving properties of any aerial. Thus, if an aerial radiates a fraction $G(\theta, \phi)\, d\Omega$ of its total radiated power into a solid angle $d\Omega$ at (θ, ϕ), the power it delivers to a matched load when plane waves of intensity S watt m^{-2} are incident from the direction (θ, ϕ) is

$$P = S\lambda^2 G(\theta, \phi) \tag{18.6}$$

where λ is the wavelength.

If an instrument has a bandwidth $\delta\nu$ its resolving time τ is of the order of $\tau \approx 1/\delta\nu$ and so the noise energy received per resolving time from a source at T is kT. Thus any measurement on the source will only yield significant results if the coherent energy transfer exceeds kT. This sets an important lower limit to the sensitivity of experimental apparatus. In many cases this lower limit is achieved with relatively simple electronic systems.

Shot Noise

In many electrical systems, e.g., a thermionic diode or a transistor, applied fields influence the motion of charged particles in regions where they are not maintained in thermal equilibrium by collisions with a lattice or heat sink. In this case the macroscopic fluctuations in the output of the system arise from the fluctuations in the number of particles which enter the active region of the system. In a later section we shall consider the relation of these fluctuations to Johnson noise and the general principles of statistical mechanics. In this section we adopt a simplified approach.

The emission of electrons from a hot cathode, or the transit of electrons across the depletion layer in a semiconductor diode, is a random process in the sense that the probability of emission of an electron in any short interval depends only on the length of the interval. The emission therefore obeys Poisson statistics and, if in some finite interval τ the mean number of electrons emitted is $\bar{N}$, individual intervals will yield values of N which fluctuate about $\bar{N}$ with a mean square deviation

$$\overline{\Delta N^2} = \overline{(N - N)^2} = \bar{N}. \tag{18.7}$$

The charge transported in a time τ is eN and so

$$\overline{\Delta q^2} = e\bar{q}.$$

The current averaged over the interval is q/τ and so

$$\overline{\Delta I^2} = \frac{1}{\tau^2}\overline{\Delta q^2} = \frac{e}{\tau}\bar{I}. \tag{18.8}$$

A theorem due to MacDonald (1949) leads directly to the power spectrum or spectral density of the fluctuations

$$\overline{dI^2} = 2e\bar{I}\,d\nu \tag{18.9}$$

and this formula is the basis of the calculations of noise and fluctuations in many types of device, e.g., diodes, transistors, photomultipliers, and microwave tubes. Equation (18.9), however, gives only the fluctuations in the current entering the active region of the device. In many cases the observed external effects are grossly altered by interactions in the active region. For example, in a space-charge-limited diode the coherent interaction of all the electrons in the region between cathode and anode completely alters the spectral distribution of the current fluctuations observed in the external leads. The low frequency components, for which ν is considerably less than the plasma frequency at the potential minimum in front of the cathode, are reduced by a factor of the order of kT/eV where T is the cathode temperature and V the anode voltage and higher frequency components are also appreciably modified. The behaviour of shot noise in electron beams in vacuum has been extensively studied since the 1930's. The two classic papers are North (1940) and Rack (1938) but there is also a general review of the field by Smullin and Haus (1959).

Noise in a Semiconductor Diode

When a forward bias V is applied to a semiconductor diode the forward current is given by an expression of the general form

$$I = I_0\left(f\left(\frac{eV}{kT}\right)-1\right) \tag{18.10}$$

where I_0 is a constant. If the charge carriers do not recombine in the depletion layer this simplifies to

$$I = I_0 \exp(eV/kT) - 1. \tag{18.11}$$

In this case the current consists of electrons injected at a rate $I_0 \exp(eV/kT)$ from one side of the junction and at a rate I_0 from the other, and we can apply the shot noise formula (18.9) to calculate the power spectrum of the fluctuations. This is

$$\overline{dI^2} = 2eI_0 \exp(eV/kT)\,d\nu + 2eI_0\,d\nu. \tag{18.12}$$

If the forward bias is appreciable the first term dominates the result, and in most practical cases we have

$$dI^2 \approx 2eI\,d\nu. \tag{18.13}$$

At zero bias the fluctuations are

$$dI^2 = 4eI_0 \, d\nu \tag{18.14}$$

and, since the conductance at zero bias is

$$g = \frac{\partial I}{\partial V} = \frac{eI_0}{kT}, \tag{18.15}$$

we can express this as

$$dI^2 = 4kTg \, d\nu. \tag{18.16}$$

This gives the short-circuit mean square current fluctuations. The open-circuit mean square voltage fluctuations are

$$dV^2 = \frac{dI^2}{g^2} = \frac{4kT}{g} \, d\nu. \tag{18.17}$$

This agrees with the Johnson noise result, as it must, for a diode at zero bias is a system in thermal equilibrium.

Noise in Transistors

The bipolar *n–p–n* transistor is essentially a device in which electrons from the *n* emitter region are injected across the depletion layer of the forward biased emitter-base junction into the *p* type base. In the base the injected electrons are minority carriers and diffuse towards the reverse biased collector-base junction where a fraction $\alpha \lesssim 1$ of the injected electrons pass across to the collector. When an electron of charge $q = -e$ leaves the emitter and crosses to the base a short pulse of total charge $-q$ flows in the emitter lead and a pulse q in the base lead as the electron crosses the depletion layer. When the electron reaches the collector-base junction a pulse $-q$ flows in the base lead and q in the collector lead. For most purposes we can regard the pulses in the base lead as instantaneous and cancelling. Fluctuations in the current in the collector lead come from electrons which appear in the mean collector current I_c; fluctuations in the base lead only from electrons which recombine in the base and contribute to the mean base current I_b. Because these processes are random and uncorrelated the base and collector current fluctuations i_b and i_c are given by

$$i_b^2 = 2eI_b \, \delta\nu,$$

$$i_c^2 = 2eI_c \, \delta\nu.$$

The emitter current fluctuations are

$$i_e^2 = (i_b + i_c)^2 = (i_b^2 + i_c^2) = 2e(I_b + I_c) \, \delta\nu$$

because i_b and i_c are uncorrelated. Since, however, $I_b + I_c = I_e$ (the mean emitter current), this is just what we should expect for i_e^2.

In circuit applications, where we can neglect high frequency effects due to charge storage or transit time in the base, the noise can be represented by the uncorrelated current generators shown in Fig. 18.1.

In a field-effect transistor the resistivity of a thin semiconductor channel forming one plate of a capacitor is controlled by the voltage on the other plate or gate. If the channel is *n* type silicon a negative voltage applied to the gate results in the removal of carriers from the channel and an increase in resistance. In this device the carriers in the channel are maintained in thermal equilibrium by collisions with the lattice and the fundamental source of

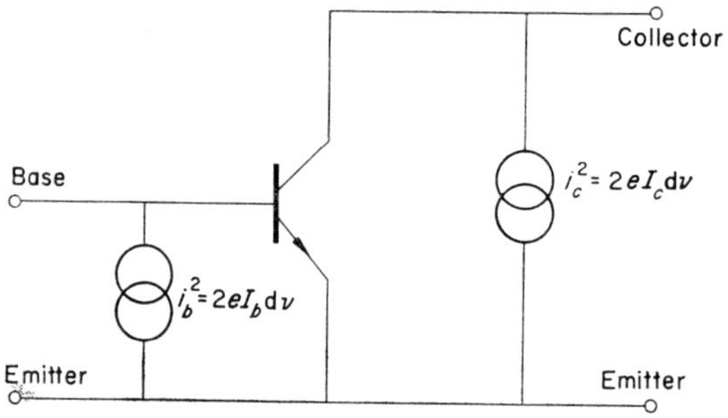

FIG. 18.1. Representation of noise by uncorrelated current generators.

noise is Johnson or thermal noise. The effects of distributed thermal noise in the channel are somewhat tedious to evaluate but at low radio frequencies can be represented by a current generator in parallel with the channel. A more convenient representation is shown in Fig. 18.2 where the noise is represented by a voltage generator in the control or gate lead.

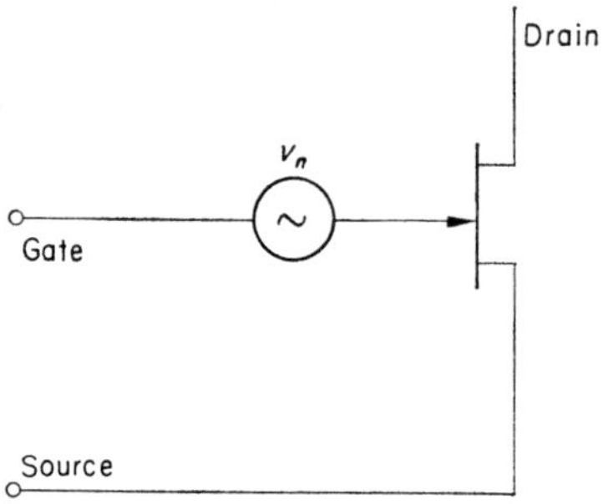

FIG. 18.2. Representation of noise by a voltage generator in the gate lead.

The mean square value of the equivalent voltage fluctuations associated with a bandwidth δv is approximately

$$v^2 = \frac{4kT}{g_m} \delta v \tag{18.18}$$

where g_m is the mutual conductance of the device. We shall return to this topic in a later section where we show that, even though practical amplifying devices themselves always introduce noise, nevertheless it is possible to observe Johnson or thermal noise fluctuations originating in a resistor or signal source. The demonstration that we can observe macroscopic fluctuations in a body belonging to a statistical ensemble at T using a device which, though not itself in thermodynamic equilibrium, is exchanging heat freely with a reservoir at T is clearly an important link in the relation between microscopic and macroscopic phenomena.

The Relation between Electrical Noise and Statistical Mechanics

The thermal or Johnson noise fluctuations in voltage across circuit elements, and the shot noise fluctuations in the number of particles in transit in active devices such as transistors, are clearly related to the statistical dispersion implicit in the notion of a thermodynamic ensemble, and it is of some interest to investigate this connection.

The simplest description of fluctuations in a thermodynamic ensemble is in the terms of the fluctuations $\langle \nu_i^2 \rangle$ in the occupation number of a state or group of states in an ensemble. If n_i is the mean occupation number then this can be expressed as

$$\langle \nu_i^2 \rangle = kT\frac{\partial n_i}{\partial \mu} \tag{18.19}$$

where μ is the chemical potential. For a system at constant temperature and pressure this can be replaced by

$$\langle \nu_i^2 \rangle = -kT\frac{\partial n_i}{\partial E_i} \tag{18.20}$$

where E_i is the energy of the state or states in question. These two results refer to a single instant in time and, in noise problems, we are either interested in the fluctuations in finite intervals or the spectral density of the fluctuations. Thus, however we make use of say (18.20) we will eventually have to perform some sort of time average. We shall not give the complete analysis (see Robinson, 1969) but merely indicate the nature of the argument. We begin by discussing Johnson noise.

If we have a conducting body with two external leads the instantaneous current in the leads can be expressed in terms of the instantaneous configuration (in phase space) of the charge carriers in the body.

Thus, $I(t)$ depends on $n_i + \nu_i(t)$. At a later time $t+\tau$ the ν_i will still be partially correlated with their values at t and, as a result, although $\langle I(t) \rangle$ and $\langle I(t+\tau) \rangle$ vanish, the ensemble average $\langle I(t)\, I(t+\tau) \rangle$ will have a functional dependence on τ determined by the dynamical equations of motion of the charge carriers. The Wiener–Khintchine theorem can be used to give the power spectrum or spectral density of $I(t)$ as

$$w(f) = 4\int_0^\infty \langle I(t)\, I(t+\tau) \rangle \cos 2\pi f\tau \, d\tau.$$

The remainder of the calculation then consists of showing that $w(f)$ is given by

$$w(f) = 4kTG$$

where GV^2 is the power dissipated in the body when an alternating voltage of frequency f and r.m.s. amplitude V is applied to the two external leads. The short-circuit current fluctuations associated with a frequency interval δf are then

$$i_n^2 = w(f)\,\delta f = 4kTG\,\delta f \tag{18.21}$$

which is equivalent to the Johnson formula. We omit the proof that $w(f) = 4kTG$. It is not trivial but it is quite general and applies equally to particles obeying quantum statistics or classical statistics.

In the case of shot noise we recognize that electrons emitted across a potential barrier come from the non-degenerate tail of a Fermi–Dirac distribution so that (18.20) yields

$$\langle \nu_i^2 \rangle = n_i. \tag{18.22}$$

Thus, if only electrons which occupy some particular group of states are emitted, the mean square fluctuations in the number emitted is equal to the mean number emitted and we immediately recover the basic shot noise result (18.9). Because equation (18.22) holds for every sub-division of the emitted electrons into, for example, energy classes, and n_i depends on the energy class and the temperature of the emitter, it is perhaps not surprising that in many devices, e.g., vacuum diodes, triodes, and microwave tubes, the irreducible minimum noise content of the electron beam generally involves the temperature of the emitter. Indeed the noise temperature of an electronic device, defined in the next section, is generally found to be proportional to the temperature of the emitter. This is as true for a bipolar transistor as for a travelling-wave tube.

Noise Figure and Noise Temperature

If we have a signal source of internal resistance R at a temperature T with an available signal power P_s, the mean square noise voltage across the source in a bandwidth δv is $V_n^2 = 4RkT\,\delta v$ and the mean square open-circuit signal voltage is $V_s^2 = 4RP_s$. The ratio of noise to signal is

$$\frac{V_n^2}{V_s^2} = \frac{kT\,\delta v}{P_s}$$

which is independent of R. If we connect this source to the input of the field-effect transistor shown in Fig. 18.2 the mean square signal voltage at the gate is $4RP_s$ but the effective noise is $4kTR\,\delta v+(4kT/g_m)\,\delta v$. The noise to signal ratio is worse by a factor known as the noise figure; in this case it is

$$F = 1+\frac{1}{g_m R}. \tag{18.23}$$

At high and low frequencies other effects must be considered, but equation (18.23) is reasonably accurate at say 1 MHz, if $R < 10^4$. Since it is quite possible for g_m to be 10^{-2}A V^{-1} we see that $F = 1{\cdot}01$. The total noise output will be 99% due to noise originating at R and so the amplifier output will predominantly be determined by the statistical ensemble to which R belongs. If the source R is at T_s and the field effect transistor at T_t we have

$$F = 1+\frac{T_t}{g_m R T_s}$$

but we could also express this as

$$F = 1+\frac{T_n}{T_s}, \tag{18.24}$$

where

$$T_n = \frac{T_t}{g_m R} \tag{18.25}$$

is known as the noise temperature of the amplifier. If an amplifier is characterized by a noise temperature T_n and is connected to a source at T_s the equivalent noise input power in δv is $k(T_s+T_n)\,\delta v$ and the threshold signal power required is of this order. We may remark electronic amplifiers and masers allow us to achieve values of T_n below 3 K at all frequencies from about 1 kHz to 20 GHz. In practice this is not worth while unless the source itself is effectively at a low temperature.

Although in this chapter we have dealt largely with electronic topics the reader should be warned that the discussion has been confined to the general principles and that, in practice, the design of low noise systems which approach the fundamental limits of sensitivity involves practical consideration which we have entirely ignored. Our purpose in couching the discussion in terms of electronics is to emphasize that the study of fluctuations is not only a topic in theoretical physics but also has applications which are easily observable in the laboratory. The consequences of (18.19) $\langle v_i^2 \rangle = kT\,\partial n_i/\partial \mu$ can, for example, be easily demonstrated with the aid of two or three transistors and a simple oscilloscope.

Problems

1. A galvanometer coil of area A and N turns with a moment of inertia I is suspended in a uniform magnetic field B in the plane of the coil. What is the mean square value of the voltage fluctuations across the terminals if the temperature of the system is T?

2. A superconducting coil, forming a resonant circuit with a superconducting capacitance C, is contained in a brass case at a temperature T. What is the r.m.s. value of the fluctuating voltage across C?

3. An amplifier of voltage gain 10^6, high input impedance and bandwidth 1 MHz has its input shorted by a resistance of $10^4\,\Omega$ at 300 K. What is the minimum possible r.m.s. noise voltage at the output?

4. A directive aerial enclosed in a plastic dome points vertically upwards. When it is used as a transmitter 1% of the power is absorbed in the dome. If it is used as a receiver what is the available noise power at the aerial terminals in a bandwidth $\delta\nu$? Assume that the dome is at 300 K and the aerial receives only noise from outer space with an effective temperature of 3 K.

5. Show that the cross-section or receiving area of an isotropic aerial is $\lambda^2/4\pi$ and independent of its physical size.

6. A temperature limited diode has a current of 1 mA which passes through a load resistance $R = 10^4\,\Omega$. Compare the magnitudes of the voltage fluctuations across R due to Johnson noise and shot noise when the resistance is at 300 K.

7. Why cannot equation (18.17) be applied to a thermionic diode?

8. A thin block of a medium of low conductivity is placed between two parallel plane electrodes of area A and separation d connected by an external wire. If the medium contains mobile carriers of charge q express the current in the external lead at any instant in terms of the microscopic velocities of the charge carriers. How can this be expressed if the relation between the energy E_i of a charge carrier and its momentum p is $E_i = E_0 + \alpha p^2 + \beta p^4$?

9. The transistor in Fig. 18.1 has a mutual conductance $g_m = eI_c/kT$. It is connected to a signal source at the same temperature with an internal resistance R. Show that the noise figure of the system has a minimum value $1 + (I_b/I_c)^{\frac{1}{2}}$ when R is chosen to be equal to $1/g_m(I_c/I_b)^{\frac{1}{2}}$. If T = 300 K, $I_b = 10\,\mu$A and $I_c = 1$ mA, what is the noise temperature of the system?

10. The fact that the effective noise temperature of outer space is known to be about 3 K is an important fact in cosmology. Given that amplifiers with a noise temperature appreciably less than 3 K do not exist, how do we know this fact?

CHAPTER 19

Résumé and Bibliography

Résumé

CHAPTERS 1–8

The macroscopic field equations in a region free from material bodies are

$$\nabla \cdot \mathbf{B} = 0, \qquad \nabla \wedge \mathbf{E}+\dot{\mathbf{B}} = 0,$$

$$\varepsilon_0 \nabla \cdot \mathbf{E} = \rho, \quad \frac{1}{\mu_0}\nabla \wedge \mathbf{B}-\varepsilon_0 \dot{\mathbf{E}} = \mathbf{J}, \tag{19.1}$$

and the expression for the force density is

$$\mathbf{F} = \rho\mathbf{E}+\mathbf{J} \wedge \mathbf{B}. \tag{19.2}$$

We regard these equations as exact in any macroscopic context and the quantities ρ and $\mathbf{J}$ as continuous variables. In the presence of matter we find it convenient to introduce new vectors $\mathbf{D}$ and $\mathbf{H}$ which reduce to $\varepsilon_0\mathbf{E}$ and $\mathbf{B}/\mu_0$ in vacuum. In terms of these vectors the inhomogeneous equations in the set (19.1) become

$$\nabla \cdot \mathbf{D} = \rho, \quad \nabla \wedge \mathbf{H}-\dot{\mathbf{D}} = \mathbf{J} \tag{19.3}$$

where new ρ and $\mathbf{J}$ refer only to free, mobile or accessible charges and currents as distinct from bound charge in matter. The difference vectors

$$\mathbf{P} = \mathbf{D}-\varepsilon_0\mathbf{E}$$

and

$$\mathbf{M} = \frac{1}{\mu_0}\mathbf{B}-\mathbf{H} \tag{19.4}$$

vanish in vacuum and can be shown to have the property that the dipole moment of a body of volume V is $\int \mathbf{P}\, \mathrm{d}V$ or $\int \mathbf{M}\, \mathrm{d}V$. The quantities $-\nabla \cdot \mathbf{P}$ and $\nabla \wedge \mathbf{M}+\dot{\mathbf{P}}$ generate fields $\mathbf{E}$ and $\mathbf{B}$ as though they represented charge and current densities

$$\rho_b = -\nabla \cdot \mathbf{P}, \quad \mathbf{J}_b = \nabla \wedge \mathbf{M}+\dot{\mathbf{P}} \tag{19.5}$$

and just as $\mathbf{J}$ and ρ satisfy the conservation equation $\nabla \cdot \mathbf{J}+\dot{\rho} = 0$ we have

$$\nabla \cdot (\dot{\mathbf{P}}+\nabla \wedge \mathbf{M})+\frac{\partial}{\partial t}(-\nabla \cdot \mathbf{P}) = 0.$$

This leads us to believe that $-\nabla \cdot \mathbf{P}$ and $\nabla \wedge \mathbf{M}+\dot{\mathbf{P}}$ in some way represent the densities of bound charge and current in matter. In an exclusively macroscopic formulation of electromagnetism this is about as far as we can go. The constitutive relations which connect $\mathbf{P}$ with $\mathbf{E}$ and $\mathbf{M}$ with $\mathbf{H}$ are to be regarded as purely empirical relations to be obtained by measurement. We might suspect that **P** or **M** could be related to the electric and magnetic dipole moment densities in matter but we cannot derive this relation from macroscopic considerations alone. Macroscopic electromagnetism is consistent with this interpretation of **P** and **M** but it does not require it. Indeed it is actually wrong for if **D** satisfies $\nabla \cdot \mathbf{D} = \rho$, where ρ represents only mobile charge, then $\mathbf{D}\varepsilon_0\mathbf{E}$ is not exactly the dipole moment density but includes terms such as $(\partial Q_{ij}/\partial r_j)$ which arise from any quadrupole moment density. In practice this is of little significance but it suffices to show that the identification of **P** with the dipole moment density is not a consequence of macroscopic theory alone.

Once we admit that matter contains sub-atomic charged particles, whose extent is certainly negligible on any macroscopic scale, we are faced with the problem of reconciling the use of (19.1) as exact equations with this knowledge. Thus, we have to show that if we adopt the equations

$$\nabla \cdot \mathbf{b} = 0, \qquad \nabla \wedge \mathbf{e} + \dot{\mathbf{b}} = 0$$

$$\varepsilon_0 \nabla \cdot \mathbf{e}(\mathbf{R}) = \sum_n q(n)\, \delta(\mathbf{r}(n) - \mathbf{R}) \tag{19.6}$$

$$\frac{1}{\mu_0} \nabla \wedge \mathbf{b}(\mathbf{R}) - \varepsilon_0 \dot{\mathbf{e}}(R) = \sum_n q(n) \dot{\mathbf{r}}(n)\, \delta(\mathbf{r}(n) - \mathbf{R})$$

as the basic field equations including both bound and mobile charges, then, in a macroscopic context, they lead to equations (19.1) and (19.3). To do this we recognize that any macroscopic problem is characterized by a finite length or resolution Λ and that its solution cannot involve variables with spatial Fourier components of frequency greater than $k_0 \approx 1/\Lambda$. Thus, if the Fourier expansions of all the variables in (19.6) are truncated at spatial frequencies near k_0 the resulting variables are still exact for all macroscopic processes. The implications of this procedure are discussed in chapter 5 and there it is shown that the truncated variables do, in fact, satisfy the field equations (19.1) and that, provided there are a reasonable number of particles in any macroscopically significant volume, ρ and **J** behave as continuous variables.

If the distinction is made between bound and mobile charges then we can also show (chapter 8) that the truncated contributions to ρ and **J**, due to bound charge, can be expressed as

$$\rho_b = -\nabla \cdot \mathbf{P} + \frac{\partial^2}{\partial r_i \partial r_j} Q_{ij} + \text{etc.}$$
$$\mathbf{J}_b = \nabla \wedge \mathbf{M} + \dot{\mathbf{P}} + \text{etc.} \tag{19.7}$$

where **P**, **M**, Q_{ij}, etc., are the truncated expressions for the various multipole moment densities in matter. The vectors **D** and **H** can then be regarded as auxiliary vectors defined by

$$\mathbf{D} = \varepsilon_0 \mathbf{E} + \mathbf{P} - \frac{\partial Q_{ij}}{\partial r_j} \text{ etc.}$$
$$\mathbf{H} = \frac{1}{\mu_0} \mathbf{B} - \mathbf{M} \text{ etc.} \tag{19.8}$$

The separation into mobile and bound charges is only helpful if the macroscopic scale Λ is greater than the dimensions of the atomic sub-units of the medium. The expression of ρ_b and $\mathbf{J}_b$ in terms of **P**, **M**, etc., is only possible if there are many sub-units in any macroscopically significant volume.

Résumé and Bibliography

The identification of **P** and **M** with the dipole moment densities establishes contact between the macroscopic field equations and the microscopic structure of a medium. The procedure is compatible not only with relativity but also with the formation of a quantum or ensemble average. Indeed the truncation process is formally rather similar to the ensemble averaging process. Thus, this formulation opens the way to an atomic theory of macroscopic electromagnetic properties. It is also sufficiently precise to deal with macroscopic fluctuations in the properties of a system which is only known to be a member of a statistical ensemble.

Not all problems, unfortunately, lead to a clear distinction between their microscopic and macroscopic aspects. This is particularly the case with the local field acting on an atom in a solid medium. Nevertheless, even if this problem cannot be solved it can at least be stated.

CHAPTERS 9–12

All the properties of the macroscopic fields are implicit in the macroscopic field equations and in these chapters they are set out together with the consequences of combining the field equations with the general conservation laws for momentum and energy.

CHAPTERS 13–14

Particular types of constitutive relation arise from particular features of the structure of a medium. In these chapters the emphasis is on the general nature of these constitutive relations and in particular on constitutive relations more complicated than the elementary, isotropic, linear relations. The almost exclusive pre-occupation of many texts with these important, but rather special, relations conveys a misleading impression that they are almost universally valid and that only a few very esoteric media display any other type of behaviour.

CHAPTER 15

Electromagnetic texts usually deal rather cursorily with thermodynamics and fail to emphasize the distinction between the free energy of the system F, with the thermodynamic identity $\delta F = -S\,\delta T + \int(\mathbf{E}\,.\,\delta\mathbf{D} + \mathbf{H}\,.\,\delta\mathbf{B})\,\mathrm{d}V$, and other functions, e.g., F^* which, in the dielectric case, has the identity $\delta F^* = -S\,\delta T - \int\mathbf{P}\,.\,\delta\mathbf{E}^*\,\mathrm{d}V$ where $\mathbf{E}^*$ is no longer the actual field in the system. When one finds that, for example, several well-known texts treat superconductors as diamagnetic media in thermodynamics, and that the power and elegance of thermodynamic arguments are rather generally ignored in electromagnetism it is apparent that a rather more extended discussion is required. In this connection we may remark that problem 2 of chapter 16 is not a joke. It appeared in a well-known physical journal.

CHAPTER 16

Any calculation or interpretation of macroscopic properties in terms of atomic structure ultimately relies on statistical mechanics. The calculation usually involves so many difficult statistical and atomic problems that little attention is paid to the electromagnetic aspect. The most important result in this section is that the free energy function F^S obtained in statistical mechanics is either the function F^* or $\tilde{F}^*$ of chapter 15. Thus the dipole moment of a macroscopic body is $-(\partial F^s/\partial E^*)_T$ or, if the statistical problem has been formulated in terms of a local field, $-(\partial F^s/\partial E^{\mathrm{local}})_T$.

CHAPTERS 17–18

Thermal radiation and Johnson noise in electrical circuits are two aspects of the same phenomenon. Both topics are amenable to thermodynamic or statistical treatment. Indeed Planck's investigation of the connection between these two types of calculation is *the* turning point in twentieth-century physics. Students often find the elementary treatment of Planck's radiation law more convincing than Nyquist's much more closely argued treatment of Johnson noise. This is to some extent because Nyquist's result leads to numerical values which are easily visualized and verified experimentally. For this reason chapter 18 emphasizes the experimental consequences of electromagnetic fluctuations. The way a macroscopic theory deals with statistical fluctuations is an important test of the consistency of the theory. Noise in electrical circuits has a more fundamental importance in physics. It represents the ultimate practical limit on our ability to perform refined and precise experiments. The numerical result of problem 9 of chapter 17 is particularly relevant to these remarks.

Bibliography

The literature of electromagnetism is so extensive that any selection of books must be to some extent arbitrary. The list that follows represents therefore a rather personal selection. It includes only books that the author has found interesting, helpful or at least not a bar to understanding.

The theory of electromagnetism is one of the outstanding intellectual achievements of mankind and its history makes compelling reading. We are fortunate that E. T. Whittaker's *History of the Theories of the Aether and Electricity* (Nelson, 1953, 2 vols.) does justice to the subject. Because it was written by a scientist who himself contributed to mathematical physics and was equipped to comprehend all aspects of the subject it is a unique contribution to the history of science.

Either the author's *Electromagnetism* (O.U.P., to be published 1974) or A. F. Kip's *Fundamentals of Electricity and Magnetism* (McGraw-Hill, 1969) would be suitable introductory texts. They are usefully supplemented by *Electricity and Magnetism* by B. Bleaney and B. Bleaney (O.U.P., 1965), a modern text giving a full discussion of the experimental aspects of the subject. At a similar level *Classical Electricity and Magnetism*, by M. Abraham and R. Becker (Blackie, 1944 and subsequent editions), is a much more formal text notable for its thorough discussion of the use of vectors.

A more advanced, but still relatively general, text is J. D. Jackson's *Classical Electrodynamics* (Wiley, 1962). This deals lucidly with the development of electromagnetic theory from roughly the point at which this book ends. Two books at a similar level but emphasizing a different aspect of the subject are S. A. Schelkunoff's monumental *Electromagnetic Waves* (Van Nostrand, 1943) and *Fields and Waves in Communication Electronics* by S. Ramo, J. R. Whinnery and T. van Duzer (Wiley, 1965). Schelkunoff's book is notable for its formal clarity while Ramo *et al.* make a remarkably happy choice of topics for extended treatment.

Three texts, J. A. Stratton's *Electromagnetic Theory* (McGraw-Hill, 1941), W. R. Smythe's *Static and Dynamic Electricity* (McGraw-Hill, 1939), and J. H. Jeans, *Electricity and Magnetism* (O.U.P., 1951), contain a wealth of information and innumerable solved problems. In this connection *Mathematical Physics* by H. and B. S. Jeffreys (C.U.P., 1956) is also useful.

A superb text covering much of the same material as this book but with an entirely different emphasis is *Electrodynamics of Continuous Media.* by L. D. Landau and E. M. Lifshitz (Pergamon, 1960). It assumes a high level of erudition in the reader but, in compensation, covers an immense range of topics.

C. Kittel's *Introduction to Solid State Physics* (Wiley, 1971) is a useful general text on the origin of constitutive relations. The classic work, however, is still J. H. van Vleck's *Electric and Magnetic Susceptibilities* (O.U.P., 1932). More recent and specialized texts are:

D. Wagner, *Introduction to the Theory of Magnetism* (Pergamon, 1972).
H. Frohlich, *Theory of Dielectrics* (O.U.P., 1958).
C. P. Smythe, *Dielectric Behaviour and Structure* (McGraw-Hill, 1955).
F. Jona and G. Shirane, *Ferro-electric Crystals* (Pergamon, 1962).
K. Hoselitz, *Ferromagnetic Properties of Metals and Alloys* (O.U.P., 1952).
R. M. Bozorth, *Ferro-magnetism* (Van Nostrand, 1951).
A. Abragam and B. Bleaney, *Electron Paramagnetic Resonance of Transition Metal Ions* (O.U.P., 1970).
A. Abragam, *Principles of Nuclear Magnetism* (O.U.P., 1961).
H. B. G. Casimir, *Magnetism and Very Low Temperatures* (Dover, 1964).
C. G. B. Garrett, *Magnetic Cooling* (Wiley, 1954).

Résumé and Bibliography

C. J. Gorter, *Paramagnetic Relaxation* (Elsevier, 1947).
N. Bloembergen, *Non-Linear Optics* (Benjamin, 1965).
V. M. Agranovich and V. L. Ginzburg, *Spatial Dispersion* (Wiley, 1964).

A brief account of thermodynamic principles is given in *Classical Thermodynamics*, by A. B. Pippard (C.U.P., 1957) but *Statistical Physics*, by L. D. Landau and E. M. Lifshitz (Pergamon, 1968), is perhaps the most generally useful text although *Thermodynamics* by P. T. Landsberg (Wiley, 1961) also gives a complete and remarkably rigorous exposition of the subject.

Classical Charged Particles, by F. Rohrlich (Addison-Wesley, 1965) deals with just those topics connected with the structure of elementary charges that we have ignored.

The classic text on the relation between microscopic and macroscopic electromagnetism is *The Theory of Electrons* by H. A. Lorentz now happily available as a reprint (Dover, 1952). A more recent account is given by L. Rosenfeld in *Theory of Electrons* (North-Holland, 1951).

The first departure from Lorentz's treatment was made by P. Mazur and B. R. A. Nijboer (*Physica* **19,** 971–986, 1953) using ideas put forward in connection with hydrodynamics by J. G. Kirkwood and J. H. Irving (*J. Chem. Phys.* **18,** 817–829, 1950). S. R. de Groot and J. Vlieger (*Physica* **31,** 254–268, 1965) generalized Mazur and Nijboer's work to include retardation and relativistic effects. There is a full account in S. R. de Groot's book *The Maxwell Equations* (North-Holland, 1969). The quantum mechanical generalization was first made by K. Schram (*Physica* **26,** 1080, 1960) and has since been treated more generally by J. M. Crowther and D. ter Haar (*Proc. Kon. Ned. Akad. Wet.* **74,** 351–357, 1971). G. Russakoff (*Am. J. of Physics* **38,** 1188–1195, 1970) discussed the nature of the Lorentz averaging procedure and greatly clarified its meaning and significance. The use of truncation was introduced by F. N. H. Robinson (*Physica* **54,** 329–341, 1971). There is also a very thorough discussion of the use of the auxiliary vectors **D, P, H,** and **M** by P. Pershan (*J. Appl. Phys.* **38,** 1482–1490, 1967).

Noise in Electrical Circuits, by F. N. H. Robinson (O.U.P., 1962) gives a brief introduction to this topic. The series of papers *Noise and Stochastic Processes*, edited by N. Wax (Dover, 1954) contains most of the classic papers on the more formal aspects of the subject.

Finally, we remind the reader that electromagnetic properties are tabulated. The series *Landolt–Bornstein* published by Springer is continually growing and contains an enormous range of data. There are also innumerable more specialized compilations, e.g., the refractive index tables by A. N. and H. Winchell (Academic Press, 1954 and 1964).

APPENDIX

Vectors

MANY of the vector relations found in the text may be unfamiliar and their proof or verification by elementary methods is often tedious. We therefore collect these relations together and give proofs where these are not obvious. We take the point of view that a vector is a quantity of physical significance having magnitude and direction and that two commensurable vectors obey the parallelogram law of addition. For a more sophisticated but less extensive treatment the reader is advised to consult either *Physical Mathematics* by C. H. Page (Van Nostrand, 1955) or *Mathematical Physics* by H. and B. S. Jeffreys (C.U.P., 1956).

The first two sections deal with definitions. The remaining sections contain the results.

1. Vectors

The simplest vector $\mathbf{r}$, the displacement of a point from the origin, can be used to represent the point itself. If ϕ is a scalar the notation $\phi(\mathbf{r})$ denotes that ϕ is to be evaluated at $\mathbf{r}$, not just the functional dependence of a function ϕ on the position of a point.

If $\mathbf{A}$ and $\mathbf{B}$ are two vectors of magnitudes A and B which make an angle $\theta \leqslant \pi$ with each other we define their scalar product as

$$\mathbf{A} \cdot \mathbf{B} = AB \cos \theta \tag{A1}$$

and their vector product as

$$\mathbf{A} \wedge \mathbf{B} = (AB \sin \theta)\mathbf{c} \tag{A2}$$

where $\mathbf{c}$ is a unit vector along that normal to the plane of $\mathbf{A}$ and $\mathbf{B}$ which represents the direction a right-handed screw advances when turned from $\mathbf{A}$ to $\mathbf{B}$ through $\theta < \pi$. (If $\theta = \pi$ then $\mathbf{A}$ and $\mathbf{B}$ do not define a plane.)

The vector derivative of $\phi(\mathbf{r})$ is $\mathbf{grad}\ \phi$ defined by

$$d\phi = (\mathbf{grad}\ \phi) \cdot \mathbf{dr} \tag{A3}$$

and is a vector in the direction of most rapid increase of ϕ.

An element of area has magnitude and, once a sign convention has been chosen, a direction, that of its positive normal. If $\mathbf{dS}$ is an element of a closed surface we take the outward normal as positive. If $\mathbf{E}(\mathbf{r})$ is a vector function of position we define its scalar derivative by

$$\text{div}\ \mathbf{E}(\mathbf{R}) = \lim_{V \to 0} \frac{1}{V} \oint \mathbf{E}(\mathbf{r}) \cdot \mathbf{dS} \tag{A4}$$

where V is the volume enclosed by a closed surface of integration containing the point $\mathbf{R}$.

If $\mathbf{S}$ is a plane area bounded, in a definite sense, by a closed path of integration, so that the positive normal to $\mathbf{S}$ is in the direction of advance of a right-handed screw as the path is traversed, we define the vector derivative of **E** as **curl E** satisfying

$$\lim_{S\to 0}\left[(\mathbf{curl\,E})\,.\,\mathbf{S}-\oint \mathbf{E}\,.\,\mathrm{d}\mathbf{r}\right] = 0. \tag{A5}$$

It is a vector normal to that orientation of a surface element of fixed magnitude that maximizes the line integral.

Apart from the addition of vectors the five operations defined above, together with integration, are the only operations required or defined in vector analysis.

2. Coordinates and Components

If we have three intersecting families of surfaces, each surface in a family being specified by a single parameter q_ν where $\nu = 1, 2, 3$, the ordered set of three numbers q_1, q_2, q_3, or more briefly q_ν, are the coordinates of the common point of intersection of the three surfaces. In crystallography we use non-orthogonal plane surfaces but here we shall only consider orthogonal systems in which the three surfaces intersect at right angles. In an orthogonal system the displacement $\mathbf{dr}$ induced by a change $\mathrm{d}q$ can be expressed as

$$\mathbf{dr} = \sum_\nu h_\nu \mathbf{a}_\nu \,\mathrm{d}q_\nu \tag{A6}$$

where the $\mathbf{a}_\nu$ are orthogonal unit vectors satisfying

$$\mathbf{a}_\nu\,.\,\mathbf{a}_\mu = \delta_{\mu\nu}. \tag{A7}$$

Both the $\mathbf{a}_\nu$ and the measures h_ν may be functions of the coordinates q_ν of the point displaced. Indeed the only orthogonal system in which all the $\mathbf{a}_\nu$ and h_ν are constant is the Cartesian system based on families of planes normal to orthogonal axes. In this system $h_\nu = 1$, $(\nu = 1, 2, 3)$.

If $\mathbf{F}(\mathbf{r})$ is a vector defined at $\mathbf{r}$ its components in a coordinate system q are

$$F_\nu(\mathbf{r}) = \mathbf{F}(\mathbf{r})\,.\,\mathbf{a}_\nu(\mathbf{r}). \tag{A8}$$

Two vectors defined at the same point are equal if their components are equal. In Cartesians alone, the vectors need not be defined at the same point. In any orthogonal system, however, their magnitudes are the same.

In an orthogonal system the property (A7) leads to the relation

$$\mathbf{F}(\mathbf{r}) = \sum_\nu F_\nu(\mathbf{r})\mathbf{a}_\nu(\mathbf{r}) \tag{A9}$$

and the scalar product of two vectors defined at the same point is

$$\mathbf{F}\,.\,\mathbf{G} = \sum_\nu F_\nu G_\nu. \tag{A10}$$

Henceforth we shall always, except when we explicitly state otherwise, imply a summation over repeated suffixes, thus (A10) can be written simply as $F_\nu G_\nu$.

We define the totally anti-symmetric quantity $\varepsilon_{\lambda\mu\nu}$ to be zero if any two subscripts are equal and to be $+1$ or -1 if the subscripts are in cyclic or anticyclic order. Its only non-zero components are

$$\varepsilon_{123} = -\varepsilon_{321} = \varepsilon_{231} = -\varepsilon_{213} = \varepsilon_{312} = -\varepsilon_{132} = 1 \tag{A11}$$

and it has the useful property (sum over k implied)

$$\varepsilon_{ijk}\varepsilon_{klm} = \delta_{il}\delta_{jm}-\delta_{im}\delta_{jl}. \tag{A12}$$

It can be used to express the components of a vector product as

$$(\mathbf{F} \wedge \mathbf{G})_\lambda = \varepsilon_{\lambda\mu\nu} F_\mu G_\nu. \tag{A13}$$

In component form the three derivatives are

$$(\mathbf{grad}\ \phi)_\nu = \frac{1}{h_\nu}\frac{\partial \phi}{\partial q_\nu} \quad \text{(no sum over } \nu\text{)}, \tag{A14}$$

$$\text{div}\ \mathbf{E} = \frac{1}{h_1 h_2 h_3}\left[\frac{\partial}{\partial q_1}(E_1 h_2 h_3) + \frac{\partial}{\partial q_2}(E_2 h_3 h_1) + \frac{\partial}{\partial q_3}(E_3 h_1 h_2)\right] \tag{A15}$$

and

$$(\mathbf{curl}\ \mathbf{E})_i = \frac{1}{h_1 h_2 h_3}\left[h_i \varepsilon_{i\lambda\mu}\frac{\partial}{\partial q_\lambda}(h_\mu E_\mu)\right] \quad \text{(no sum over } i\text{)}. \tag{A16}$$

In Cartesians where $h_1 = h_2 = h_3 = 1$ it is useful to define the vector operator ∇ by

$$\nabla = \mathbf{a}_i \frac{\partial}{\partial r_i}, \tag{A17}$$

and we can then express the derivatives as

$$\begin{aligned} \mathbf{grad}\ \phi &= \nabla\ \phi \\ \text{div}\ E &= \nabla \cdot \mathbf{E} \\ \mathbf{curl}\ \mathbf{E} &= \nabla \wedge \mathbf{E}. \end{aligned} \tag{A18}$$

This is so convenient that we regard $\nabla\phi$ as a synonym for **grad** ϕ etc. even when no coordinate system is implied. Note carefully that except in Cartesians $\nabla\phi$, $\nabla \cdot \mathbf{E}$, and $\nabla \wedge \mathbf{E}$ are given by (A14), (A15) and (A16) and that

$$\nabla\ \phi \neq \mathbf{a}_\nu \frac{\partial \phi}{\partial q_\nu},$$

$$\nabla \cdot \mathbf{E} \neq \frac{\partial E_\nu}{\partial q_\nu},$$

$$\nabla \wedge \mathbf{E} \neq \mathbf{a}_\lambda \varepsilon_{\lambda\mu\nu} \frac{\partial E_\nu}{\partial q_\mu}.$$

We also write the expression div **grad** ϕ as $\nabla \cdot \nabla\phi = \nabla^2\phi$. If A_i is a Cartesian component of a vector **A** it is a scalar, so $\nabla^2 A_i$ is a permissible expression and it is convenient to define $\nabla^2\mathbf{A}$ as $\mathbf{a}_i \nabla^2 A_i$. It should be emphasized that $\nabla^2\mathbf{A}$ is meaningless except as the aggregate of the three Cartesian component expressions $\nabla^2 A_i$. If $A_\nu = \mathbf{a}_\nu \cdot \mathbf{A}$ is a general orthogonal component of **A**, not in a Cartesian system, $\nabla^2\mathbf{A}_\nu = \nabla^2(\mathbf{a}_\nu \cdot \mathbf{A})$ and $\mathbf{a}_\nu$ as well as **A** depends on position.

If **X** is a complicated vector expression and we can show that its Cartesian components X_i are equal to the Cartesian components of a simpler or more useful vector **Y**, then $\mathbf{X} = \mathbf{Y}$ is independent of any particular coordinate system. We use this technique in the remaining sections. Note however that the relation (valid in Cartesians)

$$(\nabla \wedge (\nabla \wedge \mathbf{A}))_i = (\nabla(\nabla \cdot \mathbf{A}))_i - \nabla^2 A_i$$

cannot be carried over into more general coordinate systems because $\nabla^2 A_i = \text{div}\ \mathbf{grad}\ A_i$ has no invariant vector significance.

3. Triple Products of Vectors

There are basically two possible products $\mathbf{E}\,.\,(\mathbf{F} \wedge \mathbf{G})$, a scalar, and $\mathbf{E} \wedge (\mathbf{F} \wedge \mathbf{G})$, a vector.

$$\mathbf{E}\,.\,(\mathbf{F} \wedge \mathbf{G}) = \mathbf{F}_i\varepsilon_{ijk}F_jG_k = F_i\varepsilon_{ijk}G_jE_k = \mathbf{F}\,.\,(\mathbf{G} \wedge \mathbf{E}) \qquad \text{(A19)}$$

where the second step uses the relation $\varepsilon_{ijk} = \varepsilon_{jki}$. $(\mathbf{E} \wedge (\mathbf{F} \wedge \mathbf{G}))_i = \varepsilon_{ijk}E_j\varepsilon_{klm}F_lG_m$ and the sum over k, using (A12), gives

$$(\mathbf{E} \wedge (\mathbf{F} \wedge \mathbf{G}))_i = E_jF_iG_j - E_jF_jG_i = F_i(\mathbf{G}\,.\,\mathbf{E}) - G_i(\mathbf{E}\,.\,\mathbf{F})$$

thus

$$\mathbf{E} \wedge (\mathbf{F} \wedge \mathbf{G}) = \mathbf{F}(\mathbf{G}\,.\,\mathbf{E}) - (\mathbf{E}\,.\,\mathbf{F})\mathbf{G}. \qquad \text{(A20)}$$

To memorize this result note that **E, F, G** remain in cyclic order, as they do in (A19).

4. Differentials of Products

We need not prove the obvious results

$$\nabla(\phi\psi) = \phi\nabla\psi + \psi\nabla\phi, \qquad \text{(A21)}$$

$$\nabla\,.\,(\phi\mathbf{F}) = \phi\nabla\,.\,\mathbf{F} + \mathbf{F}\,.\,\nabla\phi, \qquad \text{(A22)}$$

or

$$\nabla \wedge (\phi\mathbf{F}) = \phi\nabla \wedge \mathbf{F} - \mathbf{F} \wedge \nabla\phi, \qquad \text{(A23)}$$

but $\nabla\,.\,(\mathbf{F} \wedge \mathbf{G})$ requires discussion

$$\nabla\,.\,(\mathbf{F} \wedge \mathbf{G}) = \varepsilon_{ijk}\frac{\partial}{\partial r_i}(F_jG_k) = \varepsilon_{ijk}F_j\frac{\partial G_k}{\partial r_i} + \varepsilon_{ijk}G_k\frac{\partial F_j}{\partial r_i}$$

$$= -\varepsilon_{jik}F_j\frac{\partial G_k}{\partial r_i} + \varepsilon_{kij}G_k\frac{\partial F_j}{\partial r_i}$$

and so

$$\nabla\,.\,(\mathbf{F} \wedge \mathbf{G}) = -\mathbf{F}\,.\,(\nabla \wedge \mathbf{G}) + \mathbf{G}\,.\,(\nabla \wedge \mathbf{F}). \qquad \text{(A24)}$$

Next consider

$$[\nabla \wedge (\mathbf{F} \wedge \mathbf{G})]_i = \varepsilon_{ijk}\varepsilon_{klm}\frac{\partial}{\partial r_j}(F_lG_m) = \frac{\partial}{\partial r_j}(F_iG_j - F_jG_i)$$

(using equation A12). We have $\partial/\partial r_j(F_iG_j) = (\mathbf{G}\,.\,\nabla)F_i + F_i\nabla\,.\,\mathbf{G}$ and so

$$\nabla \wedge (\mathbf{F} \wedge \mathbf{G}) = (\mathbf{G}\,.\,\nabla)\mathbf{F} + \mathbf{F}(\nabla\,.\,\mathbf{G}) - (\mathbf{F}\,.\,\nabla)\mathbf{G} - \mathbf{G}(\nabla\,.\,\mathbf{F}). \qquad \text{(A25)}$$

The relation

$$[\mathbf{F} \wedge (\nabla \wedge \mathbf{G})]_i = \varepsilon_{ijk}\varepsilon_{klm}F_j\frac{\partial G_m}{\partial r_l} = F_j\frac{\partial G_j}{\partial r_i} - F_j\frac{\partial G_i}{\partial r_j}$$

and the expression obtained by interchanging **F** and **G** yield

$$[\mathbf{F} \wedge (\nabla \wedge \mathbf{G}) + \mathbf{G} \wedge (\nabla \wedge \mathbf{F})]_i = \frac{\partial}{\partial r_i}(\mathbf{F}\,.\,\mathbf{G}) - (\mathbf{F}\,.\,\nabla)G_i - (\mathbf{G}\,.\,\nabla)F_i.$$

This gives us our last relation in this section:

$$\nabla(\mathbf{F}\,.\,\mathbf{G}) = \mathbf{F} \wedge (\nabla \wedge \mathbf{G}) + \mathbf{G} \wedge (\nabla \wedge \mathbf{F}) + (\mathbf{F}\,.\,\nabla)\mathbf{G} + (\mathbf{G}\,.\,\nabla)\mathbf{F}. \qquad \text{(A26)}$$

5. Double Differentials

$$\nabla\,.\,(\nabla\phi) = \nabla^2\phi = \sum_i \frac{\partial^2\phi}{\partial r_i^2}\,. \qquad \text{(A27)}$$

$$\nabla \cdot (\nabla \wedge \mathbf{F}) = \frac{\partial}{\partial r_i}\varepsilon_{ijk}\frac{\partial F_k}{\partial r_j} = -\frac{\partial}{\partial r_j}\left(\varepsilon_{jik}\frac{\partial F_k}{\partial r_i}\right) = -\nabla \cdot (\nabla \wedge \mathbf{F})$$

and so

$$\nabla \cdot (\nabla \wedge \mathbf{F}) \equiv 0. \tag{A28}$$

Similarly

$$\nabla \wedge (\nabla\phi) \equiv 0. \tag{A29}$$

The only other allowable expressions are $\nabla \wedge (\nabla \wedge \mathbf{F})$ and $\nabla(\nabla \cdot \mathbf{F})$. We have

$$[\nabla \wedge (\nabla \wedge \mathbf{F})]_i = \varepsilon_{ijk}\varepsilon_{klm}\frac{\partial}{\partial r_j}\frac{\partial F_m}{\partial r_l} = \frac{\partial^2 F_j}{\partial r_i \partial r_j} - \frac{\partial^2 F_i}{\partial r_j \partial r_j}$$

and

$$\nabla_i(\nabla \cdot \mathbf{F}) = \frac{\partial^2 F_j}{\partial r_i \partial r_j}$$

and so

$$[\nabla \wedge (\nabla \wedge \mathbf{F}) - \nabla(\nabla \cdot \mathbf{F})]_i = -\nabla^2 F_i. \tag{A30}$$

The right hand side cannot be recognized as a Cartesian component of a vector (other than the left hand side). Equation (A30) can be written as

$$\nabla \wedge (\nabla \wedge \mathbf{F}) - \nabla(\nabla \cdot \mathbf{F}) = -\nabla^2 \mathbf{F} \tag{A31}$$

only if we remember that $(\nabla^2 \mathbf{F})_i$ is defined to mean $\nabla^2 F_i$ with F_i a Cartesian component of $\mathbf{F}$.

6. Special Relations Involving r

$$\nabla \cdot \mathbf{r} = 3, \tag{A32}$$

$$\nabla \wedge \mathbf{r} = 0, \tag{A33}$$

$$\mathbf{r} = \tfrac{1}{2}\nabla(r^2), \tag{A34}$$

$$(\mathbf{F} \cdot \nabla)\mathbf{r} = \mathbf{F}, \tag{A35}$$

and if $\mathbf{F}_0$ is a constant vector

$$\nabla(\mathbf{F}_0 \cdot \mathbf{r}) = \mathbf{F}_0. \tag{A36}$$

A somewhat more complicated relation is useful in discussing multipole fields and angular momentum. This is

$$\nabla \wedge (\mathbf{r} \wedge \nabla\phi) = \mathbf{r}\,\nabla^2\phi - \nabla(\phi + \mathbf{r} \cdot \nabla\phi). \tag{A37}$$

The left hand side in component form gives

$$\begin{aligned}
[\nabla \wedge (\mathbf{r} \wedge \nabla\phi)]_i &= \varepsilon_{ijk}\varepsilon_{klm}\frac{\partial}{\partial r_j}\left(r_l\frac{\partial\phi}{\partial r_m}\right) = \frac{\partial}{\partial r_j}\left(r_i\frac{\partial\phi}{\partial r_j}\right) - \frac{\partial}{\partial r_j}\left(r_j\frac{\partial\phi}{\partial r_i}\right) \\
&= r_i\nabla^2\phi - 2\frac{\partial\phi}{\partial r_i} - r_j\frac{\partial}{\partial r_i}\frac{\partial\phi}{\partial r_j} \\
&= r_i\nabla^2\phi - \frac{\partial\phi}{\partial r_i} - \frac{\partial}{\partial r_i}\left(r_j\frac{\partial\phi}{\partial r_j}\right) \\
&= r_i\nabla^2\phi - (\nabla\phi)_i - \nabla_i(\mathbf{r} \cdot \nabla\phi)
\end{aligned}$$

which is the result (A37).

If ϕ satisfies Laplace's equation so that $\nabla^2\phi = 0$ we have

$$\nabla \wedge (\mathbf{r} \wedge \nabla\phi) = -\nabla(\phi + \mathbf{r} \cdot \nabla\phi). \tag{A38}$$

If in addition ϕ is expressed in spherical polar coordinates as a spherical harmonic $\phi = 1/r^{l+1}\,Y_{lm}(\theta, \phi)$ and $\mathbf{r}$ in (A38) is the radius vector from the origin, we have, since $\mathbf{r} \cdot \nabla\phi = r\,(\partial\phi/\partial r)$,

$$\phi + \mathbf{r} \cdot \nabla\phi = -l\phi,$$

and so, in this case,

$$\nabla \wedge (\mathbf{r} \wedge \nabla\phi) = l\,\nabla\,\phi. \tag{A39}$$

Now this particular form for ϕ is the general form for the potential due to an electric multipole moment of order l at the origin. Thus the electric field $\mathbf{E}$ of an electric multipole can be expressed either as $\mathbf{E} = -\nabla\phi$ or as $\mathbf{E} = \nabla \wedge \mathbf{A}$ where

$$\mathbf{A} = \frac{-1}{l}\,\mathbf{r} \wedge \nabla\phi. \tag{A40}$$

Notice, however, that this result uses $\nabla^2\phi = 0$ and so is not necessarily valid at the origin, and also that it cannot be applied to the potential due to a charge since then $l = 0$ and (A39) vanishes.

7. Integrals

If ϕ is a scalar function of position and x, y, z are Cartesian coordinates we can evaluate the volume integral $\int_V (\partial\phi/\partial x)\,\mathrm{d}V$ over a volume V bounded by a closed surface S by considering first cylinders of cross-section $\mathrm{d}y\,\mathrm{d}z$ parallel to the x axis. One of these cylinders contributes $(\phi\,\mathrm{d}y\,\mathrm{d}z)_2 - (\phi\,\mathrm{d}y\,\mathrm{d}z)_1$ where (1) and (2) refer to the ends of the cylinder at the surface. But $(\mathrm{d}y\,\mathrm{d}z)_2 = \mathbf{a}_x \cdot \mathrm{d}\mathbf{S}_2$ and $(\mathrm{d}y\,\mathrm{d}z)_1 = -\mathbf{a}_x \cdot \mathrm{d}\mathbf{S}_1$ where $\mathbf{a}_x$ is the unit vector parallel to the x axis and $\mathrm{d}\mathbf{S}_1$ and $\mathrm{d}\mathbf{S}_2$ are elements of the surface with positive outward normals. If a line parallel to any one of the Cartesian axes cuts the surface S a finite number of times we have

$$\int \frac{\partial\phi}{\partial x}\,\mathrm{d}V = \oint \phi\,\mathrm{d}S_x$$

or, in our earlier notation,

$$\int \frac{\partial\phi}{\partial r_i}\,\mathrm{d}V = \oint \phi\,\mathrm{d}S_i\,. \tag{A41}$$

Thus

$$\int \nabla\phi\,\mathrm{d}V = \oint \phi\,\mathrm{d}\mathbf{S}. \tag{A42}$$

All vector integral relations involving closed surfaces are based on this result. Thus, since the Cartesian components of a vector $\mathbf{F}$ considered separately are scalars we also have

$$\int \frac{\partial F_j}{\partial r_i}\,\mathrm{d}V = \oint F_j\,\mathrm{d}S_i \tag{A43}$$

and, if we set $i = j = 1, 2, 3$ and sum over i, we obtain Gauss' integral theorem

$$\int \nabla \cdot \mathbf{F}\,\mathrm{d}V = \oint \mathbf{F} \cdot \mathrm{d}\mathbf{S}. \tag{A44}$$

Also if we consider

$$\varepsilon_{ijk} \int \frac{\partial F_k}{\partial r_j}\,\mathrm{d}V = \varepsilon_{ijk} \oint F_k\,\mathrm{d}S_j = -\varepsilon_{ijk} \oint F_j\,\mathrm{d}S_k$$

we obtain

$$\int(\nabla \wedge \mathbf{F})\,\mathrm{d}V = -\oint \mathbf{F} \wedge \mathrm{d}\mathbf{S}. \tag{A45}$$

We now construct a series of more complicated relations.

If ϕ is a scalar and $\mathbf{F}$ a vector

$$\nabla(\phi\mathbf{F}) = \phi\nabla\,.\,\mathbf{F}+\mathbf{F}\,.\,\nabla\phi$$

and so

$$\int(\phi\nabla\,.\,\mathbf{F}+\mathbf{F}\,.\,\nabla\phi)\,\mathrm{d}V = \oint\phi\mathbf{F}\,.\,\mathrm{d}\mathbf{S}. \tag{A46}$$

If, in addition $F = \nabla\psi$ we obtain

$$\int(\phi\nabla^2\psi+\nabla\psi\,.\,\nabla\phi)\,\mathrm{d}V = \oint\phi\nabla\psi\,.\,\mathrm{d}\mathbf{S}. \tag{A47}$$

If we interchange ϕ and ψ and subtract the resulting equation we find

$$\int(\phi\nabla^2\psi-\psi\nabla^2\phi)\,\mathrm{d}V = \oint\phi\nabla\psi\,.\,\mathrm{d}\mathbf{S}-\oint\psi\nabla\phi\,.\,\mathrm{d}\mathbf{S}, \tag{A48}$$

a result sometimes known as Green's theorem.

If on the other hand we set $\phi = \psi$ we have

$$\int(\phi\nabla^2\phi+(\nabla\phi)^2)\,\mathrm{d}V = \oint\phi\nabla\phi\,.\,\mathrm{d}\mathbf{S}. \tag{A49}$$

If ϕ satisfies Laplace's equation, so that $\nabla^2\phi = 0$, and in addition either ϕ or $\nabla\phi$ vanishes on S, we can conclude that $\int(\nabla\phi)^2\,\mathrm{d}V = 0$ and so $\nabla\phi = 0$ and ϕ is a constant.

If $\mathbf{F}$ and $\mathbf{G}$ are two vectors

$$\frac{\partial}{\partial r_j}(F_jG_i) = (\mathbf{F}\,.\,\nabla)G_i+G_i\nabla\,.\,\mathbf{F},$$

and so

$$\int[(\mathbf{F}\,.\,\nabla)\mathbf{G}+\mathbf{G}(\nabla\,.\,\mathbf{F})]\,\mathrm{d}V = \oint\mathbf{G}(\mathbf{F}\,.\,\mathrm{d}\mathbf{S}). \tag{A50}$$

In particular if $\mathbf{F} = \mathbf{P}$ and $\mathbf{G} = \mathbf{r}$, so that $(\mathbf{P}\,.\,\nabla)\mathbf{r} = \mathbf{P}$, we obtain

$$\int\mathbf{P}\,\mathrm{d}V = -\int\mathbf{r}(\nabla\,.\,\mathbf{P})\,\mathrm{d}V+\oint\mathbf{r}(\mathbf{P}\,.\,\mathrm{d}\mathbf{S}). \tag{A51}$$

A vector $\mathbf{F}$ is said to be irrotational if $\nabla \wedge \mathbf{F} = 0$, and to be solenoidal if $\nabla\,.\,\mathbf{F} = 0$. Thus $\mathbf{F} = \nabla\phi$ is irrotational and $\mathbf{G} = \nabla \wedge \mathbf{A}$ is solenoidal. Consider then the volume integral of the scalar product of two vectors, one irrotational and the other solenoidal:

$$\int\mathbf{F}\,.\,\mathbf{G}\,\mathrm{d}V = \int G_j\frac{\partial\phi}{\partial r_j}\,\mathrm{d}V = \int\frac{\partial}{\partial r_j}(\phi G_j)\,\mathrm{d}V = \oint\phi\mathbf{G}\,.\,\mathrm{d}\mathbf{S} \tag{A52}$$

or

$$\int\mathbf{F}\,.\,\mathbf{G}\,\mathrm{d}V = \int\varepsilon_{ijk}F_i\frac{\partial A_k}{\partial r_j}\,\mathrm{d}V = \int\varepsilon_{ijk}\frac{\partial}{\partial r_j}(A_kF_i)\,\mathrm{d}V = -\oint\mathbf{F}\,.\,(\mathbf{A} \wedge \mathrm{d}\mathbf{S}). \tag{A53}$$

If *either* $\mathbf{F}$ *or* $\mathbf{G}$ vanishes on S we have

$$\int\mathbf{F}\,.\,\mathbf{G}\,\mathrm{d}V = 0. \tag{A54}$$

If $\mathbf{F} = \nabla\phi$ and $\mathbf{G} = \nabla\psi$ are both irrotational vectors,

$$\int(\mathbf{F} \wedge \mathbf{G})_i\,\mathrm{d}V = \int\varepsilon_{ijk}F_j\frac{\partial\psi}{\partial r_k}\,\mathrm{d}V = \int\varepsilon_{ijk}\frac{\partial}{\partial r_k}(\psi F_j)\,\mathrm{d}V = \oint\psi(\mathbf{F} \wedge \mathrm{d}\mathbf{S})_i$$

thus

$$\int\mathbf{F} \wedge \mathbf{G}\,\mathrm{d}V = \oint\psi(\mathbf{F} \wedge \mathrm{d}\mathbf{S}) = -\int\phi(\mathbf{G} \wedge \mathrm{d}\mathbf{S}), \tag{A55}$$

and again if *either* **F** *or* **G** vanishes on S

$$\int \mathbf{F} \wedge \mathbf{G}\, \mathrm{d}V = 0. \tag{A56}$$

If **J** is a solenoidal vector and $\mathbf{J} = \nabla \wedge \mathbf{M}$ we have

$$\int (\mathbf{r} \wedge \mathbf{J})_i\, \mathrm{d}V = \int \mathbf{r} \wedge (\nabla \wedge \mathbf{M})_i\, \mathrm{d}V = \int \varepsilon_{ijk}\varepsilon_{klm} r_j \frac{\partial M_m}{\partial r_l}\, \mathrm{d}V = \int \left(r_j \frac{\partial M_j}{\partial r_i} - r_j \frac{\partial M_i}{\partial r_j} \right) \mathrm{d}V.$$

We can rearrange this as

$$\int (\mathbf{r} \wedge \mathbf{J})_i\, \mathrm{d}V = \int \left\{ \frac{\partial}{\partial r_i}(r_j M_j) - \frac{\partial}{\partial r_j}(r_j M_i) - M_j \frac{\partial r_j}{\partial r_i} + M_i \frac{\partial r_j}{\partial r_j} \right\} \mathrm{d}V,$$

but $\partial r_j / \partial r_i = \delta_{ij}$ and $\partial r_j / \partial r_j = 3$ so that we have

$$\int \mathbf{r} \wedge \mathbf{J}\, \mathrm{d}V = \oint \mathbf{M} \,.\, \mathbf{r}\, \mathrm{d}\mathbf{S} - \oint \mathbf{M}(\mathbf{r} \,.\, \mathrm{d}\mathbf{S}) + 2\int \mathbf{M}\, \mathrm{d}V. \tag{A57}$$

We can combine the surface integrals using (A20) and write this as

$$\tfrac{1}{2}\int (\mathbf{r} \wedge \mathbf{J})\, \mathrm{d}V = \int \mathbf{M}\, \mathrm{d}V - \tfrac{1}{2}\oint \mathbf{r} \wedge (\mathbf{M} \wedge \mathrm{d}\mathbf{S}). \tag{A58}$$

If **M** vanishes on S we obtain the result used in discussing the magnetic moment of a current distribution:

$$\tfrac{1}{2}\int \mathbf{r} \wedge \mathbf{J}\, \mathrm{d}V = \int \mathbf{M}\, \mathrm{d}V. \tag{A59}$$

The couple $\boldsymbol{\Gamma}$ acting on a localized current distribution in a field **B** has components

$$\Gamma_i = \int [\mathbf{r} \wedge \{(\nabla \wedge \mathbf{M}) \wedge \mathbf{B}\}]_i\, \mathrm{d}V = \int \varepsilon_{ijk}\varepsilon_{klm}\varepsilon_{lpq} r_j \frac{\partial M_q}{\partial r_p} B_m\, \mathrm{d}V$$

and summation over l yields

$$\Gamma_i = \int \varepsilon_{ijk} r_j B_m \left(\frac{\partial M_k}{\partial r_m} - \frac{\partial M_m}{\partial r_k} \right) \mathrm{d}V.$$

We cannot do much with the expression unless **B** is a constant vector and can be taken outside the integral. In this case, with a rearrangement of the terms we have

$$\Gamma_i = B_m \int \left[\varepsilon_{ijk} \left\{ \frac{\partial}{\partial r_m}(M_k r_j) - \frac{\partial}{\partial r_k}(M_m r_j) \right\} - \varepsilon_{ijk} \left\{ M_k \frac{\partial r_j}{\partial r_m} - M_m \frac{\partial r_j}{\partial r_k} \right\} \right] \mathrm{d}V,$$

but

$$\varepsilon_{ijk} \frac{\partial r_j}{\partial r_m} = \varepsilon_{ijk}\, \delta_{jm},$$

$$\varepsilon_{ijk} \frac{\partial r_j}{\partial r_k} = 0,$$

and so

$$\Gamma_i = -B_m \int \varepsilon_{imk} M_k\, \mathrm{d}V + B_m \oint \varepsilon_{ijk} M_k r_j\, \mathrm{d}S_m - B_m \oint \varepsilon_{ijk} M_m r_j\, \mathrm{d}S_k,$$

or

$$\Gamma = -\mathbf{B} \wedge \int \mathbf{M}\, \mathrm{d}V + \oint (\mathbf{r} \wedge \mathbf{M})\mathbf{B} \,.\, \mathrm{d}\mathbf{S} - \oint (\mathbf{B} \,.\, \mathbf{M})\mathbf{r} \wedge \mathrm{d}\mathbf{S}. \tag{A60}$$

If **M** vanishes on S we have simply

$$\Gamma = -\mathbf{B} \wedge \int \mathbf{M}\, \mathrm{d}V. \tag{A61}$$

8. Poisson's Equation

If, within a volume V bounded by a closed surface S, the function ϕ satisfies Poisson's equation

$$\nabla^2\phi = -\rho$$

and, in addition, either ϕ or $\partial\phi/\partial n$, the normal component of $\nabla\phi$ is specified on S, then ϕ is determined to within a constant. The proof is simple. If ϕ_1 and ϕ_2 are two possible solutions then $\nabla^2(\phi_1-\phi_2) = 0$ and $(\phi_1-\phi_2)\,\partial/\partial n\,(\phi_1-\phi_2)$ vanishes on S. But from equation (A49)

$$\int \{\nabla(\phi_1-\phi_2)\}^2\,dV = \oint (\phi_1-\phi_2)\frac{\partial}{\partial n}(\phi_1-\phi_2)\,dS.$$

Thus $\nabla(\phi_1-\phi_2) = 0$ in V and ϕ_1 differs from ϕ_2 by at most a constant; the constant is zero if ϕ is specified on S.

The function $1/|\mathbf{R}-\mathbf{r}|$ satisfies Laplace's equation except at $\mathbf{r} = \mathbf{R}$ where it is singular. For any closed volume

$$\int \nabla^2\left(\frac{1}{|\mathbf{R}-\mathbf{r}|}\right) dV(\mathbf{r}) = \oint\left(\nabla_r\frac{1}{|\mathbf{R}-\mathbf{r}|}\right).\,d\mathbf{S} = -\oint\frac{(\mathbf{r}-\mathbf{R})\,.\,d\mathbf{S}}{|\mathbf{R}-\mathbf{r}|^3}.$$

This is -4π if S encloses $\mathbf{R}$ and zero otherwise. Thus if $\phi(\mathbf{r})$ is non-singular when $\mathbf{r} = \mathbf{R}$ we have, taking the integral over any volume which includes $\mathbf{R}$,

$$\int \phi(\mathbf{r})\nabla_r^2\left(\frac{1}{|\mathbf{R}-\mathbf{r}|}\right) dV = -4\pi\phi(\mathbf{R}). \tag{A62}$$

If in equation (A48):

$$\int(\phi\nabla^2\psi-\psi\nabla^2\phi)\,dV = \oint\phi\frac{\partial\psi}{\partial n}\,dS-\oint\psi\frac{\partial\phi}{\partial n}\,dS$$

we set $\nabla^2\phi = -\rho$ and $\psi = 1/|\mathbf{R}-\mathbf{r}|$ then, when the volume of integration includes the point $\mathbf{R}$, we obtain

$$-4\pi\phi(\mathbf{R})+\int\frac{\rho(r)\,dV(\mathbf{r})}{|\mathbf{R}-\mathbf{r}|} = \oint\phi(\mathbf{r})\frac{\partial}{\partial n}\left(\frac{1}{|\mathbf{R}-\mathbf{r}|}\right)dS-\oint\frac{1}{|\mathbf{R}-\mathbf{r}|}\frac{\partial\phi(\mathbf{r})}{\partial n}\,dS.$$

If we apply this to a case where ρ is zero except in a finite region and add the physical assumption that $\phi \to 0$ as $\mathbf{R} \to \infty$ then, by taking the surface S at infinity, we have

$$\phi(\mathbf{R}) = \int\frac{\rho(\mathbf{r})\,dV(\mathbf{r})}{4\pi\,|\mathbf{R}-\mathbf{r}|}, \tag{A63}$$

where the volume integral extends over the whole of space, or at least the region where $\rho(\mathbf{r})$ is non-zero. This is the unique solution of Poisson's equation subject to the assumption that effects due to a localized charge distribution vanish at a great distance.

9. Representation of a Vector in Terms of Its Sources and Vortices

If a vector $\mathbf{F}$ satisfies

$$\nabla\,.\,\mathbf{F} = \sigma, \quad \nabla \wedge \mathbf{F} = \boldsymbol{\omega}, \tag{A64}$$

we shall refer to σ and $\boldsymbol{\omega}$ as the sources and vortices of $\mathbf{F}$. This is a convenient nomenclature drawn from hydrodynamics.

We can express $\mathbf{F}$ as

$$\mathbf{F} = \mathbf{F}_0-\nabla\phi+\nabla \wedge \mathbf{A}, \tag{A65}$$

where $\mathbf{F}_0$ is a constant vector and

$$\nabla^2\phi = -\sigma, \quad \nabla \wedge (\nabla \wedge \mathbf{A}) = \boldsymbol{\omega}, \tag{A66}$$

and further, since $\mathbf{A}$ is arbitrary to the extent of any vector which can be expressed as a gradient, we can also insist that $\nabla \cdot \mathbf{A} = 0$. The Cartesian components of $\mathbf{A}$ then satisfy

$$\nabla^2 A_i = -\omega_i. \tag{A67}$$

If all the sources and vortices of $\mathbf{F}$ lie in a finite region of space it is reasonable to suppose that $\mathbf{F}$ and also ϕ and $\mathbf{A}$ vanish at a great distance. This implies that $\mathbf{F}_0 = 0$ and that we can express ϕ and A_i as

$$\phi(\mathbf{R}) = \int \frac{\sigma(\mathbf{r})\, \mathrm{d}V(\mathbf{r})}{4\pi|\mathbf{R}-\mathbf{r}|}, \tag{A68}$$

$$A_i(\mathbf{R}) = \int \frac{\omega_i(\mathbf{r})\, \mathrm{d}V(\mathbf{r})}{4\pi|\mathbf{R}-\mathbf{r}|}. \tag{A69}$$

In a more general orthogonal coordinate system with unit orthogonal vectors $\mathbf{a}_\nu(\mathbf{R})$ at $(\mathbf{R})$ we have

$$A_\nu(\mathbf{R}) = \mathbf{a}_\nu(\mathbf{R}) \cdot \int \frac{\boldsymbol{\omega}(\mathbf{r})\, \mathrm{d}V(\mathbf{r})}{4\pi|\mathbf{R}-\mathbf{r}|}. \tag{A70}$$

This indicates that if, quite properly, we write equation (A69) as a vector relation

$$\mathbf{A}(\mathbf{R}) = \int \frac{\boldsymbol{\omega}(\mathbf{r})\, \mathrm{d}V(\mathbf{r})}{4\pi|\mathbf{R}-\mathbf{r}|}, \tag{A71}$$

because it is a relation between vectors at different points $\mathbf{r}$ and $\mathbf{R}$ we must not expect components of $\mathbf{A}$ and $\boldsymbol{\omega}$, other than Cartesian components, to be related except by equation (A70). We leave it to the reader to verify that $\mathbf{A}$ given by (A71) does in fact satisfy $\nabla_R \cdot \mathbf{A}(\mathbf{R}) = 0$. We now have

$$\mathbf{F}(\mathbf{R}) = -\nabla_R \int \frac{\nabla_r \cdot \mathbf{F}(\mathbf{r})\, \mathrm{d}V(\mathbf{r})}{4\pi|\mathbf{R}-\mathbf{r}|} + \nabla_R \wedge \int \frac{\nabla_r \wedge \mathbf{F}(\mathbf{r})\, \mathrm{d}V(\mathbf{r})}{4\pi|\mathbf{R}-\mathbf{r}|}. \tag{A72}$$

We next look at a related problem. Let $\mathbf{F}$ represent a localized current density $\mathbf{J}$ or a field $\mathbf{E}$, which vanishes except in a finite volume V bounded by a surface S; we now investigate whether we can express $\mathbf{F}$ as $-\nabla\phi + \nabla \wedge \mathbf{A}$ where ϕ and $\mathbf{A}$ vanish outside S.

Consider first the case where $\nabla \cdot \mathbf{F} = 0$ so that we can dispense with ϕ and express $\mathbf{F}$ solely as $\nabla \wedge \mathbf{A}$. Clearly $\mathbf{A}$, which is given by

$$\mathbf{A}(\mathbf{R}) = \int \frac{\nabla_r \wedge \mathbf{F}(\mathbf{r})\, \mathrm{d}V(\mathbf{r})}{4\pi|\mathbf{R}-\mathbf{r}|},$$

does not automatically vanish outside S although $\nabla \wedge \mathbf{A}$ and $\nabla \cdot \mathbf{A}$ are both zero in this region. We can, however, add to $\mathbf{A}$ the gradient of any scalar $\nabla\psi$ without altering $\mathbf{F}$. Thus we can put

$$\mathbf{A} \rightarrow \mathbf{A}' = \mathbf{A} + \nabla\psi.$$

Outside S we have $\nabla \cdot \mathbf{A} = 0$ and so we must have $\nabla^2\psi = 0$ if $\mathbf{A}'$ is to be zero outside S. But if $\nabla^2\psi = 0$ both inside and outside S and $\nabla\psi = 0$ at infinity, ψ can only be a constant and this is useless. Thus $\nabla^2\psi \neq 0$ within S and we cannot have $\nabla \cdot \mathbf{A}' = 0$ inside S. This is, however, the only restriction and so if $\nabla \cdot \mathbf{F}$ is zero everywhere and $\mathbf{F}$ is zero outside S we can express $\mathbf{F}$ as $\nabla \wedge \mathbf{A}'$ where $\mathbf{A}'$ is also zero outside S.

If instead $\nabla \wedge \mathbf{F} = 0$ but $\nabla \cdot \mathbf{F} \neq 0$ within S the situation is rather different for, although we can dispense with $\mathbf{A}$, the solution

$$\phi(\mathbf{R}) = \int \frac{\nabla_r \cdot \mathbf{F}(\mathbf{r})\, \mathrm{d}V(\mathbf{r})}{4\pi|\mathbf{R}-\mathbf{r}|}$$

is unique (except for a trivial added constant) and does not obviously vanish outside S. Indeed not only is ϕ non-zero outside S but even $\nabla\phi$ is also non-zero and since $\mathbf{F} = -\nabla\phi$ this contradicts the assumption that $\mathbf{F} = 0$ outside S. Thus a vector $\mathbf{F}$ satisfying $\nabla \wedge \mathbf{F} = 0$ everywhere cannot vanish outside S unless it has a special form. The restriction $\nabla \cdot \mathbf{F} = 0$ in S is too strong, for then $\mathbf{F} = 0$ everywhere. The only other possibility is that all the multipole moments defined by

$$\int r^l Y_{lm}(\theta, \phi)\, \nabla \cdot \mathbf{F}\, \mathrm{d}V$$

should vanish (where Y_{lm} is a spherical harmonic and $l = 0, 1, 2$ etc.).

The special case when $l = 0$ tells us that

$$\int \nabla \cdot \mathbf{F}\, \mathrm{d}V = \oint \mathbf{F} \cdot \mathrm{d}\mathbf{S} = 0$$

which we knew already, since $\mathbf{F} = 0$ on S, but for $l > 0$ we have new conditions to be satisfied by $\mathbf{F}$.

If $\mathbf{F}$ represents an electric field $\mathbf{E}$ so that $\nabla \cdot \mathbf{F} = \rho/\varepsilon_0$, then $\mathbf{F}$ can only vanish outside S if all the multipole moments of the charge distribution associated with $\mathbf{E}$ vanish. Thus at least there must be no net charge. It then suffices if the charge distribution is spherically symmetric but this is not necessary, as we can see by considering any set of charges with overall neutrality inside a closed metal box, together with their image charges.

If $\mathbf{F}$ represents a current density $\mathbf{J}$ and we require $\mathbf{J}$ to be zero outside S then $\nabla \cdot \mathbf{J} = -\dot{\rho}$ and in this case all time derivatives of the multipole moments, including the total charge, in the region must vanish.

In general if $\mathbf{J}$ is a localized current distribution, zero outside a closed surface S, we can always express it as

$$\mathbf{J} = \nabla \wedge \mathbf{M} - \nabla\psi$$

but if we insist that $\mathbf{M}$ is also zero outside S, we must also insist that ψ is zero (or a constant) outside S. This, as we have seen, is only possible if all the multipole moments of $\nabla \cdot \mathbf{J} = -\dot{\rho}$ are zero. The relation (see A59)

$$\tfrac{1}{2}\int (\mathbf{r} \wedge \mathbf{J})\, \mathrm{d}V = \int \mathbf{M}\, \mathrm{d}V$$

is not only valid if $\nabla \cdot \mathbf{J} = 0$ but also if the weaker condition

$$\int r^l Y_{lm}(\theta, \phi) \nabla \cdot \mathbf{J}\, \mathrm{d}V = -\int r^l Y_{lm}(\theta, \phi) \dot{\rho}\, \mathrm{d}V = 0 \tag{A73}$$

is satisfied for $l = 0, 1, 2$, etc. Note that when $l = 0$ this yields

$$\int \nabla \cdot \mathbf{J}\, \mathrm{d}V = \oint \mathbf{J} \cdot \mathrm{d}\mathbf{S} = 0$$

and so no current must leave the region. This is already implied by our initial assumption that $\mathbf{J}$ was localized.

10. Representation of a Scalar as the Divergence of a Vector

In dealing with dielectric media with a specified polarization $\mathbf{P}$ the term $-\nabla \cdot \mathbf{P}$ acts as a source for the electric field $\mathbf{E}$ and it helps us to visualize the relation between $\mathbf{P}$ and $\mathbf{E}$ if we introduce an equivalent charge density

$$\rho = -\nabla \cdot \mathbf{P}. \tag{A74}$$

Now if we had no preconceived notions about the microscopic structure of a dielectric and regarded **P** as purely an empirical macroscopic vector we might wonder whether if we were given ρ we could reverse the procedure and obtain **P** from ρ. Clearly one possibility is

$$\mathbf{P}(\mathbf{R}) = \nabla_R \int \frac{\rho(\mathbf{r})\, dV(\mathbf{r})}{4\pi|\mathbf{R}-\mathbf{r}|}, \tag{A75}$$

but this is not unique since we do not know that $\nabla \wedge \mathbf{P} = 0$. On the other hand we might reasonably require that $\mathbf{P} = 0$, except within the finite region V occupied by the body. This as we shall see is not enough to make **P** unique but it does impose a restriction on the form of $\rho(\mathbf{r})$.

Outside, the body $\mathbf{P}(\mathbf{R})$ given by (A75) can be expanded in terms of a set of multipole fields using an origin within the body. We indicate this by

$$\mathbf{P} = -\nabla(\phi_0 + \phi_1 + \ldots \phi_l + \ldots)$$

where the ϕ_l are complete spherical harmonics. According to the results at the end of section 6 we can also express this as

$$\mathbf{P} = -\nabla\phi_0 - \nabla \wedge \left\{\mathbf{R} \wedge \nabla\left(\frac{\phi_1}{1} + \frac{\phi_2}{2} + \ldots \frac{\phi_l}{l} + \ldots\right)\right\}.$$

Except for the first term, $-\nabla\phi_0$, in this expansion all the terms have zero divergence and so the new vector defined by

$$\mathbf{P}'(\mathbf{R}) = \nabla_R \int \frac{\rho\, dV}{4\pi|\mathbf{R}-\mathbf{r}|} + \nabla \wedge \{\mathbf{R} \wedge \nabla\,(\phi_1 + \text{etc.})\}$$

used for all **R** within and without the body leads to the correct value of ρ. Outside the body, however, this reduces to

$$\mathbf{P}'(\mathbf{R}) \rightarrow -\nabla\phi_0$$

and this will be zero if the net charge of the body is zero. Thus, if $q = \int \rho\, dV = 0$ is zero, this is a sufficient condition for a vector **P** satisfying (A74) to exist which is also zero outside the body. It is a necessary condition, for if S is a closed surface outside, but containing the body,

$$q = \int \rho\, dV = -\int \nabla \cdot \mathbf{P}\, dV = -\int P_n\, dS$$

and if **P** is to be zero on S, then q must also be zero.

We have now proved a statement frequently made in discussion of dielectrics that if a body has no net charge we can express the charge distribution in the body as $-\nabla \cdot \mathbf{P}$ where **P** vanishes outside the body. Despite this **P** is still not unique for we can add to (A76) the curl of any vector without altering the value of ρ. Thus if we add $\nabla \wedge \mathbf{G}$ to **P** the only restriction on **G** is that $\nabla \wedge \mathbf{G}$ should vanish outside the body. Vectors with this property certainly exist; indeed, in section 9, we have shown that for any specified form of $\nabla \wedge \mathbf{G}$ within the body we can always find a vector **G** which vanishes outside the body.

Fortunately microscopic theory leads to a unique definition of the vector $\mathbf{P} = \mathbf{D} - \varepsilon_0\mathbf{E}$, which is non-zero only in matter, as the macroscopic dipole moment density of the medium. We do not have to work backwards from ρ to **P**; rather **P** is prescribed, and $\rho = -\nabla \cdot \mathbf{P}$

is only either a convenient way of visualizing the effects of **P** or a compact way of relating the macroscopic vector **P** to the macroscopic effects of the distribution of microscopic bound charges.

We conclude this section with a number of essentially trivial remarks about the consequences of the relation (A74). First of all if **P** is prescribed to be non-zero in a body and zero outside the body, the vector identity

$$\int \mathbf{P}\,\mathrm{d}V = -\int \mathbf{r}(\nabla \cdot \mathbf{P})\,\mathrm{d}V + \oint \mathbf{r}(\mathbf{P}\cdot \mathrm{d}\mathbf{S})$$

applied to a volume including the body and a surface outside the body yields

$$\int \mathbf{P}\,\mathrm{d}V = \int r\rho\,\mathrm{d}V.$$

This identifies the integral as the total dipole moment of the body but does not identify **P** as the dipole moment density. Secondly if we take S as the physical surface of the body we have

$$\int \mathbf{P}\,\mathrm{d}V = -\int \mathbf{r}(\nabla \cdot \mathbf{P})\,\mathrm{d}V + \oint \mathbf{r}(\mathbf{P}\cdot \mathrm{d}\mathbf{S})$$

but in the volume integral we must not now include in $\nabla \cdot \mathbf{P}$ the singularity at the surface. This is taken care of by the surface integral. Thirdly if **P** is a prescribed distribution of polarization in a body we can express the resulting electrostatic potential either as

$$\phi(\mathbf{R}) = -\int \frac{\nabla_r \cdot \mathbf{P}(\mathbf{r})\,\mathrm{d}V(\mathbf{r})}{4\pi\varepsilon_0|\mathbf{R}-\mathbf{r}|}$$

where the surface of the volume V lies outside the body or as

$$\phi(\mathbf{R}) = -\int \frac{\nabla_r \cdot \mathbf{P}(\mathbf{r})\,\mathrm{d}V(\mathbf{r})}{4\pi\varepsilon_0|\mathbf{R}-\mathbf{r}|} + \oint \frac{\mathbf{P}\cdot \mathrm{d}\mathbf{S}}{4\pi\varepsilon_0|\mathbf{R}-\mathbf{r}|}$$

where the surface S of V coincides with the physical surface of the body where P_n is discontinuous.

Finally the considerations of section 9 show us that we cannot insist that **P** vanish outside the body and simultaneously require that $\nabla \wedge \mathbf{P} = 0$ unless all the multipole moments of the body vanish, in which case the result is useless. Thus the vector **P** used in dielectric problems does not normally have a non-zero curl.

11. Total and Partial Time Derivatives

If θ is a scalar quantity associated with matter in a medium moving with a velocity **v**, the total time derivative $\mathrm{d}\theta/\mathrm{d}t$ is the rate of change of θ at a point moving with the medium and it is related to $\partial\theta/\partial t$ the partial derivative at a fixed point in space by

$$\frac{\mathrm{d}\theta}{\mathrm{d}t} = \frac{\partial\theta}{\partial t} + \mathbf{v}\cdot\nabla\theta = \frac{\partial\theta}{\partial t} + v_j\frac{\partial\theta}{\partial r_j}. \tag{A76}$$

The value of $\mathrm{d}\theta/\mathrm{d}t$ is determined by the rate of change of variables which describe the local state of the medium, for example the temperature, but in what follows we shall only consider those changes which result directly and inescapably from the motion. These are changes in the local orientation of the medium and the local strain.

If **l** is an infinitesimal vector fixed in the medium then, in a displacement **u**(**r**) of the medium the components of **l** are changed from l_i to $l_i + l_j\,\partial u_i/\partial r_j$ and the square of the length of the vector is changed by

$$\left(l_i+l_j\frac{\partial u_i}{\partial r_j}\right)^2-l^2 \approx 2l_il_j\frac{\partial u_i}{\partial r_j}.$$

The quantity $\partial u_i/\partial r_j$ can be decomposed into a symmetric part and an anti-symmetric part thus:

$$\frac{\partial u_i}{\partial r_j} = \tfrac{1}{2}\left(\frac{\partial u_i}{\partial r_j}+\frac{\partial u_j}{\partial r_i}\right)+\tfrac{1}{2}\left(\frac{\partial u_i}{\partial r_j}-\frac{\partial u_j}{\partial r_i}\right) = S_{ij}-w_{ij}, \tag{A77}$$

and we see that the change in l^2 is simply $2l_il_jS_{ij}$. Thus the symmetric combination S_{ij} describes the local strain. A uniform rotation of the medium through an infinitesimal angle $\boldsymbol{\phi}$ changes the point $\mathbf{r}$ to $\mathbf{r}+\mathrm{d}\mathbf{r} = \mathbf{r}+\boldsymbol{\phi} \wedge \mathbf{r}$ and so

$$\mathrm{d}r_i = \varepsilon_{ijk}\phi_j r_k.$$

This rotation leads to $S_{ij} = 0$, but we have

$$w_{ij} = -\varepsilon_{ijk}\phi_k, \tag{A78}$$

and so w_{ij} describes a local rigid rotation. For small values of the velocity $\mathbf{v}$ the quantities $\mathrm{d}S_{ij}/\mathrm{d}t$ and $\partial S_{ij}/\partial t$ are negligibly different and can be written as

$$\dot{S}_{ij} = \tfrac{1}{2}\left(\frac{\partial v_i}{\partial r_j}+\frac{\partial v_j}{\partial r_i}\right) \tag{A79a}$$

and similarly

$$\dot{w}_{ij} = \tfrac{1}{2}\left(\frac{\partial v_i}{\partial r_j}-\frac{\partial v_i}{\partial r_i}\right). \tag{A79b}$$

If θ is a true scalar property, e.g., the temperature, it is independent of orientation and then we must have

$$\frac{\mathrm{d}\theta}{\mathrm{d}t} = a\dot{S}_{jj} = a\nabla\cdot\mathbf{v} \tag{A80}$$

where a is a scalar constant. The situation is different if θ is the component of a vector for then we shall have, say, for a component M_i of a permanent magnetization,

$$\frac{\mathrm{d}M_i}{\mathrm{d}t} = \dot{w}_{ij}M_j+\gamma_{ijk}\dot{S}_{jk} \tag{A81}$$

where the first term is due to rotation, and

$$\gamma_{ijk} = \frac{\mathrm{d}M_i}{\mathrm{d}S_{jk}} \tag{A82}$$

is a tensor coefficient to be derived from the equation of state. Since S_{jk} is symmetric γ_{ijk} is necessarily symmetric in j and k and so we also have

$$\frac{\mathrm{d}M_i}{\mathrm{d}t} = \tfrac{1}{2}\left(\frac{\partial v_i}{\partial r_j}-\frac{\partial v_j}{\partial r_i}\right)M_j+\gamma_{ijk}\frac{\partial v_j}{\partial r_k}. \tag{A83}$$

Rotation transforms the components of a tensor such as the dielectric constant ε_{ij}, like the product of two vector components a_i and b_j, and so in this case

$$\frac{\mathrm{d}\varepsilon_{ii}}{\mathrm{d}t} = \varepsilon_{ik}\dot{w}_{jk}+\varepsilon_{kj}\dot{w}_{ik}+\alpha_{ijkl}\dot{S}_{kl}. \tag{A84}$$

Here α_{ijkl} is the tensor coefficient

$$\alpha_{ijkl} = \frac{\mathrm{d}\varepsilon_{ij}}{\mathrm{d}S_{kl}} \tag{A85}$$

to be obtained from the equation of state. Again it must be symmetric in k and l. We therefore have

$$\frac{\mathrm{d}\varepsilon_{ij}}{\mathrm{d}t} = \tfrac{1}{2}\,\varepsilon_{ik}\left(\frac{\partial v_j}{\partial r_k}-\frac{\partial v_k}{\partial r_j}\right)+\tfrac{1}{2}\,\varepsilon_{kj}\left(\frac{\partial v_i}{\partial r_k}-\frac{\partial v_k}{\partial r_i}\right)+\alpha_{ijkl}\frac{\partial v_k}{\partial r_l} \tag{A86}$$

and the partial derivative is

$$\frac{\partial \varepsilon_{ij}}{\partial t} = \frac{\mathrm{d}\varepsilon_{ij}}{\mathrm{d}t}-v_k\frac{\partial \varepsilon_{ij}}{\partial r_k}. \tag{A87}$$

Two scalar properties, the charge density ρ and the mass density τ, satisfy conservation equations of the form

$$\frac{\partial \rho}{\partial t}+\nabla\,.\,(\rho\mathbf{v}) = 0 \quad \text{and} \quad \frac{\partial \tau}{\partial t}+\nabla\,.\,(\tau\mathbf{v}) = 0, \tag{A88}$$

and so, for example,

$$\frac{\mathrm{d}\tau}{\mathrm{d}t} = -\nabla\,.\,(\tau\mathbf{v})+\mathbf{v}\,.\,\nabla\tau = -\tau\nabla\,.\,\mathbf{v} = -\tau\dot{S}_{jj}. \tag{A89}$$

If this is used to eliminate $\dot{S}_{jj}$ in equation (A80) there results

$$\frac{\mathrm{d}\theta}{\mathrm{d}t} = -\frac{a}{\tau}\frac{\mathrm{d}\tau}{\mathrm{d}t} \tag{A90}$$

and we see that the coefficient a is given by

$$a = -\tau\frac{\mathrm{d}\theta}{\mathrm{d}\tau}. \tag{A91}$$

The partial time derivative of a scalar is therefore

$$\frac{\partial \theta}{\partial t} = -\tau\frac{\mathrm{d}\theta}{\mathrm{d}\tau}\nabla\,.\,\mathbf{v}-(\mathbf{v}\,.\,\nabla)\theta. \tag{A92}$$

In an initially isotropic solid the rotational terms in equation (A86) cancel and leave

$$\frac{\partial \varepsilon_{ij}}{\partial t} = \alpha_{ijkl}\frac{\partial v_k}{\partial r_l}-v_k\frac{\partial \varepsilon}{\partial r_k}\delta_{ij}. \tag{A93}$$

Since there is no preferred direction in the unstrained solid the most general relation possible between ε_{ij} and S_{kl} is

$$\mathrm{d}\varepsilon_{ij} = a_1 S_{ij}+a_2 S_{kk}\delta_{ij}. \tag{A94}$$

The tensor coefficient is therefore reduced to

$$\alpha_{ijkl} = a_1\delta_{ik}\delta_{jl}+a_2\delta_{ij}\delta_{kl}, \tag{A95}$$

and so

$$\frac{\partial \varepsilon_{ij}}{\partial t} = a_1\frac{\partial v_i}{\partial r_j}+a_2\nabla\,.\,\mathbf{v}\delta_{ij}-(\mathbf{v}\,.\,\nabla)\varepsilon\delta_{ij}. \tag{A96}$$

Appendix

For isotropic, hydrostatic strain we have $\partial v_i/\partial r_j = \frac{1}{3}\nabla \cdot \mathbf{v}\delta_{ij}$ and, in this case,

$$\frac{\partial \varepsilon_{ij}}{\partial t} = \{(\tfrac{1}{3}a_1+a_2)\nabla \cdot \mathbf{v}-(\mathbf{v} \cdot \nabla)\varepsilon\}\delta_{ij}. \tag{A97}$$

If this is compared with (A80) or (A91) we see that $\frac{1}{3}a_1+a_2$ corresponds to $a = -\tau\,(\mathrm{d}\varepsilon/\mathrm{d}\tau)$.

In the text $\partial\varepsilon_{ij}/\partial t$ occurs only in the combination $\frac{1}{2}\varepsilon_0 E_i E_j\,(\partial\varepsilon_{ij}/\partial t)$ and this reduces i and j to the status of dummy tensor indices. If use is also made of the symmetry $\varepsilon_{ij} = \varepsilon_{ji}$ equation (A87) leads to

$$\tfrac{1}{2}\varepsilon_0 E_i E_j \frac{\partial \varepsilon_{ij}}{\partial t} = \tfrac{1}{2}\varepsilon_0\left\{\varepsilon_{kj}E_iE_j-\varepsilon_{ij}E_jE_k\right\}\frac{\partial v_i}{\partial r_k}-T_{ij}^{\text{es}}\frac{\partial v_i}{\partial r_j}-\tfrac{1}{2}\varepsilon_0 v_i E_j E_k\frac{\partial \varepsilon_{jk}}{\partial r_i}, \tag{A98}$$

where the electrostrictive tensor is

$$T_{ij}^{\text{es}} = -\tfrac{1}{2}\varepsilon_0\alpha_{klij}E_kE_l. \tag{A99}$$

For an initially isotropic solid medium this is

$$T_{ij}^{\text{es}} = -\tfrac{1}{2}\varepsilon_0(a_1E_iE_j+a_2E^2\delta_{ij}) \tag{A100}$$

and, for a fluid, in which ε remains isotropic and behaves like a scalar

$$T_{ij}^{\text{es}} = \tfrac{1}{2}\varepsilon_0\tau\frac{\mathrm{d}\varepsilon}{\mathrm{d}\tau}E^2\delta_{ij}. \tag{A101}$$

We now turn to a rather different problem and consider the total rate of change $\mathrm{d}\psi/\mathrm{d}t$ of the flux ψ of a vector $\mathbf{F}$ linking an infinitesimal plane element of area $\mathbf{A}$ in a moving medium. As the medium moves it sweeps out volume at a rate $\mathbf{A} \cdot \mathbf{v}$ and so captures new lines of $\mathbf{F}$ at a rate $(\mathbf{A} \cdot \mathbf{v})\nabla \cdot \mathbf{F}$. At the same time an element $\mathrm{d}\mathbf{l}$ of the perimeter of $\mathbf{A}$ sweeps out area at a rate $\mathbf{v} \wedge \mathrm{d}\mathbf{l}$ and so crosses lines of $\mathbf{F}$ at a rate $\mathbf{F} \cdot (\mathbf{v} \wedge \mathrm{d}\mathbf{l}) = -(\mathbf{v} \wedge \mathbf{F}) \cdot \mathrm{d}\mathbf{l}$. Thus the total rate of change of flux is

$$\frac{\mathrm{d}\psi}{\mathrm{d}t} = \mathbf{A} \cdot \frac{\partial \mathbf{F}}{\partial t}+(\mathbf{A} \cdot \mathbf{v})\nabla \cdot \mathbf{F}-\oint(\mathbf{v} \wedge \mathbf{F}) \cdot \mathrm{d}\mathbf{l}.$$

Since $\mathbf{A}$ is infinitesimal the line integral is $\mathbf{A} \cdot \{\nabla \wedge (\mathbf{v} \wedge \mathbf{F})\}$ and therefore

$$\frac{\mathrm{d}\psi}{\mathrm{d}t} = \mathbf{A} \cdot \left\{\frac{\partial \mathbf{F}}{\partial t}+\mathbf{v}(\nabla \cdot \mathbf{F})-\nabla \wedge (\mathbf{v} \wedge \mathbf{F})\right\}. \tag{A102}$$

If this is applied to a magnetic field $\mathbf{B}$, for which $\nabla \cdot \mathbf{B} = 0$, the flux change is related to the line integral of $\mathbf{E}$ round a path moving with the medium, and so the curl of $\mathbf{E}$ evaluated in a moving coordinate system is related to $\partial\mathbf{B}/\partial t$ at a fixed point in space by

$$\nabla^m \wedge \mathbf{E} = \frac{\partial \mathbf{B}}{\partial t}-\nabla \wedge (\mathbf{v} \wedge \mathbf{B}). \tag{A103}$$

Problems

1. Two particles at $\mathbf{r}_1$ and $\mathbf{r}_2$ at $t = 0$ travel in straight lines with constant velocities $\mathbf{v}_1$ and $\mathbf{v}_2$ and eventually collide. Show that $(\mathbf{r}_1-\mathbf{r}_2) \wedge (\mathbf{v}_1-\mathbf{v}_2) = 0$.

2. If $\mathbf{E} = -\nabla\phi$ show that $\mathbf{E}$ is normal to a surface of constant ϕ.

3. The thermal capacity per unit volume of a fluid is C, a constant, and its thermal conductivity, also a constant, is K show that the temperature T obeys the differential equation $K\nabla^2 T = C\dot{T}$. How is this equation modified if K and C are functions of position?

4. A bucket containing water rotates at an angular velocity ω about its axis of symmetry carrying the water with it. If $\mathbf{v}(\mathbf{r})$ is the velocity of the water at a radius $\mathbf{r}$ what is the form of $\nabla \wedge \mathbf{v}(\mathbf{r})$?

5. What meaning, if any, can we attach to the components, in spherical polar coordinates, of a general displacement vector $\mathbf{r}_2 - \mathbf{r}_1$?

6. If $\mathbf{A}$, $\mathbf{B}$, and $\mathbf{C}$ are vector displacements from the origin what is the geometric significance of $\mathbf{A} . (\mathbf{B} \wedge \mathbf{C})$? What meaning, if any, can we attach to the sign of the triple product?

7. Show that

$$(\mathbf{A} \wedge \mathbf{B}) . (\mathbf{C} \wedge \mathbf{D}) = (\mathbf{A} . \mathbf{C})(\mathbf{B} . \mathbf{D}) - (\mathbf{A} . \mathbf{D})(\mathbf{B} . \mathbf{C}).$$

8. Express $\nabla\phi$, $\nabla . \mathbf{F}$, $\nabla \wedge \mathbf{F}$ and $\nabla^2\phi$ in cylindrical polar coordinates (ρ, θ, z). Compare the ρ components of $\nabla(\nabla . \mathbf{F}) - \nabla \wedge (\nabla \wedge \mathbf{F})$ and $\nabla^2 F_p$.

9. If p is the hydrostatic pressure in a fluid and V is a volume bounded by a closed surface S what is the physical interpretation of the equation

$$\int \nabla p \, \mathrm{d}V = \oint p \, \mathrm{d}\mathbf{S}?$$

10. An incompressible fluid is confined in a rigid fixed container which it fills. Show that the total linear momentum of the fluid is zero.

11. Show that if the fluid in problem 10 is in motion within the container the motion cannot be irrotational.

12. A vector $\mathbf{F} = -\nabla\phi$ is zero on and outside a closed surface S which bounds a volume V. Show that

$$\int r_i r_j \nabla . \mathbf{F} \, \mathrm{d}V = k\delta_{ij}$$

where k is a constant. (Hint: consider $(\partial/\partial r_k)(r_i r_j F_k)$.)

13. Show that if $\mathbf{p}$ is a constant vector

$$\nabla(\mathbf{p} . \nabla\phi) + \nabla \wedge (\mathbf{p} \wedge \nabla\phi) = \mathbf{p}\nabla^2\phi.$$

Hence show that for any closed surface

$$\oint \nabla\left(\mathbf{p} . \nabla\left(\frac{1}{r}\right)\right) . \mathrm{d}\mathbf{S} = 0.$$

14. If ϕ is a scalar function of position which depends only the distance r from the origin show that

$$\nabla^2\phi = \frac{1}{r^2}\frac{\partial}{\partial r} r^2 \frac{\partial\phi}{\partial r} = \frac{1}{r}\frac{\partial^2}{\partial r^2}(r\phi).$$

15. If $\mathbf{p}$ is a constant vector show that for any closed surface $\oint \mathbf{p} . \mathrm{d}\mathbf{S} = 0$.

16. Show that the volume enclosed by a closed surface is $\frac{1}{3}\oint \mathbf{r} . \mathrm{d}\mathbf{S}$.

References

ABRAGAM, A. (1961) *Principles of Nuclear Magnetism*, O.U.P.
AGRANOVICH, V. M. and GINZBURG, V. L. (1966) *Spatial Dispersion in Crystal Optics*, Interscience.
AMBLER, E., HAYWARD, R. W., HOPPES, D. D., and HUDSON, R. P. (1957) *Phys. Rev.*, **105**, 1413.
BLOEMBERGEN, N. (1965) *Non-Linear Optics*, Benjamin.
BORN, M. and WOLF, E. (1959) *Principles of Optics*, Pergamon Press.
CADY, W. G. (1962) *Piezo-Electricity*, Dover.
CASIMIR, H. B. G. (1948) *Proc. Kon. Ned. Akad. Wet.* **51**, 793.
CROWTHER, J. M. and TER HAAR, D. (1971) *Proc. Kon. Ned. Akad. Wet.* B, **74**, 351.
FAULKNER, E. A. (1969) *Introduction to the Theory of Linear Systems*, Chapman & Hall.
LE FEVRE, R. J. W. (1965) *Advances in Physical Organic Chemistry*, **3**, 1.
FIERZ, M. (1960) *Helv. Phys. Acta*, **33**, 855.
FROHLICH, H. (1958) *Theory of Dielectrics*, O.U.P.
GOLDSTEIN, H. (1959) *Classical Mechanics*, Addison-Wesley.
DE GROOT, S. R. (1969) *The Maxwell Equations*, North-Holland.
DE GROOT, S. R. and VLIEGER, J. (1965) *Physica*, **31**, 254.
HOSELITZ, K. (1952) *Ferromagnetic Properties of Metals and Alloys*, O.U.P.
IRVING, J. H. and KIRKWOOD, J. G. (1950) *J. Chem. Phys.*, **18**, 817
JACKSON, J. D. (1962) *Classical Electrodynamics*, Wiley.
JEFFRIES, C. D. (1963) *Cryogenics*, **3**, 41.
JEGGO, C. R. and BOYD, G. D. (1970) *J. Appl. Phys.*, **41**, 2741.
JOHNSON, J. B. (1928) *Phys. Rev.*, **32**, 97.
JONES, R. V. and RICHARDS, J. C. S. (1954) *Proc. R. Soc.*, A **221**, 480.
KITTEL, C. (1971) *Introduction to Solid State Physics*, Wiley.
KLEINMAN, D. A. (1962) *Phys. Rev.*, **126**, 1977.
LANDAU, L. D. and LIFSHITZ, E. M. (1960) *Electrodynamics of Continuous Media*, Pergamon Press.
LANDAU, L. D. and LIFSHITZ, E. M. (1968) *Statistical Physics*, Pergamon Press.
LANDAU, L. D. and LIFSHITZ, E. M. (1970) *Theory of Elasticity*, Pergamon Press.
MACDONALD, D. K. C. (1949) *Phil. Mag.*, **40**, 561.
MAZUR, P. and NIJBOER, B. R. A. (1953) *Physica*, **19**, 971.
MESSIAH, A. (1961) *Quantum Mechanics*, vol. 1, North-Holland.
MILLER, R. C. (1964) *Appl. Phys. Lett.*, **5**, 17.
NORTH, D. O. (1940) *R.C.A. Review*, **4**, 441.
NYE, J. F. (1957) *Physical Properties of Crystals*, O.U.P.
NYQUIST, H. (1928) *Phys. Rev.*, **32**, 110.
ONSAGER, L. (1936) *J. Am. Chem. Soc.*, **58**, 1456.
PERSHAN, P. (1963) *Phys. Rev.*, **130**, 919.
PERSHAN, P. (1967) *J. Appl. Phys.*, **38**, 1482.
PLATT, J. R. (Senior Author) (1964) *Systematics of the Electronic Spectra of Conjugated Molecules*, Wiley.
RACK, A. J. (1938) *Bell Syst. Tech. J.*, **17**, 592.
RAMO, S., WHINNERY, J. R., and VAN DUZER, T. (1965) *Fields and Waves in Communication Electronics*, Wiley.
ROBINSON, F. N. H. (1963) *Physics Letters*, **4**, 180.
ROBINSON, F. N. H. (1965) *Proc. R. Soc.*, A **286**, 525.
ROBINSON, F. N. H. (1967) *Bell. Syst. Tech. J.*, **46**, 913.
ROBINSON, F. N. H. (1969) *Int. J. Electronics*, **26**, 227.
ROHRLICH, F. (1965) *Classical Charged Particles*, Addison-Wesley.
ROSENFELD, L. (1951) *Theory of Electrons*, North-Holland.
SCHELKUNOFF, S. A. (1943) *Electromagnetic Waves*, Van Nostrand.
SCHRAM, K. (1960) *Physica*, **26**, 1080.
SIMON, A., ROSE, M. E., and JAUCH, J. M. (1951) *Phys. Rev.*, **84**, 1158.
SMULLIN, L. D. and HAUS, H. A. (1959) *Noise in Electronic Devices*, Wiley.
SMYTHE, C. P. (1955) *Dielectric Behaviour and Structure*, McGraw-Hill.
SPAARNAY, M. J. (1958) *Physica*, **24**, 751.

References

STONER, E. C. (1949) *Phil. Mag.*, **36**, 803.
STOPES-ROE and WHITWORTH (1971) *Nature*, **234**, 31.
STRATTON, J. A. (1941) *Electromagnetic Theory*, McGraw-Hill.
VAN VLECK, J. H. (1932) *Theory of Electric and Magnetic Susceptibilities*, O.U.P.
VAN VLECK, J. H. (1937) *J. Chem. Phys.*, **5**, 320.
WEYL, H. (1911) *Math. Ann.*, **71**, 441.

Answers to Problems

CHAPTER 2

1. No. 2. No. 4. Consider the limited mobile charge density in the silicon.

CHAPTER 3

1. Let the loop be the circle $r = a$ in the plane $z = 0$, then

$$A_\phi = \frac{\mu_0 I}{4\pi r}\int_0^{2\pi} \frac{a\cos\theta\, d\theta}{(a^2+r^2+z^2-2ar\cos\theta)^{\frac{1}{2}}}$$

which may be evaluated in terms of complete elliptic integrals.

2. $E_x = -j\omega A \exp j(\omega t-\beta z)$, $B_y = \beta E_x/\omega$, $\overline{(\mathbf{E}\wedge\mathbf{B})_z} = \frac{1}{2}\omega\beta A^2$.
3. At R the scalar potential changes from $q/4\pi\varepsilon_0 R$ to $q/4\pi\varepsilon_0|\mathbf{R}-\mathbf{r}|$ during the interval $0 < t < \tau$. During this interval $\mathbf{A}$ also changes, but the combination of the unretarded and retarded parts of $\mathbf{A}$ produces changes in the fields $\mathbf{B} = \nabla\wedge\mathbf{A}$ and $\mathbf{E} = -\nabla\phi-\dot{\mathbf{A}}$ which occur during the interval $R/c < t < \tau+R/c$. After this interval has elapsed $\mathbf{B}$ is zero, and $\mathbf{E} = -\nabla(q/4\pi\varepsilon_0|\mathbf{R}-\mathbf{r}|)$. Thus the initial and final fields are the usual electrostatic fields but the transient behaviour is much more complicated.

CHAPTER 4

1. Use $\nabla\wedge\mathbf{E} = 0$ in conjunction with equation (4.8).
3. $\alpha\varepsilon_0 Ar^2$, radially outwards.
4. $\pi r^2 I$. 5. Zero. 6. $-E(qr_0+\frac{1}{2}qE/k)$, $\frac{1}{2}qE^2/k$.

CHAPTER 5

1. Near the anode the electron density is $N = [(4\varepsilon_0 V)/(9ed^2)] \approx 3\times10^{15}$ m^{-3}. Thus $k_0 < 3^{1/3}\times10^5$ m^{-1}.
2. We require $N > 1/\lambda^3$ and also, for no attenuation, $Ne^2/\varepsilon_0 m < (2\pi\nu)^2$. This implies $\lambda > e^2/\varepsilon_0 mc^2$.
4. If, for example, we set $k_0^2 = 10^{11}$ cm^{-2} we require $10^{-\nu/2\nu} < 10^{-6}$, i.e. $\nu \geqq 5$. If $\nu = 5$ a Fourier component with $k = 10^6$ cm^{-1} is reduced by $\exp(-10^4)$.
5. $\varepsilon+4\pi\times10^{-3}$ where ε is the value for the plastic alone. Somewhat less than 3×10^{10} Hz. At higher frequencies there will be diffraction effects.

CHAPTER 6

1. 1 mA corresponds to 6×10^{15} electrons sec^{-1}, 1μC to $3{\cdot}6\times10^{-2}$ electrons per μsec.
2. If the mean free path is not much less than a wavelength λ, the continuum equations do not describe momentum transfer across distances of macroscopic significance. The relevant pressure is about 10^{-4} atm.

CHAPTER 7

Let n_i be the mean number and $n_i+\nu_i$ be the actual number of electrons of velocity v_i crossing the plane in a particular interval τ. For this interval

$$\overline{\delta v^\tau} = \frac{\Sigma_i(n_i+\nu_i)v_i}{\Sigma_i(n_i+\nu_i)} - \frac{\Sigma_i n_i v_i}{\Sigma_i n_i} \approx \frac{1}{\Sigma_i n_i}\left\{\Sigma_i\nu_i v_i-\bar{v}\Sigma_i\nu_i\right\}.$$

The ensemble average is zero but

$$\langle(\overline{\delta v^\tau})^2\rangle = \left\{\frac{1}{\Sigma_i n_i}\right\}^2\{\Sigma_i\langle\nu_i^2\rangle v_i^2-2\bar{v}\Sigma_i\langle\nu_i^2\rangle v_i+\bar{v}^2\Sigma_i\langle\nu_i^2\rangle\}$$

and, since $\langle\nu_i^2\rangle = n_i$ and $\Sigma_i n_i = \tau N_0$, we obtain

$$\langle(\overline{\delta v^\tau})^2\rangle = \frac{1}{\tau N_0}\{\bar{v}^2-(\bar{v})^2\}.$$

Finally $\overline{v^2} = 2kT/m$ and $(\bar{v})^2 = \pi kT/2m$, which gives the required result.

Answers to Problems

CHAPTER 8

1. $f = \int \Pi_n \Pi_e \{\delta(\mathbf{R}(n) - \mathbf{R}(n, 0) - \mathbf{s})\delta(\mathbf{r}(n, e) - \mathbf{r}(n, e, 0) - \mathbf{s})\} \phi(\mathbf{s})\, d^3\mathbf{s}$,
   $g_n = \int \delta(\mathbf{R}(n) - \mathbf{R}(n, 0) - \mathbf{s}) \Pi_e \{\delta(\mathbf{r}(n, e) - \mathbf{r}(n, e, 0) - \mathbf{s})\} \phi(\mathbf{s})\, d^3\mathbf{s}$.
2. See Appendix, equation (A50).
3. **B** is the same in both cases. In the solenoid $\mathbf{H} = \mathbf{B}/\mu_0$, in the cylinder **H** is opposed to **B**.
4. $\mathbf{H} = -\dfrac{\mathbf{m}}{(\mu+2)a^3}, \quad \mathbf{B} = \dfrac{2\mu_0 \mathbf{m}}{(\mu+2)a^3}$.
5. $\mathbf{p} = 4\pi a^3 \varepsilon_0 \mathbf{E}$ in weak fields. In strong fields one part of the sphere will be completely depleted of mobile carriers and **p** will increase more slowly with **E**.
6. If $\mathbf{J} = \nabla \wedge \mathbf{I}$, then $\nabla . \mathbf{J} = 0$ and $\oint J_n\, dS = 0$.
7. The charge is zero.

CHAPTER 9

2. $I = \omega\pi r^2 B \sin \omega t/R$. No.
3. $\mathbf{B} = \mu_0 \mathbf{M}, \mathbf{H} = 0$.
4. This follows from $\oint D_n\, dS = q = 0$.
5. This follows from $q = \oint D_n\, dS = 0$.

7, 8, 9. Consider the changing magnetic flux linked by a closed path which follows the beam from cathode to collector and then returns to the cathode outside the entire system.

10. $B = -2\mu_0 H = \dfrac{2\mu_0 H_r H_c}{(H_r^2 + 4H_c^2)^{\frac{1}{2}}}$
11. Consider the consequences of $\nabla \wedge \mathbf{H} = 0$.
13. Zero.
14. About 200 A. Slightly less than 1 tesla.
15. The dipole moment is $4\pi a^3 \varepsilon_0 E$ and the surface charge density at colatitude θ is $3\varepsilon_0 E \cos\theta$.
16. $\mathbf{m} = -2\pi a^3 \mathbf{B}_0/\mu_0$. The surface current at colatitude θ is $I_\phi = -\dfrac{3B_0 \sin\theta}{2\mu_0}$.
17. Inside the sphere $H_s = -M/3$. In the hole $H = H_s + M/2 = M/6$. The force per unit length is $\mu_0 MI/6$.
18. If the radii of the body and the cavity are b and c
$$E = \frac{9\varepsilon E_0}{(2+\varepsilon)(2\varepsilon+1) - 2(\varepsilon-1)^2(c/b)^3}.$$
21. Approximately 1 cm.
22. $E_\phi = -\frac{1}{2}\omega r \mu_0 n I \sin \omega t$.
23. $H_\phi = \frac{1}{2}\varepsilon\varepsilon_0 r \dot{V}/d$.
24. Regard the air velocity **u** as the analogue of **E**. The whole system is a dipole.
25. $\frac{3}{4}$ tesla.
26. Numerically equal to ε_0.

CHAPTER 10

7. The radiated power is about 4×10^6 W m^{-2} and the pressure 10^{-2} N m^{-2} or 10^{-7} atm.
9. Consider the phases of the primary and secondary currents and the magnetic field in the hollow interior of the yoke.
13. The attractive force is $\frac{1}{2}\varepsilon_0 A \phi^2/d$.
14. For $r > a, p = \dfrac{\varepsilon - 1}{2\varepsilon_0 \varepsilon^2}\left(\dfrac{q}{4\pi r^2}\right)^2$.
15. $h = \frac{1}{2}\dfrac{\varepsilon_0(\varepsilon-1)E^2}{\tau g}$.
16. $\dfrac{3\varepsilon}{1+2\varepsilon} qE^*$ and $\dfrac{1-\varepsilon}{1+2\varepsilon} qE^*$. The sum is qE^*.
17. IB^*.
19. For $t > 0, -\frac{1}{2}(\varepsilon_0 E_0^2 + B^2/\mu_0) \sin^2 \omega t$; for $t < 0$, zero.
20. If the z axis is the outward normal to the surface and y is parallel to **B**,
    $B_y = 2B \cos(kz \cos\theta) \exp j(\omega t - kx \sin\theta)$,
    $E_x = -2jcB \cos\theta \sin(kz\cos\theta) \exp j(\omega t - kx \sin\theta)$,
    $E_z = -cB_y \sin\theta$.
    The mean energy density is $[(2B^2)/(\mu_0)]\{1 + \sin^2\theta \cos(2kl\cos\theta)\}$ and $T_{zz} = -B^2 \cos^2\theta/\mu_0$.

Answers to Problems

CHAPTER 11

1. Wave in (a) a coaxial cable; (b) in an electron linear-accelerator; (c) in a rectangular wave-guide; (d) in a travelling-wave tube or outside a light-pipe.
2. See J. D. Jackson, *Classical Electrodynamics* (Wiley, 1962), pp. 268–277.
3. A superposition of plane waves diverging from the z axis.
4. $\frac{1}{2}I\left(\frac{n^2-1}{n^2+1}\right)^2$.
6. About -2×10^{-10}.
7. 10^{-5}.

CHAPTER 12

3. The point group 3 is acentric, $\bar{3}$ is not. The presence of either piezo-electricity or a linear electro-optic effect indicates 3.
4. Only the polar point group 6 mm can display a pyro-electric effect.
5. By the rapid reversal of a strongly enough applied field.
6. Very small differences in absorption coefficients produce large effects in optically thick specimens.

CHAPTER 13

1. Magnetic: $\theta \approx 0{\cdot}03$ K; electric: $\theta \approx 300$ K.
6. See J. F. Nye (1957).
7. About 1 part in 10^{10}.
10. $\chi_e^* = 0{\cdot}95$; $\chi_i^* = 0{\cdot}925$. Approximately twice as large.
13. The point group must be acentric.
14. The point group possesses a polar axis.

CHAPTER 14

4. Equation (14.8) gives $-1{\cdot}5\times 10^{-9}$. The experimental value is -4×10^{-9}.
5. $B \approx 2$ tesla.
6. $C/3T$.
8. Approximately 10^{-8} sec.
9. $\Delta Q \approx -Q_0 v\chi''/V \approx -10^{-2}$.
10. $[8\pi/27]\mu_0 M^2 r^3$ where M is the magnetization.

CHAPTER 15

4. $F = F_0+\frac{1}{2}\frac{q^2 x_0}{\varepsilon\varepsilon_0 A}\left(1-\frac{q^2}{4\varepsilon\varepsilon_0 A^2 Y}\right)$.
5. 5 joule.
6. $\frac{1}{2}\gamma\phi^2$.
7. $\pi\gamma\tau\phi^2/2T$.
8. Consider the interaction between the dipole moment of the sphere and its images in the two electrodes.
10. 1250 joule.
11. Approximately 3×10^{-3} K.
12. Sphere: $\mu_0(\mu+2)\ \mathfrak{M}^0 \wedge \mathbf{H}^*/3$; disc: $\mathfrak{M}^0 \wedge \mathbf{B}^*$.
13. Yes.

CHAPTER 16

2. The entropy removed on magnetization cannot exceed R ln 2 per mole of cerium ions. The corresponding proton nuclear entropy is $24R$ ln 2.
3. $\frac{2\times I\pi^2 N}{3\hbar^2\varepsilon_0 V}$.

CHAPTER 17

1. Stefan's constant: $5{\cdot}67\times 10^{-8}$ W m^{-2} deg^{-4}.
2. 1·7 mm, 17 μm, 8300 Å.
3. $3{\cdot}3\times 10^{-6}$ atm.
4. Nuclear: (R ln 4)/23 per cm^3; lattice: $4\times 10^{-6}R$ per cm^3; electronic: $2\times 10^{-5}R$ per cm^3; radiation: $1{\cdot}5\times 10^{-22}R$ per cm^3.
9. Approximately 800 V m^{-1}.

Answers to Problems

CHAPTER 18

1. $kT(ABN)^2/I$. 2. $(kT/C)^{\frac{1}{2}}$. 3. About 12 V.
4. $kT_e\,\delta\nu$ where $T_e = 300/100+3(1-1/100) \approx 6$ K.
6. 1/200.
7. Is a thermionic diode in thermal equilibrium?
8. $I = \Sigma_i q v_i / d$. Replace v_i by $[(\partial E_i)/(\partial p)]$.
9. 30 K.

APPENDIX

1. If they collide at t, $\mathbf{r}_1+\mathbf{v}_1 t = \mathbf{r}_2+\mathbf{v}_2 t$ and so $\mathbf{r}_1-\mathbf{r}_2 = t(\mathbf{v}_2-\mathbf{v}_1)$, thus $\mathbf{r}_1-\mathbf{r}_2$ is parallel to $\mathbf{v}_2-\mathbf{v}_1$.
2. If $\mathbf{dr}$ is a displacement on a surface of constant ϕ, then $\mathbf{dr} \cdot \nabla\phi = 0$.
3. The heat flux $\mathbf{J}$ obeys $\nabla \cdot \mathbf{J} = C\dot{T} = 0$ and $\mathbf{J} = -K\nabla T$, and so $\nabla \cdot (K\nabla T) = C\dot{T}$.
4. 2ω parallel to the axis.
5. Virtually none.
6. The volume of the parallelepiped defined by the three vectors. The sign implies a conventional ordering of **A**, **B**, and **C**.
8. Impossible; $\nabla^2\mathbf{F}$ is not a vector.
10. Set $\mathbf{P} = \rho\mathbf{v}$ in equation (A51).
11. The motion is solenoidal; if it is also irrotational equation (A52) gives $\int v \mathrm{d}^2\, V = 0$.

Author Index

Subject Index

Subject Index